Engineering and Governing the Climate

Key Issues in Climate Change and Sustainability Ethics, Politics and Policy

Series Editors: Idil Boran, Xavier Landes, and
Morten Ebbe Juul Nielsen

This series provides quality, research-based monographies that address specific issues raised by climate change (e.g., immigration, geoengineering, and corporate social responsibility). The contributions come from various horizons within the social sciences and humanities. The series provides the knowledge and conceptual tools for understanding the challenges climate change poses to our societies. The ultimate objectives are to connect natural sciences, social sciences, and public policy to inform about the various aspects of climate change and contribute to foster mitigation and adaptation. In sum, *Key Issues in Climate Change and Sustainability: Ethics, Politics and Policy* provides the conceptual and practical tools to navigate the challenges of a more uncertain world.

Engineering and Governing the Climate

Ethical and Political Issues

Xavier Landes

ROWMAN & LITTLEFIELD
Lanham • Boulder • New York • London

Published by Rowman & Littlefield
An imprint of The Rowman & Littlefield Publishing Group, Inc.
4501 Forbes Boulevard, Suite 200, Lanham, Maryland 20706
www.rowman.com

86-90 Paul Street, London EC2A 4NE, United Kingdom

British Library Cataloguing in Publication Information Available

Library of Congress Cataloging-in-Publication Data Available

ISBN 9781538145609 (cloth) | ISBN 9781538145616 (paperback)
| ISBN 9781538145623 (ebook)

To the memory of Paulette and René Miquel.

Contents

Acknowledgments

Writing a book is never a totally individual endeavor. It often owes a lot to the people with whom we share intimacy, work, or friendship. These pages constitute no exception. They are the result of the contribution and support of many. Among those, I would like to express my gratitude to Liene Judeika, Anne-Marie and Francis Landes, Christel, François, Quentin and Axel Haschard, as well as all my close friends who endured the peculiarities and vagaries of an author who is embarked upon such a daunting journey. Through his attentive reading, Christopher Rieber provided an invaluable assistance in the last stages of the process. People at Rowman & Littlefield have been incredibly patient, indulging for my delays and changes of plan. The administration at the Stockholm School of Economics in Riga has been extremely supportive and accommodating. In particular, Anders Paalzow, SSE Riga rector, offered me a rare opportunity in modern academia: the privilege of writing a book while not being overburdened by my professional duties.

Abbreviations and Acronyms

AFOLU: Agriculture, Forestry and Other Land Use
BECCS: bioenergy with carbon capture and sequestration/storage
CBA: cost–benefit analysis
CBD: Convention on Biological Diversity
CCS: carbon capture and sequestration/storage
CCT: cirrus cloud thinning
CDR: carbon dioxide removal
CE: climate engineering
CFCs: chlorofluorocarbons
CNSA: China National Space Administration
COP: Conference of the Parties (United Nations)
DACS: direct air capture with sequestration/storage
EJ: exajoules
EMT: environmental modification techniques
ENSO: El Niño–Southern Oscillation
EOR: enhanced oil recovery
ETS: Emissions Trading System
FAO: Food and Agriculture Organization (United Nations)
GEO: geosynchronous orbit
GeoMIP: Geoengineering Model Intercomparison Project
GHG: greenhouse gas
GLENS: Geoengineering Large Ensemble Project
GSSP: Global Boundary Stratotype Section and Points
Gt: gigatons
GWP: global warming potential
HCFCs: hydrochlorofluorocarbons
HFCs: hydrofluorocarbons

HSRC: Haida Salmon Restoration Corporation
IAM: integrated assessment model
IEA: International Energy Agency
IPCC: Intergovernmental Panel on Climate Change (United Nations)
LEO: low Earth orbit
MCB: marine cloud brightening
MRV: measurement, reporting, and verification
NASA: National Aeronautics and Space Administration
            (United States)
NDCs: nationally determined contributions
NETs: negative emissions technologies
NRC: National Research Council
OECD: Organisation for Economic Co-operation and Development
OIF: ocean iron fertilization
RCP: representation concentration pathway
R&D: research and development
REDD+:   Reducing Emissions from Deforestation and Forest
            Degradation in Developing Countries (United Nations)
SAI: stratospheric aerosol injection
SCoPEx: Stratospheric Controlled Perturbation Experiment
SCS: soil carbon sequestration/storage
SEL: Sun-Earth Lagrange point
SPICE: Stratospheric Particle Injection for Climate Engineering
SRM: solar radiation management
SSI: stratospheric sulfur injection
STDR: social time discount rate
TRL: technological readiness level
UHI: urban heat island
UNEP: United Nations Environment Programme
UNFCCC: United Nations Framework Convention on Climate Change
USB: urban surface brightening

Introduction

Climate Change and Alteration in the Anthropocene

In the heat of a summer afternoon, aircrafts are crossing a milky blue sky somewhere over Canada. In the open sea, automated ships are spraying salt-water droplets in the air, leaving white plumes in their wake. In the Gulf of Guinea, boats plough the surface, sprinkling iron dust. Far above, at an altitude of 2,000 kilometers, orbiting satellites constantly readjust their mirrors, following the course of the sun. Somewhere among the Icelandic volcanic landscapes, commanding metallic structures equipped with large fans are sucking up atmospheric carbon before injecting it in the deep geological layers of the Earth. Those images offer a glimpse into what could turn out to be a "not-so-distant" future if humanity or some countries decide to engineer the climate. Whereas some interventions might appear improbable, others will reach maturity in the coming decades or are even already researched or deployed, such as urban surface brightening, afforestation/reforestation, carbon capture and sequestration/storage (CCS), and bioenergy with carbon capture and sequestration/storage (BECCS). Moreover, due to the constant rise of greenhouse gas (GHG) concentrations and the inefficacy of international politics to curb emissions, climate alteration is becoming more attractive.

TEXTBOX 0.1 GREENHOUSE GAS AND GLOBAL WARMING POTENTIAL

Carbon dioxide (CO_2) is not the only or even the most common or potent GHG. Water vapor (H_2O) is more common but less potent. Methane (CH_4), nitrous oxide (N_2O), ozone (O_3), chlorofluorocarbons (CFCs), hydrofluorocarbons (HFCs), hydrochlorofluorocarbons (HCFCs), and other gases are more potent but less common. Moreover, the lifetime of these substances varies. Some GHGs remain in the atmosphere for a short period, such as CH_4 (12 years); others reside much longer like fluorinated gases (CFC_8, HCFCs, and HFCs), whose lifetimes are comprised of between hundreds of thousands of years. By convention, GHGs' potency is evaluated in terms of global warming potential (GWP), which expresses the relative positive radiative forcing of a gas compared to CO_2, and within a given time horizon (usually 100 years). Radiative forcing is

> the net change in the energy balance of the Earth system due to some imposed perturbation. It is usually expressed in watts per square meter averaged over a particular period of time and quantifies the energy imbalance that occurs when the imposed change takes place. (Myrhe *et al.* 2013, 664)

CO_2 constitutes the benchmark (unity 1), and the GWPs of other GHGs are evaluated relative to it (cf. table 0.1). In the words of the Intergovernmental Panel on Climate Change (IPCC) GWP is

> [a]n index measuring the radiative forcing following an emission of a unit mass of a given substance, accumulated over a chosen time horizon, relative to that of the reference substance, carbon dioxide (CO_2). The GWP thus represents the combined effect of the differing times these substances remain in the atmosphere and their effectiveness in causing radiative forcing. (IPCC 2014, 124)

Table 0.1   **Global Warming Potential (GWP) of Some Greenhouse Gases (GHG)**

	Lifetime (year)	Cumulative forcing over 20 years	Cumulative forcing over 100 years
Carbon dioxide (CO_2)	Variable	1	1
Methane (CH_4)	12.4	84	28
Nitrous oxide (N_2O)	121.0	264	265
Carbon tetrafluoride (CF_4)	50,000.0	4,880	6,630

The table reads that one unit of CH_4 warms the atmosphere over 100 years as much as 28 units of CO_2. Based on IPCC 5th Assessment Report (IPCC 2014, 87).

Geoengineering, which is the most popular designation, or climate engineering (CE), which is the term favored for this book, is an umbrella term and a controversial concept used to refer to a variety of methods for countering climate change or some of its effects, primarily global warming. They share the characteristic of promising to alter the climate by lowering GHG emissions/concentrations or attenuating solar radiation. Denominations such as interventions, initiatives, techniques, and methods seem more appropriate than technologies, considering that CE designates a broad array of actions ranging from low (e.g., planting trees, brightening urban surfaces) to high technicity (e.g., carbon capture from the ambient air and space reflectors).

The landmark 2009 report by the British Royal Society provides the canonical definition of geoengineering/CE as the "deliberate large-scale manipulation of the planetary environment to counteract anthropogenic climate change" (Shepherd *et al.* 2009, 1). In addition, it lays down a key, sometimes contested, distinction between *carbon dioxide removal* (*CDR*) which "address [es] the root cause of climate change by removing greenhouse gas from the atmosphere," and *solar radiation management* (*SRM*), which "attempt[s] to offset the effects of increased greenhouse gas concentrations by causing the Earth to absorb less solar radiation" (Shepherd *et al.* 2009, ix). The present book follows this classification. Through either GHG concentrations or incoming radiation, both interventions aim at lowering radiative and climate forcing, that is the energy flowing into the Earth system.[1]

CDR intends to capture and sequestrate CO_2 and can be divided into two broad categories. The first encompasses the enhancement of land-based (soils, forests, and wetlands) and ocean-based (e.g., accelerated weathering and fertilization) carbon sinks. The second assembles full-fledged engineering processes such as direct air capture with sequestration/storage (DACS) (carbon is trapped from the ambient atmosphere) or carbon capture and sequestration/ storage (CCS) (CO_2 is filtered out from the flue gas, e.g., at a coal plant, and injected underground). SRM intends to enhance Earth's albedo, which is the Earth's capacity to reflect shortwave incoming solar radiation. Interventions range from innocuous (e.g., painting roofs and streets in white or cultivating brighter crops) to controversial (e.g., dispersing sulfur in the stratosphere) and unrealistic (e.g., putting in orbit space shades). What those methods have in common is to not tackle carbon emissions and concentration, at the origins of climate change. Instead, the focus is on taming its main consequence: thermic increase. SRM does not roll back ocean acidification and the catastrophic impact on marine life like corrals, seashells, or crabs.

Due to the growing concern about climate change combined with the failure of nations to undertake proper mitigation, CE is increasingly discussed as a set of possible solutions. No serious researcher defends CE as *the* only climate strategy though. The priority rests on mitigation, supported by adaptive measures. Nonetheless, the more the climate crisis will unfold through extreme events like heatwaves, droughts, storms, floods, and sea-level rise, the more CE will appear tempting to policymakers. Because many CE options will become more accessible as time is passing by and because emergencies might lure governments or private actors into deployment, each citizen is entitled to a modicum of adequate knowledge about climate alteration. The consequences could prove so drastic that any decision of researching and deploying CE needs to pass public scrutiny. At least, citizens have the moral right to understand the stakes behind spraying sulfur in the stratosphere, stimulating marine stratocumuli, capturing carbon through burning biomass in powerplants, or fertilizing the oceans.

This introduction begins with a discussion of the concept of geoengineering and its limited adequacy for qualifying intentional, global climate alterations for countering anthropogenic climate change. CE is defended as a more descriptively accurate concept. Next, the reasons for why climate alteration increasingly appears to be a legitimate solution are presented. Obviously, they have to do with the perceived climate emergency, which is not properly tackled by international politics. The push from part of the research community is also key to understand the rising legitimacy of CE. Then, the projects of carbon capture and albedo enhancement are replaced under the deeper time horizon of the Anthropocene. In that respect, CE could be interpreted less as opening a new era than as pursuing human interference in the climate system, which does not constitute a justification for engineering the climate. In addition, it is shown how different debuts for the Anthropocene carry different injunctions about how to engineer the climate. This chapter closes on the general presentation of the place of CE within climate policy, along with mitigation and adaptation, followed by a brief exposition of the rationale and structure of this book.

HOW ACCURATE IS GEOENGINEERING DENOMINATION?

The geoengineering label is frequently criticized for various reasons (Cairns 2014; Heyward 2013; National Research Council [NRC] 2015a). Following Jay Michaelson (2013), three are important to mention. First, geoengineering would be too lax, bundling fundamentally irreducible interventions (e.g., putting under the same label afforestation with stratospheric sulfur injection [SSI]). Second, it would be inaccurate, with the prefix *Geo* being misleading. In general, CDR and SRM primarily deal with air, water, and solar radiation, not with land or geography, save exceptions (cf. chapters 3 and 4). Third, it conveys an unfortunate science-fiction flavor, a regrettable reminiscence of the post-war technocratic faith. Hence, Michaelson advocates for replacing geoengineering with "climate management."

The first objection has some teeth. Geoengineering indeed serves to qualify a wide array of interventions, some seemingly alien to others, turning it into a catch-all concept (Hulme 2014, 5). A possible response is to underline that those interventions share enough for being lumped together, primarily their objective of *modifying the climate through the alteration of radiative forcing*.

The second issue lies in the denomination being descriptively inaccurate. As a substitute, Michaelson proposes climate management, which gets rid of the *geo-* and *-engineering* altogether. It could, however, be argued that *Geo* etymologically originates in Gaia (γαια), the Greek divinity incarnating Earth. Hence, it refers not only to the land, but to Earth in its globality, including, and not limited to, its geology or geography. Indeed, climate interventions target not only the Earth's surface (e.g., through reforestation, sequestration, or surface albedo enhancement), but also the aerosols, clouds, and space too. Then, it would be difficult to see why the atmospheric layers fall outside the concept of *Geo* standing for the entire planet. Nonetheless, geoengineering does not encompass drastic alterations of the Earth's geophysical features as in the case of terraforming. The crux of the methods is to change the composition of the atmosphere or to block part of the incoming solar radiation, even if it could produce some visible surface alterations (e.g., canopy). In conclusion, Michaelson's reticence toward the prefix *Geo* possesses some validity.

The use of *engineering* appears problematic as well for descriptive and moral reasons. First, siding with Michaelson, the US NRC judges that geoengineering, as its variations like CE, "implies a greater level of precision and control than might be possible" (NRC 2015a, viii). The US NRC then advises using "climate intervention" as a more fitting concept since it expresses "an action intended to improve a situation" (NRC 2015a, viii). Second, the reservation voiced by the NRC possesses a moral dimension, which echoes Michaelson's third criticism of the hubristic undertone carried by geoengineering. This criticism is advanced by other authors (e.g., Hamilton 2013a; Jamieson 2013). In sum, the use of engineering would be misleading, by suggesting that CE interventions could achieve more control than realistically possible, while exuding the objectionable anthropogenic arrogance of dominating nature.

However, a global hubristic intent does not necessarily have to drive climate interventions. More modest ambition could be at work, such as protecting urban populations from heatwaves and shielding the Arctic Sea ice. Thus, the descriptive part of the objection seems difficult to uphold. It could be replied that climate alteration is nonetheless inherently hubristic, independently of the contextual, more modest, goals, as it is founded on a delusional sense of mastery. However, in this case, the problem is less with the possible intention, but the very nature of the activity itself (cf. chapter 2).

For the descriptive aspect, engineering is the study and use of physical laws and materials for the benefit of humankind. In the case of CE, such methods aim at preventing or alleviating the damages incurred by catastrophic climate change. Any discussion attributing to CE some potential efficacy needs, as a matter of fact, to assume some potency over variables such as temperatures, solar radiation, jet streams, and ocean currents. Although it is unlikely that the climate will ever be fully "controlled" in the sense implied by Hulme's analogy to a "global thermostat" (Hulme 2014), specific initiatives can still qualify as limited attempts at influencing some parameters. For instance, CDR methods such as CCS (carbon capture from the flue gas of power plants and other industrial facilities), BECCS (capture from biomass power plants), and DACS are heavily engineered infrastructure. Even the apparently more "natural," that is, less interfering, methods such as afforestation or reforestation require a modicum of (forest) engineering. Therefore, the engineering suffix appears descriptively accurate.

In sum, climate interventions demand a dose of engineering that does not necessitate adhering to a conception of humanity that does or should completely control natural forces. CE could remain modest. For sure, proponents may infuse the whole project with Promethean arrogance, but it does not have to be. In any case, returning to the terminology, this book uses CE for characterizing methods of albedo enhancement (SRM) and CDR. By doing so, CE keeps the engineering suffix, but acknowledges that the prefix *geo-* is to some extent misleading.

INTENSIFYING CLIMATE CONCERNS

After the definitional aspect, another interrogation emerges: why has CE been gradually gaining traction among scientists (e.g., it has been reluctantly considered by economic Nobel Prize William Nordhaus [2019]) and a broader public starting from the mid-2000s, in particular since Nobel Prize winner Paul Crutzen's article about SSI (Crutzen 2006)?

First, concerns have been building up, supported by pessimistic models, about mounting GHGs concentration and their implications for present and future generations, especially for "prevent[ing] dangerous anthropogenic interference with the climate system" as stipulated by the United Nations Framework Convention on Climate Change (UNFCCC). The multiplication of initiatives like marches, protests, and strikes manifest such worries. The harmful consequences of unabated climate change are becoming more difficult to ignore due to the increased frequency of catastrophes (heatwaves, droughts, floods, wildfires, etc.) and more extensive media coverage, although the causation link running from climate change to singular weather events is arduous to establish. Reports about the remaining carbon budget, ocean acidification, loss of biodiversity, rising sea levels, food insecurity, and exceptional floods fill the news, often in a sensationalist manner (Allen 2019). The surge of environmental activism, incarnated by public figures such as Greta Thunberg, alarmist articles, and documentaries, illustrate and feed anxieties.

Second, in this context of urgency, the international efforts for significantly cutting down GHG emissions are still scant despite optimistic stances from political and business leaders. Global temperatures are on a trajectory for soaring above the 2015 Paris Agreement's target of a 2°C (preferably 1.5°C) increase in 2100 by comparison to preindustrial levels (National Academies of Sciences, Engineering, and Medicine 2019, 34). The mitigation gap, that is, the difference between the carbon cut necessary to achieve Paris target and current abatement, is even more worrisome when one realizes that the net carbon emissions should reach zero as soon as possible. The difficulty is acute when considering that many climate scenarios used, for instance by the IPCC, include negative emissions technologies (NETs), another term for CDR, to respect the Paris goal (Smith *et al.* 2023).

Save drastic changes in international politics, mitigation will remain insufficient owing to technological and political gridlocks, with the most influential protagonists (industrialized countries, emerging markets, main energy, and industrial corporations) being reluctant to substantially cut their emissions (Michaelson 2013). This impotence has been discernible since the 1992 Earth Summit in Rio and the ensuing UNFCCC Conferences of the Parties (e.g., the Kyoto Protocol and the COP15 in Copenhagen).[2] Moreover, the world economies are trapped into using fossil energies, which is designated as "carbon lock-in" (Unruh 2000). Any additional investment in gas, oil, and coal powerplants commits humanity to further emissions, reinforcing the perspective of an overshoot scenario (Cui *et al.* 2019) in which "CO_2 concentrations in the atmosphere temporarily exceed some predefined, 'dangerous' threshold (before being reduced to non-dangerous levels)" (Nusbaumer and Matsumoto 2008, 164).

Finally, CE would be promoted by "invisible colleges" made of a small pool of researchers supported by a handful of institutions in developed countries (Möller 2021). Those researchers would be particularly vocal in their support to CE at the expense of more sustainable alternatives. By flooding the field with their publications, their ideas would have made their way into the IPCC reports and built up the scientific legitimacy of climate alteration. In short, the growing prominence of CE as an answer to climate change would result from the activism of a tiny community.

WOULD CLIMATE ENGINEERING ANNOUNCE A NEW ERA?

The rise of CE in the scientific and, to a lesser extent, public spheres has generated heated debates over the last couple of decades. A common view is to interpret it as a rupture corroborating the "end of nature" (McKibben 1970), which will come with fading boundaries between the natural and human orders (Carr *et al.* 2012, 176; Corner *et al.* 2013, 940–941). CE would instigate a new perilous era (Hamilton 2013a) in which humanity, or a few nations, would tweak the global thermostat. The criticism usually contains two components: an opposition to the climate manipulation per se and concerns about potential disastrous consequences.

The next chapters will elaborate on the outcomes of CE research, development, and deployment. For now, let us consider the most fundamental objection that could be framed as follows: *if undertaken, CE would represent a drastic and unprecedented shift.* Humanity would venture into the uncharted and perilous territories of global climate manipulation.[3] To assess this *exceptionalism claim,*[4] CE ought to be placed in a broader timeline. Against the view of CE as a new stage of planetary interference, it is worth noticing that anthropogenic influence on the climate, as well as on the land and biodiversity, predates industrialization.

After the Last Glacial Maximum (LGM), 21,000 years ago, and the ensuing Younger Dryas, ca. 11,700 years BP, temperatures soared, the several-kilometers-thick continental glaciers in the North hemisphere receded, and biodiversity deepened. Human groups spread on the quasi-entirety of the emerged lands, helped by milder conditions and technical breakthroughs (e.g., animal and plant domestication). Anthropogenic activities produced a significant increase in GHGs' concentrations, in the form of methane (because of ruminant cattle and rice cultivation) and carbon dioxide (due to forest clearing and biomass burning). Moreover, those activities directly affected vegetal and animal life on a noticeable scale (Ellis *et al.* 2021). The Industrial Revolution accentuated the trend by ramping up the use of fossil fuels (coal, oil, and natural gas) and expanding agriculture and animal husbandry, which in turn supported unprecedented population growth.

The expansion following the end of the LGM carried global impacts thousands of years before industrialization. The argument that CE, or industrialization, constitutes a drastic shift in Earth's history needs to be moderated. Humanity has tampered with the climate long before and on a global scale, although probably not at a comparable pace as nowadays. Of course, the fact that human beings have altered the climate in the past does not justify climate alteration nowadays. However, realizing the far reach of the anthropogenic influence calls into question the exceptionalism claim according to which CE would represent a radically novel phase in the history of Earth. If anything, CE epitomizes more of the latest stage of escalating anthropic interference than a rupture.

A last point. The climate is not immutably fixed on the current conditions. It has been evolving, shaped by factors such as changes in Earth's dynamics (eccentricity, obliquity, and precession), volcanic eruptions, meteorites, and so on. For example, when the so-called Great Dying took place (ca. 252 million years ago), the Earth warmed by 10°C over a period of ca. 50,000 years. The Younger Dryas saw the temperatures dropping between 3°C and 10°C depending on the latitudes within a few decades (Lieberman and Gordon 2018, 33). Moreover, the Earth has already experienced much hotter climates than today, for instance, during the Cretaceous (between 145 and 66 million years BP). Those glimpses into past climates help to put the current Anthropocene at his place: a dramatic change for living beings, but within a blink at the geological scale.

ANTHROPOCENE CONTROVERSIES

Although the climate is not static or immutable, even on a decadal or centennial basis, two specificities set apart the present situation from most previous global changes: the velocity of GHG's atmospheric accumulation and the ensuing rise in climate forcing and response (Nehrbass-Ahles *et al.* 2020), as well as the fact that the engine is a self-reflective species who could still steer the evolution it initiated in a less harmful direction. Moreover, because of this conscious influence, this species could be ultimately held accountable.

The anthropogenic interference runs so deep that Nobel Prize-winning physicist Paul Crutzen and ecologist Eugene Stoermer proposed in 2000 to rename the current geological period from the *Holocene* (in Ancient Greek: *holo-*, whole, -*kainos,* recent) to the *Anthropocene* (*anthropo-*, human being, -*kainos*, recent). This would acknowledge that humanity has become a "significant geological, morphological force" (Crutzen and Stoermer 2000, 17) by massively releasing GHGs, diverting or taming rivers, leveling mountains, clearing forests, undermining biodiversity, and polluting land and oceans with plastics, radionuclides, and chemicals (Lewis and Maslin 2018). Thus, this "current interval of anthropogenic global environmental change" (Zalasiewicz *et al.* 2008, 4) would require amending the *Geologic Time Scale* (cf. textbox 0.2).

Combining the Greek words for "humans" and "recent time," scientists have named this period the Anthropocene (Lewis and Maslin 2018, 5). It marks the moment when Homo sapiens transforms into a geological superpower, setting the Earth onto a novel path. The proposition triggers controversies. Although the magnitude of the anthropic influence is difficult to deny, two debates are nonetheless raging. The first, epistemic, bears on determining whether the Anthropocene refers to a genuine geological concept, a "political statement" (Finney and Edwards 2016, 4), or a feature of pop culture (Autin and Holbrook 2012). The second bears on when Anthropocene begun. Beyond factual interest, isolating a starting point generates implications for anthropogenic responsibility, especially when it comes to tackling the Anthropocene's adverse consequences through engineering the climate.

What kind of concept is the Anthropocene? Two views split the geological community. A conservative view asserts that the Anthropocene is inadequate for characterizing a new epoch (Autin and Holbrok 2012; Finney and Edwards 2016; Santana 2019). It does not respect the criteria set by the International Commission on Stratigraphy used for determining the *Geologic Time Scale*.[5] The criticism is that the Anthropocene's delineations are not rooted in stratigraphic records (sedimentation and rocks), but in direct observations of contemporary events (GHG concentrations, radionuclides fallout, biodiversity decline, global warming, etc.) (Finney and Edwards 2016). While geological periods are usually bounded by Global Boundary Stratotype Section and Points (GSSP) or golden spikes, that is, stratigraphic markers detectable in various locations across the world, the Anthropocene does not satisfy this condition. A neat basal boundary, that is, a lower frontier containing "lithologic, fossil, mineral, chemical, or geological signatures" is missing (Autin and Holbrook 2012, 60). It is also argued that current changes (e.g., in the atmosphere, landscape, and pollution) are too recent and most likely transient in the scale of the Earth's history for assuring that they will remain discernible by future geologists.

TEXTBOX 0.2 CLASSIFICATION, GOLDEN SPIKES, AND DATING[6]

With its *Geologic Time Scale,* the International Commission on Stratigraphy offers a widely accepted segmentation of Earth's Past into five hierarchical levels: eons (longest), eras, periods, series/epochs, and ages (shortest). Each upper division is partitioned further (e.g., eons are split into eras, then eras into periods, and so forth). This taxonomy usually rests on the identification of GSSPs or golden spikes, like mass extinction or the colossal release of specific components (e.g., oxygen), *which* help to pinpoint a basal boundary. These spikes are used for marking the end of a given time segment as well as the beginning of the next one. For instance, the spike between the Mesozoic and Cenozoic eras is a bolide impact that led to the extinction of non-avian dinosaurs.

On the other hand, a reformist view expresses a more open, laxer, approach to classification (Zalasiewicz *et al.* 2008; Lewis and Maslin 2016, 2018). The claim is that anthropogenic activities are imprinting "a global stratigraphic signature distinct from that of the Holocene or of previous Pleistocene interglacial phases, encompassing novel biotic, sedimentary, and geochemical change" (Zalasiewicz *et al.* 2008, 4). The contention is twofold: human interference is already noticeable *on geological grounds,* and human interference will remain identifiable by future scientists, for instance through plastic, chemical, or radionuclide (especially plutonium 239 and 240) deposits in the rock and sediment layers.

When did the Anthropocene start? While the first controversy fully belongs to geology, the second expands beyond it. The determination of the Anthropocene's golden spike carries normative implications, most notably in terms of anthropic responsibility.[7] Such implications are blatant when CE ambition is defined as "deliberate restorative planetary management" (Keith 2017, 73), annulling or "counter (acting) anthropogenic climate change" (Shepherd *et al.* 2009, 1), or "return(ing) the climate system to its 'original' state before humans began to affect it" (Jamieson 1996, 325). Then it begs the question of what should be canceled out or restored, that is, the baseline or boundary to be chosen.[8]

The discussion of the Anthropocene, and its debut, exceeds the frontiers of Earth sciences and ventures into ethics. For instance, it underscores the backward-looking responsibility for past GHG emissions and the forward-looking responsibility for tackling climate change (Fahlquist 2019, 29–54). While the former is concerned with assessing causal relations and moral blame for past actions and events, the latter deals with the consequences of anthropogenic emissions in terms of redress, remedy, compensation, restoration, or prevention of future harms. Forward-looking responsibility thus carries implications for potential, desirable or permissible, climate alterations (Buck 2017).

In conclusion, clarifying the nature of Anthropocene as a concept and identifying its start are two distinct issues. The Anthropocene may not be a proper geological epoch, but it may still be apt at capturing the magnitude of negative human interference as well as ensuing collective responsibility, whether that responsibility has to do with causation and blame, or remedy and redress. In the context of CE, the history of allegedly unintentional anthropic climate change is pertinent for evaluating proposals for deliberate, large-scale, CE interventions.

THE ONSET OF THE ANTHROPOGENIC INFLUENCE

In the literature, several dates compete as markers for the beginning of the Anthropocene (Lewis and Maslin 2016, 2018), and each proposal has ramifications for climate manipulation.

Megafauna extinctions (ca. 50,000–10,000 years BP). The disappearance of large-bodied mammals (e.g., mammoths, dire wolves, and sabertooth cats) happened across the globe over a 40,000-year span, during which human predation was pivotal. As a basal boundary, it would imply that the *Holocene* (beginning ca. 11,700 years BP) would be absorbed by the Anthropocene, and the Pleistocene (the epoch just before the Holocene) would be shortened.

The end of the Younger Dryas, start of the Neolithic Revolution (ca. 11,700 years BP). The end of the last glacial period coincides with the advent of plant and animal domestication, observable in fossilized pollen for instance. Even though domestication was slow to reach a global scale, the warming at the twilight of the Younger Dryas marks the dawn of human expansion. Such a debut would imply replacing the Holocene by the Anthropocene in the Geologic Time Scale.

The early anthropogenic hypothesis (Ruddiman 2003, 2006). This hypothesis situates the Anthropocene's inception ca. 5,000 years ago, when a continuous increase in CO_2 and methane significantly diverges from the natural trend. Ruddiman speculates the origin to be the diffusion of rice culture in southern Asia. This starting point leaves open the question of what to do with the Holocene: keep it for 5,000 years, or merge it within the previous epoch (the Pleistocene)?

The Orbis Spike (1610). The marker is a noticeable dip in atmospheric CO_2 likely due to the severe depopulation of the Americas caused by the "Columbian Exchange" during which the Europeans colonized the continent, brought in unknown diseases such as measles, smallpox, cholera, and typhus, leading to massive mortality among the indigenous peoples (Lewis and Maslin 2016, 174–175).[9] As a result, vast areas rewilded and the biomass expanded (e.g., woods and forests), capturing and sequestrating large amounts of carbon.

The Industrial Revolution (starting somewhere between the end of the eighteenth century and the mid-nineteenth century). As seen above, the proponents of the reformist view argue that humans have been leaving clear stratigraphic signatures. Such markers are often, but not always, presented as synchronous with the Industrial Revolution (Crutzen 2002; Zalasiewicz *et al.* 2008): the steep rise of GHG emissions and pollution. In addition, climate change is directly caused by the massive recourse to fossil energies initiated at this period.

The Great Acceleration (post 1945). Many debuts are offered between the end of the Second World War and nowadays, usually in relation to nuclear activities. For instance, the Working Group on the Anthropocene identifies "the artificial radionuclides spread worldwide by the thermonuclear bomb tests from the early 1950s."[10] The starting date of 1964 is sometimes proposed since it corresponds to the peak of the radionuclides' fallout (^{14}C, backed by other isotope markers such as $^{240}Pu/^{239}Pu$) from atomic weapons testing (Lewis and Maslin 2016, 176).

From the perspective of CE, the beginning of the Anthropocene is worth mentioning for two reasons. First, it deepens the picture in which climate alteration appears less as a stark rupture than the continuation of the anthropic influence that cannot be circumscribed to modern times (Ellis *et al.* 2021), which does not imply that the principle of climate intervention is acceptable. From that point of view, CE could be seen as a by-product of human expansion on the surface of Earth.[11] Second, when framed in terms of rectification or regeneration, such as taking responsibility for, remedying to, canceling out anthropogenic change, or restoring nature, justifications for CE require a reference point.[12] This raises the following questions: *which nature at what period should be restored?* Or, more broadly, *how much to engineer the climate* to return to this reference point (Keith 2013, 53–54)?

The implications may go beyond CE for including *biological* interventions such as rewilding or genetic retro-engineering of vegetal and animal species. For instance, the choice of the megafauna extinction (50,000–10,000 years BP) as a baseline could imply that humans are responsible for restoring part of those distant climate conditions as well as the species that disappeared (e.g., large mammals like mammoths). Retaining the Neolithic revolution or the early anthropogenic hypothesis could mean that, in addition to atmospheric parameters, humanity bears some responsibility for animal and plant rewilding or (de-)domestication. Should original species be revived, and under which arrangement? According to the same line of thought, should animal domestication be phased out for biological and climate restoration reasons?

The discussion stimulated by the Anthropocene is not limited to GHG emissions and their impacts on human societies. Engaging into such a discussion presupposes clarifying the contours of the morally relevant community and the inclusion of part of the living realm. Such a community may encompass entities, for which it has been demonstrated that they have legitimate claims upon humans. Thus, humanity may be considered responsible not only for tackling past and current harms to those entities, but also for preventing future damages. In the case of climate change, a strong intuition is that the morally relevant community must include the animals, plants, and ecosystems that are affected by GHG emissions. If supported, this intuition implies that evaluating the Anthropocene raises issues of backward-looking and forward-looking responsibility toward various creatures and environments.

Then, the reasons why humans may be responsible for the well-being or, more generally, conditions of nonhuman entities need to be elucidated. Are they accountable for those entities because (and to the extent to which) the latter have an instrumental value (e.g., economic services provided by ecosystems), or because they possess an intrinsic value or inherent worth?[13] According to the former, nonhuman living beings and ecosystems would simply constitute moral proxies for duties toward other humans, present and future. In accord with the latter, nonhuman beings and ecosystems per se would be the beneficiaries of anthropic obligations. Without delving deeper into those questions for the moment, it is enough to notice that discussing the Anthropocene is to venture into environmental ethics, defined as

> the moral relations that hold between humans and the natural world. The ethi-
> cal principles governing those relations determine our duties, obligations, and
> responsibilities with regard to the Earth's natural environment and all the ani-
> mals and plants that inhabit it. (Taylor 1986, 71–76)

MITIGATION, ADAPTATION, AND CLIMATE ENGINEERING

The Anthropocene highlights the intricate question of past, present, and future anthropic responsibility, especially in the case of significant harms to humans, other living beings, and ecosystems. Beyond the allocation of liability, it raises the practical issues of what should be done, by whom and how, which is the realm of climate policy understood as the *strategies and initiatives crafted by various institutions for anticipating, preparing for, and tackling climate issues.*

International climate texts ranging from the UNFCCC (1992) to the Paris Agreement (2015) rest on a number of core principles that are directly relevant for climate alteration. The first is the prevention, or minimization, of catastrophic harms, which is described by the UNFCCC as "dangerous anthropogenic interference with the climate system" (article 2). Such injunction should be fulfilled while respecting other principles, such as the precautionary principle, that is, "[w]here there are threats of serious or irreversible damage, lack of full scientific certainty shall not be used as a reason for postponing cost-effective measures to prevent environmental degradation" (UNFCCC article 3 paragraph 3).[14] Another principle is the right to sustainable development (UNFCCC article 3 paragraph 4), which includes poverty alleviation (Callies and Moellendorf 2021). Moreover, countries share this anthropogenic, collective climate responsibility to varying degrees as captured by the UNFCCC reference to "common but differentiated responsibilities" (UNFCCC article 3 paragraph 1). First, those responsibilities reflect countries' divergences in past emissions. Second, they are further modulated according to countries' capabilities. Finally, countries should enhance GHG sinks and reservoirs (UNFCCC article 3 paragraph 3; article 4 paragraph 1d), which show that climate policy is not simply an issue of curbing emissions, but also of improving the efficiency of the carbon cycle too.

TEXTBOX 0.3 CLIMATE SCIENCE AND POLICY

Robust science is a prerequisite to solid climate policy. Decision-makers need evidence, data, facts, models, and projections to determine the vulnerabilities and future risks to which populations are, and will be, exposed. Controversies have been raging about assumptions, methods, and results mobilized by climate science. Many have no scientific basis and are simply fueled by energy and industry lobbies (Mann 2012, 2021; Oreskes and Conway 2011). Overall, a few key elements are widely agreed upon among researchers. First, GHG concentrations have been climbing and the climate has been rapidly warming for the past 150 years. Second, the largest portion of such changes cannot be explained by natural factors such as natural cycles, solar activity, Earth's eccentricity, obliquity, precession, and so forth. Moreover, they fall outside the range of natural variability over such a short period of time (a couple of centuries). Third, considering the sheer volume of GHGs released by human beings since the eighteenth century, it can be safely claimed that the cause of rising GHGs concentration and climate change is overwhelmingly anthropogenic.

Despite this quasi-consensus, climate science remains challenging for policymakers owing to the prominence of uncertainty. In that respect, a distinction should be drawn between backward-looking and forward-looking scientific enquiries. The former focuses on past evolutions with the purpose of descriptive accuracy and robust explanations of past dynamics. This was traditionally the scope of climatology with the investigation for instance of the last ice age (Heymann and Achermann 2018). The latter is probabilistic by nature. It aims to support plausible projections and inferences, that is, building the predictive force of models. It is challenging because it implies forecasts about highly complex systems with irreducible (deep) uncertainty.[15] Such uncertainty stems from the nonlinearity of and singularities in climate responses (e.g., the famous "tipping points"). More importantly, they result from the future GHG emission pathways, which depend on present and forthcoming technological and political decisions. Notwithstanding such challenges as well as the limitations of evidence-based policy in terms of moral neutrality (Finkel 2018), public policymaking nonetheless needs science, with as little intrusion as possible from particular interests (e.g., corporations and political forces). The imperative is even more pressing that humanity is potentially dealing with global interference in the Earth system, i.e. CE, the effects of which will be compounded by the Anthropocene.

The two major climate strategies are mitigation and adaptation. *Mitigation* is defined by the IPCC as the "human intervention to reduce the sources or enhance the sinks of greenhouse gases" (IPCC 2015, 125), which refers to energy conservation and efficiency, along with decarbonization (Keith 2013, xix–xx). The intent is to slow, preferably halt or reverse, dangerous changes. Examples cover conventional carbon abatement (cut) as well as the development of fossil-free energies. Mitigation is the crux of policy, especially since past GHG emissions may have already committed Earth to hazardous climates and tipping points (Lenton 2019).[16] As a result, it requires more than just lowering or stopping GHG emissions and includes "net negative emissions" (the active extraction of GHGs from the atmosphere), the reason why CDR is often presented as a mitigation measure. The NETs[17] are essential, some would claim necessary (Callies and Moellendorf 2021; EASAC 2018), for remaining within the 2°C target of global average temperature increase by comparison to preindustrial times while guaranteeing the right to self-development, as stipulated in the Paris Agreement (article 2).

Adaptation is "the process of adjustment to actual or expected climate and its effects" (IPCC 2015, 118), which features, for example, building seawalls, relocating coastal and insular populations, or designing new crops resistant to heat, droughts, or exotic pests. If mitigation tackles the origin of climate change, adaptation deals with its consequences. SRM techniques can, controversially, be classified as adaptive, since enhancing Earth albedo softens global warming without addressing its cause. Due to the failure of international mitigation initiatives deployed since the 1992 Rio Earth Summit, climate policy is progressively leaning toward adaptation. This does, and should, not mean that mitigation has become irrelevant. The rise of adaptation as a strategy illustrates the difficulties experienced by states, private organizations, and citizens for coordinating in order to significantly abate carbon. Moreover, the now prominent role played by adaptation, after having been viewed as a distraction from mitigation, manifests doubts about the possibility of reversing such inertia and overcoming cooperation issues.

The barriers to robust mitigation mainly stem from international collective action problems (Morrow 2020b, 33). Collective action problems, or social dilemmas, designate situations where entities fail to collaborate owing to the pursuit of a short-term strategy of maximizing parochial interests at the expense of long-term, more sustainable, collective ones.[18] An emblematic stylized formulation is Garrett Hardin's Tragedy of the Commons (G. Hardin 1968) that describes a hypothetical case in which herders use shared pasture for cattle grazing. Each herder has an incentive to put as many animals as possible on those fields for optimizing his income. However, since every herder faces the same incentive structure, and has no interest to exert restraint if the others do not, overgrazing appears as the rational strategy, leading to the decline of the common resources, soil erosion, and diminished land carrying capacity. Therefore, by indulging in overgrazing each herder cashes in immediate and small benefits instead of longer and more sustainable ones.

Similarly, dangerous climate change stems from an international tragedy of the commons (Callies 2019, 3). Effective mitigation requires that the key GHG emitters, that is, the most advanced countries, the major emerging markets (e.g., China and India), and the main private polluters (e.g., energy, cement, and transportation companies), cooperate for drastically reducing their emissions. However, emissions continue to pile up in the atmosphere due to the massive short-term benefits of still relying on cheap fossil energies (natural gas, coal and oil) for profits, socioeconomic development, and poverty alleviation. Put differently, each entity (multinational corporations and states) is interested in letting other agents take care of the problem.[19] The influence of national and private interests is compounded by international disagreements on the attribution of responsibility for past emissions, the distribution of adaptation costs, and technological transfers for fossil-free alternatives. The main obstacle is the "fragmentation of agency" (Gardiner 2011, 24) at the international level, which impedes cooperation by generating a prisoner's dilemma, that is, a situation in which cooperation is difficult even in the presence of mutually advantageous course of action.[20]

In complement to mitigation and adaptation, CE is often presented as a third type of climate policy (e.g., Incropera 2016; Long 2017), which is contested (Heyward 2013). Succinctly put and before elaborating further in the next chapter, CE designates engineering processes applied to the climate motivated by the goal of altering radiative forcing through carbon sinks stimulation and albedo enhancement. Despite being defended (and opposed) for various reasons (cf. chapter 5), the research into and deployment of CDR and SRM appear as legitimate instruments because of the pessimistic view that mitigation will be insufficient and too late. Thus, CE will be a necessary supplement to adaptation. No serious researcher advocates for CE as the sole component of climate policy. CDR and SRM are almost always considered as elements of a hybrid or portfolio policy (e.g., Keith 2013, xix; NRC 2015a, 2015b; Svoboda 2017).

RATIONALE AND STRUCTURE OF THE BOOK

This book aims at different audiences. First, the readers who are not familiar with climate alteration can discover what those methods are, how advanced they are, and which promises and challenges they carry. The book offers an accessible synthesis of the major techniques and their state of advancement. Yet, the objective is not only descriptive. It is also to recap some of the political and ethical problems triggered by interfering with Earth's radiative forcing. This normative component may be appealing to a second audience who, while acquainted with CE, is willing to explore the large array of issues related to research, development, and deployment. More broadly, the goal is to build bridges between natural sciences and engineering, on the one hand, and social sciences, humanities, and public policy on the other hand. This cross-disciplinary strategy explains why many forthcoming presentations and analyses may appear to some climate specialists or trained ethicists as lacking some depth. This shortcoming is the consequence of an editorial choice. This book has been conceived for remaining accessible to an as-large-as-possible audience, with a particular attention to themes of direct relevance for public policy. The result is a monograph that recaps and synthetizes the state of the knowledge on climate alteration.

In sum, three major motivations actuate this book: (a) to offer a comprehensible and methodical *tour d'horizon* of the available knowledge on different CE methods, (b) to review and structure the main moral and political problems based on actual and updated scientific inputs, and (c) to diffuse information that is essential for democratic debates and decision-making. Such ambition justifies a modest dose of repetition. The five chapters have been written to remain, to some extent, independent from each other. Readers who are only interested in some aspects of climate alteration could directly go to the appropriate part. Such a choice faces limitations, with no chapter being completely isolated from the rest. However, they can be read separately. The structure of the book is as follows:

Chapter 1 is mostly conceptual. The task is to iron out a characterization of CE that is as robust as possible. For doing so, The Royal Society's canonical definition—"deliberate large-scale manipulation of the planetary environment to counteract anthropogenic climate change" (Shepherd *et al.* 2009, 1)—is exposed and challenged. This widely accepted definition appears in need of clarification. The chapter closes by setting CE apart from traditional climate policy, more specifically by differentiating between CDR and mitigation as well as between SRM and adaptation. These pages could be helpful for readers who want to understand better what CE *qua* concept is, how it is defined, with which consequences, and what are the main traits of the techniques assembled under this label.

Chapter 2 ventures further into moral and political analyses. The goal is to iron out the evaluative criteria to be applied to CE methods. The task could be summed up by a simple question: how to decide whether to initiate a given CE research or deployment project? The chapter collects the criteria that could be found in the literature and groups them into three categories: feasibility, permissibility, and preferability. The first refers to the material possibility of engaging in specific climate alterations. If the first category is predominantly, but not uniquely empirical, the second is normative. It captures two dimensions: that acceptability also matters, especially for public policy, and that it comes in various shades (moral, political, and legal). The third dimension skims the issue that, in a world of scarcity, climate strategies come with opportunity costs, that is the costs of renouncing to some alternative options induced by a given action. Thus, choices should be made and options need to be compared. This chapter is addressed to readers who are already somewhat familiar with CE methods and whose interest lies in embarking upon extended assessments or deciding for themselves whether some technique is worth pursuing.

The third and fourth chapters elaborate on CE techniques. They are made for readers who want to deepen their knowledge on the various methods of climate alteration while going beyond the misleading uniformity imposed by the CE label. Again, those chapters have been written to be to some extent self-contained. Chapter 3 presents CDR methods. It shows how they are making their way into climate models and scenarios that respect the Paris Agreement (2015) targets of 2°C, preferably 1.5°C, global warming in comparison to preindustrial times. A taxonomy is offered, based on two overarching categories: natural sink enhancement and fully engineered sinks. Each method is examined separately, presenting how it works, what the potential for CCS is, as well as the promises and challenges. The last section concludes with a general assessment of CDR following the evaluative criteria spelled out in chapter 2.

Chapter 4 mirrors this structure and intent for SRM. The techniques are divided into four categories, depending on the level of intervention: surface, troposphere, stratosphere, and outer space. Each method is reviewed separately. It shows that stratospheric aerosol injection (SAI) holds a prominent place within SRM interventions, and more generally climate alteration. Due to being, apparently, cheap, fast, and efficient, the method attracts most of the attention when it comes to albedo enhancement and CE. An immediate undesirable effect is to distort the discussions about solar CE and climate alteration. The chapter closes on a general assessment of SRM based on chapter 2's criteria.

The fifth and last chapter consists of an overview of the normative landscape, that is, the arguments mobilized at different stages for supporting or opposing climate alteration. The idea is to gather at one place the most notable claims and controversies that surround CE as a policy option. This chapter can be read independently or in conjunction with the other chapters (e.g., 3 and 4), especially because most of the evidence for different claims are exposed in the previous chapters. Chapter 5 opens with the necessity argument that asserts that *some* radiative forcing alteration would be unavoidable because of the insufficient carbon cuts (the so-called mitigation gap). Then, the common claim that research would be distinct from deployment is assessed. The conclusion is that no matter how one judges the whole CE project, it appears that some governance and regulation are already needed. The next section elaborates on the main justifications for CE as stand-alone intervention, part of a portfolio policy, or as an emergency solution. The ultimate lengthy part structures into categories the main objections that have been addressed to CE. Chapter 5 has been crafted for readers who are more inclined toward normative reflections within a public policy framework. However, it remains introductory and, as such, sets up the stage for future, more refined, moral and political analyses.

NOTES

1. "Climate *forcings* are external factors that influence climate" (Keith 2013, 47), to which the Earth reacts through climate *responses*. GHGs and incoming solar radiation contribute to radiative forcing while global warming constitutes a climate response.

2. Although CE proponents often underscore the insufficient carbon cuts (e.g., Crutzen 2006; Frumhoff and Stephens 2018), the majority asserts that mitigation should remain the first best, CE being suppletive in the short run for SRM (e.g., Keith 2013, 2017) and longer for CDR (cf. chapter 5). The Royal Society's report expresses this widespread agreement:

> the safest and most predictable method of moderating climate change is to take early and effective action to reduce emissions of greenhouse gas. No geoengineering method can provide an easy or readily acceptable alternative solution to the problem of climate change. (Shepherd *et al.* 2009, ix)

One of the most adamant defenders of SAI, Crutzen (2006, 217), indicates that "the very best would be if emissions of the greenhouse gases could be reduced so much that the stratospheric sulfur release would not need to take place."

3. Various living organisms durably changed the climate in the past, for example, during the Great Oxygenation Event caused by cyanobacteria ca. 2 billion years ago. The claim that CE would constitute an unprecedented interference overinflates the importance of the industrial civilization and humanity in general.

4. A lengthier discussion of the exceptionalism claim can be found in chapter 1.

5. This position could receive two interpretations. The first is to consider that the Anthropocene is not a geological concept at all, missing the defining features for being included into the Geologic Timeline. The second is that if the Anthropocene could eventually qualify as such a concept, it would still not be fit for characterizing an epoch, but a more transient stage such as a geological age (Santana 2019).

6. For a more exhaustive discussion, the reader can refer to Lewis and Maslin (2018, 45–78).

7. The idea that the Anthropocene conveys a collective, indiscriminate, responsibility borne by all human beings, equally, is controversial. Critics point out that most of the anthropogenic emissions result from the conjunction of a specific economic model, capitalism, and European colonialism (Bonneuil and Fressoz 2016; Davis *et al.* 2019; Moore 2017).

8. CE can of course be defended as a means, not to rectify climate change, but to fend off climate catastrophes. This claim is discussed at length in this book.

9. The "Columbian Exchange" (Crosby 1972) was not only about slavery, killing, exploitation of natural resources (e.g., silver and gold), and imported germs, but the exchange of goods, artifacts, and ideas between the two sides of the Atlantic Ocean (Nunn and Quian 2010).

10. http://quaternary.stratigraphy.org/working-groups/anthropocene/

11. The particularity of industrialized economies does not lie in releasing carbon *per se*. Previous civilizations were using biomass for heating and deforesting for agricultural purposes, leading to significant emissions. The particularity resides in the scale of emissions as well as of the sources from which fossil fuels are extracted. Nowadays, "dead" carbon is freed from stable geological sequestration while traditional societies mostly burnt biological, "living," carbon that was part of the carbon cycle.

12. Conservationist or preservation is not the only justification for CE. Technological prowess, the project of dominating or regulating the climate, could also motivate calls to climate alteration (Hulme 2014, 29) as exposed in chapters 2 and 5.

13. Intrinsic value, inherent value, and inherent worth could be considered as distinct concepts (Taylor 1986, 71–76), a point that is not debated in this book.

14. The precautionary principle can receive various characterizations, many being unsatisfactory, internally inconsistent, mutually contradictory, and so on (Sunstein 2005; Steel 2015).

15. In this book, uncertainty and deep uncertainty are treated as equivalent, that is, referring to the same phenomenon of adverse event which are unknown or for which probabilities cannot be calculated.

16. A tipping point is a threshold beyond which "a small amount of extra climate forcing, usually linked to global warming—for example, greenhouse gas forcing— triggers a qualitative change in part of the climate system" (Lenton 2021, 325). Such change is assumed to be globally detrimental, even potentially catastrophic, for humanity and ecosystems. Those points could be triggered by events like the melting of important ice bodies (e.g., West Antarctica and Greenland ice sheets), permafrost thawing, the alteration of the Atlantic Meridional Overturning Circulation, and massive diebacks of boreal and tropical forests.

17. Methods that aim at capturing GHGs and storing them away from the atmosphere (EASAC 2018, 3). The US NRC defines them as "technologies that take CO_2 out of the atmosphere and put it back into geologic reservoirs and terrestrial ecosystems" (NRC 2019, xi). The NETs are closely related, if not assimilable, to CDR methods (Carton *et al.* 2020).

18.   A general definition of a social dilemma is a situation in which a group of N-persons ($N \geq 2$) must choose between maximizing selfish interests and maximizing collective interests. It is generally more profitable for each person to maximize selfish interests, but if all choose to maximize selfish interests, all are worse off than if all choose to maximize collective interests. (Komorita and Parks 1996, 8)

19. Darrel Moellendorf (2015) claims that the difficulties in coordinating might stem from a misperception of the incentives structure. Due to the steep decrease in the prices of renewable energies and the significant short-term costs of burning fossil fuels (e.g., in terms of pollution and public health issues), immediate and robust carbon abatement may carry greater co-benefits than assumed. The net balance of an ambitious climate policy may be positive, undermining the view that it has *necessarily to be* entangled into a collective action problem.

20.   The Prisoner's Dilemma gets its name from the story of two prisoners who are separately interrogated. Naturally, there is insufficient evidence to convict them of the crime that the police suspect them of having committed. Unless they confess, the worst conviction they risk is for illegal possession of firearms, for which they would each be sentenced to a year in jail. But the police and the prosecuting attorney are devious, and they offer each prisoner the following deal: you can turn state's witness to help us put your partner away for ten years, and we'll let you off free. The only hitch is that, if both of you confess, we'll convict both of you of armed robbery and ask the judge for a lenient sentence of only six years for each of you. To confess or not to confess—that is your dilemma. If you are narrowly self-interested, you are better off confessing no matter what your partner does. Since you both must see the issue this way, you may both spend six years in jail. But if you could act together as a group with a single mind, you would act in the group's interest and hold out so that you both would spend only a year in jail. If, however, you reason from the fallacy of composition while your partner acts from self-interest, you will rest ten years in jail. It would be a painful lesson in logic. (R. Hardin 1982, 2)

Chapter 1

What Is Climate Engineering?

Human attempts at influencing the weather abound in history, ranging from magical rituals to controlled forest fires and cloud seeding (Fleming 2007, 2010). However, the ambition has always been to affect the local meteorological conditions for specific purposes. For instance, Charles Hatfield, a famous "rainmaker," was hired in 1915 by the San Diego Wide Awake Improvement Club for replenishing the Morena Dam Reservoir (Fleming 2010, 38–40). During the Vietnam War, the US Air Force seeded the clouds to disrupt the Ho Chi Minh trail in Cambodia (operation Popeye).[1] Climate engineering (CE) promises to ramp up the interference with the climate in a more enduring manner at the regional or global scale.

As seen in the introduction, geoengineering (as well as CE) is an umbrella term. It originates in a 1977 article by the Italian physicist Cesare Marchetti titled "On Geoengineering and the CO_2 Problem" (Keith 2000, 247; Morton 2015, 137). However, the idea of modifying the climate has been around for longer, around the time of the discovery of the greenhouse effect. At the opening of the twentieth century, scientists like Arrhenius and Callendar contemplated releasing enough CO_2 to avert the next glacial period(s) (Wilby 2017, 30). After the Second World War, proposals for altering the climate and weather proliferated (Fleming 2010; Keith 2000; Schneider 1996). The 1965 report to US President Johnson, *Restoring the Quality of Our Environment*, suggested albedo enhancement for counterbalancing higher CO_2 concentrations. In his works, Joseph Fletcher (1968, 1969) envisioned the possibility of influencing global temperatures, ocean currents, and tackling climate change through ambitious, but elusive, interventions. Early CE discussions often participated in or were attached to the Promethean project of modifying the Earth according to human needs. For instance, in the United States, the Lawrence Livermore National Laboratory, which played a key role in the 1990s for advancing the idea of dimming solar radiation, for example, through sun scatterers (Teller *et al.* 1997), was also involved after 1945 in controversial proposals, such as using nuclear explosions for civil engineering purposes (Project Plowshare). The fact that Livermore's scientists like Lowell Wood and Edward Teller have been pivotal for the development of CE research (Hamilton 2014; Morton 2015, 148–151) reinforces the hubristic undertone oft decried by CE opponents.

The United States was not alone in striving to modify the climate and weather. The Soviet Union also aspired to model the world and the climate (Keith 2000, 250–251), with not less than nineteen research institutions publishing on the topic during the Cold War (Schneider 1996, 291–292). CE was part of a broader enterprise of dominating natural forces, initiated by Joseph Stalin in 1948 (Fleming 2010, 197–198). Soviet climatologist Mikhail Budyko stressed the inevitability of human interference for preserving the conditions for life on Earth (Budyko 1974, 493). Thirty years before Crutzen's landmark article, Budyko was considering injecting particles in the stratosphere through aircraft (Budyko 1977, 239–245).

Thus, the project of modifying the climate as well as more specific interventions such as albedo enhancement have been advanced for at least half a century (Jamieson 1996; Keith 2000). CE cannot be reduced to the fantasy of some isolated scientists. The very idea originates from highly technological societies convinced that nature can be modeled for the greatest good. To a certain extent, those proposals embody the faith in human reason and the capacity, or duty, to control nature, which is judged as hubristic (Hamilton 2013a; Jamieson 2013). Put differently, CE could be seen as the self-conscious manifestation of the Anthropocene.

In 2009, the Royal Society offered what has since become the canonical definition for geoengineering: "deliberate large-scale manipulation of the planetary environment to counteract anthropogenic climate change" (Shepherd *et al.* 2009, 1). Variations populate the literature, from Clive Hamilton (2013, 1) "deliberate, large-scale intervention in the climate system design to counter global warming or offset some of its effects" to David Keith (2013, 48) "intentional manipulation of climate forcings with the goal of counteracting undesirable climate change." According to Oliver Morton (2015, 26), geoengineering consists of "any deliberate technological intervention in the earth system on a global scale, not just those aimed at countering, or ameliorating, the changes that people are making to the climate without deliberation."

Those definitions present CE as possessing three attributes: deliberate, large scale, and aiming at counteracting climate change (Keith 2000, 247)—each will be scrutinized over the next three sections. It will be shown that while descriptively accurate, the intentional dimension carries mixed moral implications. Moreover, it tends to draw too of a sharp dividing line between the ongoing Anthropocene period and possible climate alteration. Next, the large-scale trait will be discussed. Defining CE as necessarily large scale, especially if it is understood as global, could be misleading, as many interventions may also be localized or targeted, for instance on the Arctic region or specific cities. The third characteristic, which encapsulates the aim of CE—"counteract[ing] anthropogenic climate change"—is shown to be, again, too reductive, since presumably climate modification could be, theoretically at least, pursued for various purposes, for example, as a weapon against antagonistic countries (even if such a perspective appears unlikely as shown in chapter 5). The last two sections of the chapter distinguish, on the one hand, carbon dioxide removal (CDR) from mitigation and, on the other hand, solar radiation management (SRM) from adaptation.

DELIBERATE

Within a definition of CE, the deliberate attribute performs a triple function. First, it insists on the role played by intentions in any engineering project. Since CE stems from the human willingness to modify the climate, it cannot be interpreted as accidental or mere side effects of other activities (e.g., economic development). The intentional aspect is particularly salient when CE is contrasted with the climate change conveyed by the Anthropocene (Morrow 2014). This highlights the second function: to draw a sharp line between two forms or moments of anthropogenic climate change. On the one hand, there is the Anthropocene, in which greenhouse gas (GHG) emissions are unintended consequences of socioeconomic development. On the other hand, CE constitutes a new, qualitatively different, stage in which the climate change will be deliberate. The third function, moral, is to point out human responsibility regarding the possible results of CE. Intentions play a significant role in attributing and modulating responsibility for adverse outcomes (Jamieson 2013, 534–535). Harm brought willfully is usually judged as less excusable or justifiable than as harm caused inadvertently. Killing due to negligence (e.g., inattentive driving) is considered less wrong than assassinating someone. Criminal law in many countries reflects this difference by distinguishing manslaughter from murder.

It is important to focus on the first and third functions. By definition, any engineering project is intentional, therefore CE also should be. As stressed by Jane Long (2017, 81) "geoengineering is fundamentally *engineering*—that is, a solution designed to solve a problem." In other words, it represents a *purposeful* activity, which consists of carefully weighing goals, means, and possible outcomes. CE requires humans to act rationally in an instrumental manner (Kolodny and Brunero 2023), that is, to consciously select means for achieving specific ends, namely taming climate change or its consequences, or to recycle United Nations Framework Convention on Climate Change (UNFCCC) phraseology, "to prevent dangerous anthropogenic interference with the climate system." This self-reflective, deliberate, aspect is underscored by the literature, especially when it comes to distinguishing CE interventions from past emissions.

> [T]he notion of deliberation matters; to the earth system, a change made in passing may be no different to one made on purpose, but in the human world there is a difference between the changes you make and those that you plan, between having an effect on the future that you can foresee and having an effect that you intend. (Morton 2015, 26)

If intentions appear descriptively apt, they also carry moral implications. *Ceteris paribus*, for qualifying as permissible, climate interventions should not follow any sort of motives, but only good, just, and fair, that is, morally acceptable ones. CE ought to be led by an authentic will of doing good, contributing to a fairer state of affairs, or at least reducing harm (which introduces consequentialist elements). Such evaluation is primarily *deontological*,[2] namely it focuses on the rules, principles, and intentions that preside over agents' actions.

More generally, intentions fulfill a pivotal role in many definitions of (good) engineering, for example, "the profession in which knowledge of the mathematical and natural sciences, gained by study, experience, and practice, is applied with judgment to develop ways to use, economically, the materials and forces of nature *for the benefit of mankind*" (Oakes and Leone 2018, 2; emphasis added). Such benefit is the centerpiece of *morally acceptable* CE. Harmful or even criminal engineering is always a possibility, like the Nazi-orchestrated extermination of Jewish people with the complicity of the German industry. The reference to benevolent intentions voices more normative constraints than descriptive criteria. In other words, for being justifiable, engineering projects *must* deliberately pursue collective benefits.[3]

If the presence of intentions applies to any engineering enterprise, good intentions only characterize ethically justified engineering projects. One basic requirement of acceptability is to be guided by benevolent intents, from the research and development (R&D) stage,[4] as *per* the deontological evaluation. Yet, good intentions, even from a deontological perspective, may not be enough for securing permissibility. The maximization of collective benefits (public goods) and/or minimization of collective harms (public bads) are also necessary. At least, and under the risks and uncertainty that characterize climate interventions, any engineering enterprise ought to offer reasonable probabilities of delivering a modicum of positive impacts by lessening average global temperatures or GHG concentrations. Any project, even conducted with laudable motives, but which makes the life of people worse in comparison to the default position (no action) or equally feasible alternative, cannot be justified.[5]

At first blush, mainstream CE proposals appear driven by benevolent intentions, in line with the United Nations Framework Convention on Climate Change (UNFCC) injunction to "prevent dangerous anthropogenic interference." They offer to reverse catastrophic climate change through carbon extraction and sequestration or to shield populations against global warming by enhancing the albedo. However, permissibility assessments must go beyond screening intentions by including other morally relevant dimensions, such as potentially disastrous consequences, for example, termination shock[6] from stratospheric aerosol injection (SAI), or deadly leaks from carbon reservoirs. Special attention also needs to be paid to deep uncertainty, that is, situations in which "probability distributions are unknown" (Frisch 2018, 438).[7] Risks and (deep) uncertainty are the parameters by which CE, or some methods, is often deemed as unacceptable.

Intentions also play a crucial role in distinguishing CE from other forms of anthropogenic activities with noticeable climate implications, such as burning fossil fuels, clearing forests, or agricultural practices. They are often used to contrast the "accidental Anthropocene" (Lewis and Maslin 2018, 25) with "intentional climate change" (Jamieson 1996).[8] In the same vein, Mike Hulme considers that there is "some value in retaining a distinction between inadvertent and intentional experimentation" (Hulme 2014, 109). Such value is not only descriptive, but moral. As the argument goes, by burning fossil fuels, past generations did not search to alter the climate. They did not wish future humans to be exposed to catastrophes. Besides being descriptively apt, intentionality carries ethical implications in respect to blaming. For instance, although past generations are materially responsible for the current concentrations of GHGs, their moral responsibility may be lessened by what Simon Caney labels as "excusable ignorance" (Caney 2015, 380). The lack of knowledge about the long-term effects of GHG emissions would rule out the fact that the Anthropocene was deliberately engendered and, thus, lower responsibility, especially when compared with potential CE projects.

Among CE critics and partisans, the emphasis on intentions supports the *claim of exceptionalism*, defined by Clare Heyward (2015, 137) as asserting that "the development of geoengineering will lead to unprecedented situations, either good or bad." CE would constitute a drastic rupture in the history of anthropogenic climate change, for the better or the worse, and such a rupture would have been initiated by voluntary large-scale interventions. There would be a clear-cut dividing line *before* and *after* CE, even when taking into consideration the fact that humans have modified the climate and their environment for millennia (cf. introduction). "This ability to influence the climate represents a major paradigm shift, arguably on the order of the Copernican Revolution" (Donner 2007, 233).

The exceptionalism claim is grounded in contrasting an "accidental Anthropocene" with "intentional climate change." While roughly accurate, such a distinction nurtures too simplistic a view. It neglects that human beings have been conscious of their influence on the climate for more than a century. In 1856, Eunice Foote described the basic mechanism of the greenhouse effect (Foote 1856). In 1896, Svante Arrhenius advanced that carbon emissions could lead to surging global temperatures (Arrhenius 1896). In 1938, Guy Callendar contemplated the possibility that anthropogenic emissions would be enough for postponing the next glacial period(s) (Callendar 1938). In the following decades, this knowledge was deepened and spread beyond scientific circles. In 1965, a report to the US President Lyndon Johnson, *Restoring the Quality of Our Environment,* identified CO_2 as a "pollutant" that could cause the Antarctic ice cap to melt, sea levels to rise, and (fresh) water to become more acidic.[9] Ten years later, Manabe and Wetherald (1975) predicted that a doubling of CO_2 would increase the average global temperature by 2.93°C (Wilby 2017, 41), which is characterized as the "climate sensitivity."[10] Their study came after nearly three decades of scientific debates on the topic.

As those examples show, scientists, decision-makers, and a part of the public knew from a quite early stage of the industrialization that the release of massive amounts of carbon was akin to a planetary experiment of climate manipulation. Moreover, many warmly welcomed such changes. Debates have been raging for more than a century about using carbon emissions for securing a more hospitable planet. In consequence, how strong is the claim that the advent of CE would radically transform anthropogenic interference, especially from a moral point of view, and represent a paradigmatic shift? For sure, any climate alteration project would increase human responsibility and represent a climate intervention that *could* turn out to be unprecedented in scale (depending on the CE methods). But could such accrued responsibility in regard to CE consequences be judged as something radically new, some kind of a paradigmatic shift?

LARGE SCALE

The Royal Society's definition presents CE as being about "*large-scale* manipulation of the planetary environment" (emphasis added). Large scale could apply to two dimensions—the deployment and the impacts—which do not neatly overlap. The scale of deployment corresponds to the area and resources mobilized (land, oceans, and atmosphere). For instance, effective forest management (reforestation, afforestation, and deforestation halting) will require extensive areas of land. In contrast, technological innovation might lead to extremely efficient direct air capture with sequestration/storage (DACS) with a minimal land footprint. So, the deployment may remain limited in extension, but the effects could be planetary. Therefore, large scale could also refer to the scope of the impacts, like in the case of high-performance DACS facilities.

In the discussions on CE, SAI constitutes the archetype of the large-scale methods since the objective is to lower average global temperatures, nothing less. The technology is often interpreted as being truly global with aircraft spraying particles all over the world. Furthermore, the challenges SAI generates from the perspective of governance and justice and risk and uncertainty, with the creation of novel climates (McLaren 2018), contribute to the perception of CE as being by definition large scale, in the planetary sense.[11] However, this perception is partly misleading or at least reductive. By definition and as a matter of fact, large-scale methods are not forcibly global in the sense of impacting the whole Earth.[12] They could foster regional effects, which could be judged as being large scale, but not global.[13] Presenting CE as necessarily global reinforces concerns over humanity becoming locked into a worldwide experiment (Robock 2008a, 17), threatening a "millennial dependence risk" (National Research Council [NRC] 2015b, 60). Such worries are strong in relation to SRM, especially SAI, as illustrated by the heated debates about termination shock (Halstead 2018; Hamilton 2013a, 65; McKinnon 2019; Parker and Irvine 2018; Preston 2013, 32; Trisos *et al.* 2018).

In addition, if CE *can* be large scale, it does not have to be (Hamilton 2013a). Interventions could be limited, contained, or encapsulated in their deployment and/or impacts (Hulme 2014, 10–11). Smaller deployments are exemplified by the "soft engineering" approach defined by Robert Olson (2012, 30) as a form of local CE, which displays eight traits: (1) it "can be applied locally," (2) it is potentially "scalable to larger areas," (3) it creates "low or no anticipated negative impacts on ecosystems or society," (4) it can be quickly reversed in case of problems, (5) it "has multiple benefits, beyond impacts on climate," (6) it is "analogous to natural processes," (7) its "effects are large enough soon enough to be worthwhile," and (8) it is "cost-effective with mature technologies deployed at moderate scale." In sum, CE could be "localized" (Hamilton 2013a, 20–21) or "targeted" (Moore *et al.* 2021) even though its consequences *could* be vaster.

Soft or targeted CE could be mobilized for the urgent objective of preserving the cryosphere, which is constituted of frozen regions that exert a massive influence on the climate through albedo, thermocline,[14] and so on. A possible intervention is to erect barriers for protecting offshore ice sheets from warming currents, which melt them from underneath. Some of those interventions are not "climatic," in the sense of targeting biochemical processes, but rather mechanical, "geotechnical geoengineering solutions" focused on slowing down glaciers' velocity (Lockley *et al.* 2020, 402).

The *Arctic Ice Project*,[15] formerly known as *Ice911* (Moore *et al.* 2021; Olson 2012), offers to blanket open waters or vulnerable snow and ice cover with light-colored material such as tiny silica balls for enhancing the albedo. Applied to the permafrost layer, this intervention could help keep in check the temperature's surge since the top three meters of frozen soil contain the equivalent of double the carbon currently stored in the atmosphere (Moore *et al.* 2021, 112). The *Refreeze the Arctic Project* from the Center for Climate Repair at the University of Cambridge[16] investigates methods susceptible to increase the extent of the Arctic Sea ice through marine cloud brightening (MCB), that is, the stimulation of low-lying marine clouds, or pumping and spilling seawater on top of the existing sea ice. MCB has also been the subject of an experiment for protecting Australia's coral reef, the *MCB for the Great Barrier Reef* (Low *et al.* 2022). Another example of local CE is cirrus cloud thinning (CCT) over the poles (Mitchell and Finnegan 2009, 6) or to limit SAI deployment to subpolar and polar zones (Smith *et al.* 2022; Walker *et al.* 2021) (cf. chapter 4).

A final illustration is called "regional land radiative management" (Seneviratne *et al.* 2018), the goal of which is to enhance the albedo of cityscapes (urban surface brightening [USB]) and crops ("bio-geoengineering" [Ridgwell *et al.* 2009]). The influence on temperatures and precipitations would be limited to specific areas, with unnoticeable large-scale effects (Davin *et al.* 2014; Singarayer *et al.* 2009). The idea is to raise the reflectance of urban surfaces by brightening them or of plants by selecting shinier varieties or species. The local frequency and intensity of heat waves could be reduced, particularly during summers. The positive impact would be an accrued protection for vulnerable populations within cities, and higher agricultural yields, especially cereals such as wheat or barley in temperate areas (cf. chapter 4). In sum, targeted CE does not necessarily aim at stabilizing the global climate or fully reversing climate change, but at impacting specific areas for reducing the risk of crossing tipping points (Moore *et al.* 2021).

As the forthcoming chapters will demonstrate, soft or targeted CE is not free from moral concerns. Nonetheless, the possibility of more modest interference has methodological implications, namely any evaluation should avoid sweeping generalizations and proceed with cautious distinctions. CE encompasses a broad array of methods, each carrying its specific technical, ethical, or political specificities and challenges. Simplifications about CE being necessarily large scale, or even global, have contributed to polarizing the debate. Critics overly focus on stratospheric sulfur injection (SSI) as representative of the hazards of CE as a whole, or to suggest that SSI shortcomings apply to other methods through something that looks like the guilt by association fallacy.[17] Any solid evaluation nonetheless requires taking seriously the diversity and complexity of the different CE methods, which is the task undertaken in the third and fourth chapters. Of more immediate interest, it clearly appears to be the case that CE refers to actions and initiatives that aim to tackle or address climate change both locally and globally, leaving open interrogations about how "tackle," "address," or any other formulation can or should be understood.

COUNTERING ANTHROPOGENIC CLIMATE CHANGE

This brings the third characteristic of CE, which is teleological, that is, it bears on its finality. The Royal Society posits that CE interventions are devoted to "counteract anthropogenic climate change," which is close to formulations such as Hamilton's ("counter global warming or offset some of its effects") and Keith's ("counteracting undesirable climate change"). Among the definitions already seen, Morton offers the broadest. CE includes more than intentional climate manipulation "aimed at countering, or ameliorating, the changes that people are making to the climate without deliberation" (Morton 2015, 26). Accordingly, any interference, even undertaken for reasons different than countering climate change would qualify as CE. For instance, cloud seeding for facilitating rainfall or techniques for dissipating storms could be counted as CE.

Even if taming climate change is the main motivation for researching, developing, and deploying CE, it is perfectly conceivable that other economic or political rationales might come into play. Although the prospect remains hypothetical, CE methods could theoretically be weaponized (Stephens and Surprise 2020).[18] Excluding such activities from the definition may be too restrictive, especially because the genuine motives behind CE might remain difficult to decipher. Still, the third characteristic might be interpreted as formulating a necessary normative condition. For being considered as a legitimate *CE* initiative, a given intervention should be devoted to address anthropogenic climate change. If it does not, then it qualifies as something else.

Another question begging for clarification is the meaning to be attributed to "counteracting," "countering," or "offsetting" climate change. CE designates initiatives whose objectives are to tackle either the cause (GHGs' emissions and concentrations) or the consequences of climate change (thermal increase and ice thawing). The first category includes methods designed for capturing and storing the GHGs (mostly carbon). The bulk of the second category is made of interventions aiming at reducing local or global temperatures by enhancing the albedo. Nonetheless, addressing the consequences of GHG concentrations could take other forms, such as protecting ice sheets from warm currents, or anchoring them to the adjacent shore or onto the seafloor.

Countering climate change could be designated to either tackle the cause or the consequences. Furthermore, the range of interpretation of what could count as countering is large. As presented above, CE interventions may be tasked with fully reversing climate change, for example, by removing all the carbon emitted since a predefined point in time (which underlines the centrality of the Anthropocene debate), or blanketing the stratosphere with enough sulfate for completely offsetting global warming. The objective could also be more modest, like halving the thermal surge caused by a doubling of CO_2 atmospheric concentration (Irvine *et al.* 2019).

A potential goal could be to capture GHGs in proportions that significantly reduce the risks of "dangerous anthropogenic interference with the climate system." However, determining a threshold beyond which the climate becomes (too) dangerous or raises the likelihood of catastrophes to an unacceptable extent is arduous for several reasons. A major problem is the relative uncertainty and nonlinearity of climate responses (i.e., temperatures and adverse events) to soaring radiative forcing (i.e., GHG concentrations). A widely agreed on but imperfect frontier is a 2°C rise by 2100. The Paris Agreement stipulates in its article 2, paragraph 1 (a), that the objective is of "holding the increase in the global average temperature to well below 2°C above pre-industrial levels and pursuing efforts to limit the temperature increase to 1.5°C above pre-industrial levels." The accrued risk of crossing tipping points (e.g., collapse of Western Antarctica ice cap, the accelerated meltdown of Greenland glaciers, and permafrost thawing), beyond which positive feedback loops are anticipated, rendering the climate much more unpredictable and dangerous, justifies using 2°C as a policy goal (Lenton 2021; Lenton *et al.* 2019).

In sum, the characterization of CE as the "deliberate large-scale manipulation of the planetary environment to counteract anthropogenic climate change" suffers from shortcomings. First, if the intentional aspect is hard to dispute, it does not draw a clear-cut divide with the Anthropocene. Second, presenting CE's scope as large scale, potentially global, seems to exclude local projects, for example, those dedicated to preserve parts of the cryosphere or microclimates. Third, the qualification of CE as countering climate change is too general, leaving out of the picture other possible motives for interference while not delineating precisely enough the different objectives. *In fine*, CE does not intend to manipulate the whole planetary climate, but variables linked to radiative forcing such as carbon concentration and the albedo.

In conclusion, two alternative definitions could be offered. A wide characterization stipulates that *CE interventions in the climate system aim at significantly influencing the radiative forcing locally or globally in order to gauge specific climate responses (e.g., thermic or precipitation variation, ocean circulation changes).* This characterization accommodates methods initiated for different motives than curbing climate change, such as weaponization and increased agricultural productivity. The rest of the book will be grounded on a more restrictive definition that closely tracks how CE methods are presented and debated in the literature, that is, as techniques directly addressing anthropogenic climate change. Despite this restriction, it does not imply that CE methods cannot be mobilized for other purposes. It simply reflects the common usage according to which CE is worth considering only as far as it promises to tame dangerous climate change. Thus, the narrow definition posits that *CE characterizes anthropogenic interventions in the climate system that predominantly aim at altering radiative forcing locally or globally to slow down, stop, or reverse climate change and/or related adverse events.*

IS CLIMATE ENGINEERING REDUCIBLE TO CONVENTIONAL CLIMATE POLICY?

Before proceeding to the criteria for evaluating CE proposals and a more detailed description of CDR and SRM methods, an issue is pending. Could CE be subsumed under the traditional categories of climate policy? More precisely, does CDR constitute a form of mitigation and SRM an instance of adaptation (Möller 2021, 31)? In the affirmative, CE would simply offer variants of existing policies, perhaps more technologically loaded or ambitious in some respect, but not intrinsically different from conventional mitigation (e.g., carbon abatement and energy efficiency) in the case of CDR, or adaptation measures (e.g., building seawalls and switching to drought-resistant crops) in the case of SRM. As this section shows, assimilating CE interventions to traditional climate policy is problematic. There are solid arguments to keep the four categories separated (Heyward 2013; Keith 2000), or to invent original classifications (Boucher *et al.* 2014). CDR is not reducible to conventional mitigation since it does not prevent GHG emissions, while SRM significantly differs from sustainable adaptive measures.

CDR and Mitigation

At first sight, CDR may appear as an instance of mitigation. The Intergovernmental Panel on Climate Change (IPCC) defines mitigation as the "human intervention to reduce the sources or enhance the sinks of greenhouse gases" (IPCC 2015, 125), which is compatible with the description of some or most CDR interventions (Minx *et al.* 2018, 3; Schäfer *et al.* 2015, 20).[19] From forest management to DACS, the objective is to trap and store carbon away from the atmosphere, that is, to enhance GHG sinks. Based on international treaties such as UNFCCC and Paris Agreement, Honegger *et al.* (2021) argue that CDR is a form of mitigation. On the surface, there is no contradiction between what many CDR methods (promise to) accomplish and how mitigation is defined. Additionally, the fact that the UNFCCC (forestry under Kyoto's Clean Development Mechanism) and the IPCC explicitly consider CDR interventions (forestry, bioenergy with carbon capture and sequestration/storage [BECCS], and DACS), usually under the appellation of negative emissions technologies (NETs), as mitigation, or include them into the "mitigation portfolio" (National Academies of Sciences, Engineering, and Medicine 2019, 4), supports classifying CDR under the mitigation label (Cox *et al.* 2018).

TEXTBOX 1.1 GREENHOUSE GAS (GHG) SOURCES, SINKS, AND RESERVOIRS

It is important to grasp the distinct stages of the carbon and GHG cycles when apprehending CDR: sources, sinks, and reservoirs. The first two refer to fluxes, that is, generation or withdrawal of GHGs at a given time, whereas the third refers to a stock, that is, accumulated result of past fluxes. According to the UNFCCC Article 1 (1992), a GHG source characterizes "any process or activity which releases a greenhouse gas, an aerosol or a precursor of a greenhouse gas into the atmosphere." Sources can be natural (e.g., decaying biomass, wildfires, and volcanoes) or anthropogenic (e.g., emissions from cars, planes, or factories).

In the same article, the UNFCC defines a sink as "any process, activity or mechanism which removes a greenhouse gas, an aerosol or a precursor of a greenhouse gas from the atmosphere." For carbon, natural absorption occurs in the oceans, on lands (biomass and soils), and in the atmosphere. Natural absorption involves living organisms (e.g., through photosynthesis) or chemical reaction (e.g., through weathering). CDR offers to stimulate natural processes (e.g., by enhancing the phytoplankton intake through ocean iron fertilization [OIF]), or to engineer artificial sinks (e.g., by direct capture of the ambient CO_2 with DACS).

Still in the Article 1, the UNFCCC defines a GHG reservoir as "a component or components of the climate system where a greenhouse gas or a precursor of a greenhouse gas is stored." It designates locations where carbon dioxide, methane, and other GHGs are durably sequestrated, like pockets in the lithosphere, sediments at the bottom of the oceans, and so on. CDR relies on two kinds of reservoirs: biological (e.g., forests or phytoplankton) and geological (e.g., deep aquifers). On average, the latter guarantees more enduring storage (from several millennia up to millions of years) than the former (from a few decades to a couple of centuries).

This extensive conception is nonetheless at odds with a more restrictive, and commonly endorsed, view (Keith 2000). Mitigation is usually understood as, first, abstaining from emissions and, second, excluding *ex post* removal. For instance, Keith (2013, xix) characterizes mitigation as composed of energy conservation (refraining to use it), energy efficiency (reduced use for a given process), and decarbonization (switching to nonfossil sources). The restrictive conception or "conventional mitigation" (Keith 2000, 264) focuses on shutting down GHG sources, whereas *net* mitigation, which bundles conventional mitigation with "negative emissions," refers to "intentional human efforts to remove CO_2 emissions from the atmosphere" (Minx *et al.* 2019, 3).[20]

Except for a couple of methods like deforestation halting or no-/low-tillage, CDR consists in capturing and storing carbon *after* its release whereas carbon abatement seeks to *prevent its release* (Cox *et al.* 2018). Through its extensive understanding of mitigation, the IPCC combines interventions on sources (conventional mitigation) and sinks (CDR) in a single definition. This obfuscates fundamental differences, but also conceals practical issues such as the fact that CDR/NETs lack the maturity and scalability for curbing the increases of GHG emissions as projected in climate scenarios and policy pathways (Lenzi *et al.* 2018) (cf. chapter 3).

In addition, blending conventional mitigation with CDR, for instance into the concept of net negative emissions,[21] suggests that the latter could substitute for the former. However, their respective energy balances demonstrate that they are not equivalent. Burning fossil fuels followed by ulterior capture is *ceteris paribus* more energy-demanding than using energy for a given purpose alone (electricity generation, heating, etc.). For example, CDR imposes an energy premium (penalty) on biomass power plants (BECCS) and facilities run on fossil fuels (CCS). Similarly, fertilizing the ocean for stimulating phytoplankton bloom or crushing rocks for scattering them on land or at sea will mobilize a vast amount of, possibly fossil, energy.

Besides being descriptively inaccurate, assimilating CDR into mitigation could lessen the impetus for abatement by nurturing a false sense of security that positive emissions could be compensated for by "negative" ones (i.e., capture). However, conventional mitigation directly grapples with the origin of anthropogenic climate change, namely fossil combustion, whereas CDR tackles the issue obliquely while not requiring giving up fossil energies. A possible consequence could be to exacerbate the current carbon lock-in (Anderson and Peters 2016).[22] Moreover, CDR methods such as CCS or

BECCS owe their raison d'être to carbon emissions since they are attached to fossil-powered facilities. Thus, their extensive use could worsen the current carbon lock-in (Asayama 2021; Vergragt *et al.* 2011). In sum, strong reasons exist for keeping conventional mitigation and NETs/CDR apart when assessing their "targets, timetables, accounting methods and incentives" (McLaren *et al.* 2019, 2). Propositions have been advanced for attributing CDR a specific, "interim," status under international law (Lin 2019, 574–577).

Despite those objections, CDR may constitute an indispensable component for successful hybrid strategies. It could help to hasten net mitigation by enhancing carbon sinks and reservoirs and act as a bridging technology while waiting for decarbonization. The argument rests on the common assumption that aggressive mitigation would be either unfeasible or insufficient for averting dangerous change. Such an assumption is now frequently crafted into climate scenarios. Under those circumstances, CDR could prove crucial for narrowing this "mitigation gap" (Buylova *et al.* 2021; Mace *et al.* 2021).

Without assimilating CDR into mitigation, it could be argued that the former is essential for implementing the latter in a morally acceptable manner. A common objection against aggressive mitigation is that it condemns vulnerable populations to poverty, imposing unnecessary suffering due to socioeconomic decline. CDR would then be instrumental for lessening the dilemma between UNFCC's two central imperatives: "avoid catastrophic climate change" and secure the right to sustainable development (Callies and Moellendorf 2021). It could support robust mitigation in the long run while not impeding socioeconomic progress in the short run. That outcome could be achieved by either hastening the transition toward a state of affairs in which GHG sinks outperform sources (net-zero pathways) or by indulging humanity into temporary excessive emissions for investing into a technological breakthrough for decarbonization (end-of-the-century pathways).

The net-zero pathways or "net-zero budget scenarios" "assume climate policies that limit the remaining cumulative CO_2 emissions until carbon neutrality (net-zero CO_2 emissions) is reached" (Riahi *et al.* 2021, 1064). In those scenarios, humanity does not overspend the budget for staying under the Paris targets (Drouet *et al.* 2021; Riahi *et al.* 2021). Such pathways usually transition to carbon neutrality around mid-century. In practice, net-zero can designate two very different strategies. The first is genuinely carbon-free in the sense that it relies mainly on conventional mitigation for bringing emissions down as fast as possible. Fully decarbonizing sectors like aviation and the cement industry in due time will, however, prove extremely difficult, which could justify using CDR for residual, incompressible, emissions. Moreover, as seen above, achieving these goals could come with significant suffering in terms of foregone economic development and poverty reduction. The second approach is to achieve a net-zero state with more substantial help from CDR. That option is genuinely *net*-zero since carbon neutrality is accomplished by matching sinks to emissions. Although the first strategy may turn to carbon capture at the margin (e.g., for decarbonizing aviation), the second requires CDR to be employed at a level that may match ongoing emissions. The problem is if CDR does not compensate for indulged, accumulated, present and past emissions, a failing *net*-zero pathway will unleash an unbridled overshoot scenario in which "CO_2 concentrations in the atmosphere temporarily exceed some predefined, 'dangerous' threshold" (Nusbaumer and Matsumoto 2008, 164).

The other route is formed of the end-of-the-century pathways that rely on overshoot scenarios (Huntingford and Lowe 2007). A temporary overshoot could provide more resources, through cheap fossil fuels, for investing into technologies that will support future mitigation and adaptation (Nordhaus T. 2018). The end-of-century pathways rest on overshoot scenarios that promise a return under the carbon budget by 2100,[23] postponing the goal of mid-century carbon neutrality. An increasing number of integrated assessment models (IAMs)[24] feature overshoot pathways, partly due to the mounting influence of carbon budgets in international talks (Geden 2015). The growing view that the carbon budgets imposed by the Paris targets will be overstretched tends to support CDR (IEA 2023).

Contrary to genuine fossil-free approaches that are mostly rooted on conventional mitigation, end-of-century and truly *net*-zero pathways incorporate large amounts of CDR (de Coninck *et al.* 2018, 342), mainly, but not only, under the form of forestry management and BECCS. Again, the difference between conventional mitigation and CDR matters because the promise of capture might not materialize. Since appropriate CDR for realizing *net*-zero and end-of-the-century pathways depends on external factors like a resilient political will to decarbonize, future emissions, existing carbon lock-ins, and CDR technological breakthroughs, it generates higher uncertainty than strategies based on conventional mitigation alone. Everything else being equal, nothing guarantees that CDR will be deployed fast enough for achieving a net-zero balance while keeping emissions within the budget or at a scale that will rapidly extract carbon enough for canceling the overshoot. Also, the indulged emissions may not be used for developing fossil-free energies.

Studies show that, *ceteris paribus*, overshoot scenarios will yield lower economic growth and other benefits by 2100 than drastic, immediate, mitigation (Riahi *et al.* 2021).[25] More importantly, the cooling effect of one unit of extracted carbon might not compensate for the warming impact of one unit of released carbon. There is a "possible non-equivalence of warming and cooling caused by overshoot emissions and recapture removals" (Brander *et al.* 2021, 705). Finally, even if the warming gauged by the overshoot is fully balanced by the cooling of the subsequent capture, other climate variables might not properly adjust, such as sea-level rise, externalizing harms to future generations.

Descriptively, CDR cannot therefore be considered as analogous to conventional mitigation, and at the end it may impede carbon abatement. This nonequivalence has moral implications. Conventional mitigation is about abstaining from an act that, while carrying benefits, is also known to cause predictable and unpredictable harms, that is, risks and uncertainty, that will follow nonlinear patterns, affecting future generations who will have had no voice in the initial arbitrage between mitigation and CDR. In addition to issues of impaired efficiency and increased uncertainty, the distinction between conventional mitigation[26] and CDR raises the deeper question of whether refraining from harm is morally equivalent to committing an act that could prove harmful in the future and to intervening later for attempting to cancel the damage. The latter may be judged as *morally inferior* to the former. Not only the agent who perpetrates the initial act engages in self-serving behaviors (deontological reason), but by doing so, he exacerbates the probabilities of catastrophic outcomes for innocent people, owing to the nonlinearity of climate response, especially when tipping points are crossed (consequentialist reason). In short, the two main ethical conceptions offer arguments for not treating mitigation and CDR as equivalent (Cox *et al.* 2018).

SRM and Adaptation

Equating SRM to adaptation can also be challenged. To begin with, the IPCC defines adaptation in a manner that does not seem to exclude SRM, namely "the process of adjustment to actual or expected climate and its effects" (IPCC 2015, 118). SRM's objective is to enhance the local or global albedo, thus reducing radiative forcing, and *cool a targeted area*. SRM could be used for lowering global average temperatures, the major consequence of climate change, which supports classifying it as an adaptation measure (e.g., Minx *et al.* 2018, 4).

SRM could, nonetheless, be interpreted not so much as adapting to, but as masking climate change by precluding some effects (e.g., temperatures and loss of ice sheets and glaciers). Whereas adaptation implies reactive strategies to unfolding changes, SRM consists of artificially preventing those changes from fully manifesting. In addition, since SRM is effective only as long as it is actively pursued (cf. chapter 4), contrary to the introduction of new species or the building of seawalls that endure after the initial investments, it may not count as *resilient* adaptation. For example, SAI or MCB, the two most discussed SRM interventions (NRC 2015b), require aircraft to continuously spray aerosols in the stratosphere (SAI) or vessels to constantly feed marine cloud formations (MCB). In comparison, many adaptation instruments can be implemented in a less demanding manner or without requiring constant efforts. For instance, migrating is, on the human life scale, a circumscribed decision. Adapting crops or trees to novel conditions through crossbreeding or genetic modification imposes initial, possibly heavy, investments, but after a while the new species could thrive and be self-sustainable if the selection and acclimatization processes were properly undertaken. Moreover, further financial efforts into R&D might not be needed. Erecting seawalls and protection necessitate initial investments as well as constant maintenance, but the whole intervention does not demand to be taken *de novo*—seawalls do not have to be continuously rebuilt.[27]

Worse, SRM methods such as MCB, CCT, or SSI undermine deeper resilience. They do not simply hide climate change. They allow risks and uncertainties to accumulate. As soon as those interventions end, the temperatures will rapidly surge (Matthews and Caldeira 2007; Jones *et al.* 2013; Shepherd 2009, 24), catching up over a short period with what they should have been. Various studies point at the danger of the thermic rebound being larger than the initial thermal increase masked by the intervention (NRC 2015b, 63). Such a sudden surge will not only harm human beings (Baum *et al.* 2013), but also animal life and ecosystems (Trisos *et al.* 2018), which will have a reduced opportunity window for adapting (for instance through migration or evolution) by comparison with a pathway without SRM characterized by steadier warming. Known as the termination shock (e.g., Callies 2019a, 56; McKinnon 2019; Parker and Irvine 2018), this risk illustrates the maladaptive nature of SRM.

Depending on the future emissions, SRM will not prepare humanity to adjust to greater GHG concentrations. A method like SAI will simply conceal some of the thermic increase, which may result in a delusional sense of security or normality. Such an effect may emerge from the SRM R&D stage. The simple prospect of dodging certain effects of global warming could be sufficient for creating what has been qualified as a moral hazard ("situations where an agent, once being insured, adopts a riskier behaviour, this inflection not being intended or foresighted by the insurer" [Landes 2015]) or as risk compensation ("increase or decrease in risk-taking once an individual perceives that risk to be lower or higher, respectively" [Reynolds 2015, 177]). The curtailed sense of emergency might impede mitigation efforts, although studies suggest that the very perspective of deploying SRM methods could nurture extra motivation for ambitious carbon abatement (Reynolds 2015).

Further, a method like SAI, and possibly MCB, CCT, and space CE, will establish novel climates characterized by new temperatures, precipitation distribution, and so forth (McLaren 2018).[28] If enough aerosols are dispersed in the stratosphere for balancing global warming since the preindustrial times, local temperatures will still display diverging patterns in comparison to this baseline (Irvine *et al.* 2010). Novel regional climates will emerge, with potential negative outcomes such as the weakening of the African and South-East Asian monsoons and increasing frequency and/or intensity of droughts and famines (Burns 2016, 11; Nalam *et al.* 2018; Robock *et al.* 2008). Since populations will become more exposed to climate risks as a direct result of the intervention, SRM methods cannot be equated to effective adaptation.

To summarize, SRM could be seen as an adaptation measure, at least on the surface. However, when digging deeper, there are also strong reasons to contest the idea that it would represent a resilient or acceptable form of adaptation. Independent of those taxonomic issues, nothing precludes considering SRM a useful instrument of climate policy, for instance as a bridging technology (Callies and Moellendorf 2021). SRM deployment could constitute an imperfect, potentially harmful, but still necessary tool in the race against time for the proper decarbonizing of the economy and perhaps for engaging into ambitious GHG capture and sequestration.

NOTES

1. The Ho Chi Minh trail was a network of routes and paths connecting North to South Vietnam through Laos and Cambodia and utilized for supplying the Viet Cong rebellion.

2. Deontological ethics is one of the three main families of moral theories (Alexander and Moore 2021), along with consequentialism (Sinnott-Armstrong 2022) and virtue ethics (Hursthouse and Pettigrove 2022). It is based on duties. The moral acceptability of actions depends on the rules followed by the agents. The main difference with the other theories is that consequentialism evaluates the outcomes of the decisions, while virtue ethics considers the moral character of the agent (cf. Svoboda [2017, 95–122] for an example of virtue ethics applied to CE).

3. What counts as a benefit? For whom? How to account for risk and uncertainty? Although those questions will be discussed in the coming sections, it is worth noting that usually no mention is made of the benefits for nonhuman animals, plants, other living organisms, and ecosystems. Even if engineering appears at first as an anthropocentric activity, that is, an activity only or mainly undertaken for humans' benefits, it does not have to be.

4. Beneficence is crucial in experiments involving human subjects, which CE may become if experimented at large scale or deployed (Morrow *et al.* 2009).

5. Intentions and overall consequentialist considerations (benefit–harm ratio) do not exhaust the ethical calculus. There might be situations where the collective welfare is improved by an engineering project, but such a project is not permissible, for example, in reason of the distribution of the negative consequences or of the magnitude of such adverse consequences (McGinn 2018, 32–33).

6. The consequences of a sudden halt or failure of the geoengineering system. For SRM approaches, which aim to offset increases in greenhouse gases by reductions in

absorbed solar radiation, failure could lead to a relatively rapid warming which would be more difficult to adapt to than the climate change that would have occurred in the absence of geoengineering. SRM methods that produce the largest negative forcings, and which rely on advanced technology, are considered higher risks in this respect. (Shepherd *et al.* 2009, 35)

7. A convention is to distinguish risk, that is, adverse outcomes which odds can be calculated, from (deep) uncertainty, that is, adverse outcomes that are unknown or which odds cannot be calculated (Knight 1921). Uncertainty characterizes contexts where

analysts do not know, or the parties to a decision cannot agree on, (1) the appropriate conceptual models that describe the relationships among the key driving forces that will shape the long-term future, (2) the probability distributions used to represent uncertainty about key variables and parameters in the mathematical representations of these conceptual models, and/or (3) how to value the desirability of alternative outcomes. (Lempert *et al.* 2003, xii)

8. Dale Jamieson is critical of drawing too ambitious moral implications from the distinction between intentional and non-intentional climate change.

9. The 1965 report to US President Johnson was also one of the first official documents to evoke geoengineering.

10. In IPCC reports, equilibrium climate sensitivity (units: °C) refers to the equilibrium (steady state) change in the annual global mean surface temperature following a doubling of the atmospheric equivalent carbon dioxide (CO_2) concentration. (IPCC 2015, 120–121)

11. For instance, the ETC Group, one of the most vocal advocacy groups opposed to CE, defines it as "a set of proposed technologies to deliberately intervene in and alter Earth systems on a large—i.e., *planetary*—scale" (emphasis added).

12. Michael E. Mann offers an illustration of the assimilation of large into global when defining geoengineering as "schemes that employ *global-scale* technological intervention with the planet in the hope of offsetting the warming effects of carbon pollution" (Mann 2021, 154, emphasis added).

13. On that aspect, the analysis presented here diverges from other interpretations of CE as carrying necessarily, or by definition, global effects (e.g., Heyward 2013, 24).

14. Comprised between 200 and 1,000 m, the thermocline is an ocean "layer in which temperature decreases and density increases (. . .) markedly with depth"; furthermore, it "acts as a barrier between the warmer surface water and the colder deep-layer water" (Barry and Chorley 2010, 191).

15. https://www.arcticiceproject.org/

16. https://www.climaterepair.cam.ac.uk/refreeze-arctic

17. The guilt by association fallacy is the erroneous logic that just because someone/something A is associated with someone/something B, that someone/something A has or accepts all of the qualities of someone/something B. This fallacy permeates society, from social groups, to political campaigns, to business relationships, to the court system. (Kolb 2019, 351)

18. The 1977 United Nations Convention on the Prohibition of Military or Any Other Hostile Use of Environmental Modification Techniques (EMT) (ENMOD) bans "military or any other hostile use of environmental modification techniques having widespread, long-lasting or severe effects as the means of destruction, damage or injury to another State party" (article I).

19.   This definition subsumes all those NETs that focus on natural sink enhancement such as afforestation and reforestation (AR), soil carbon sequestration (SCS), ocean fertilization (OF), biochar (BC) or enhanced weathering (EW) as an integral part of mitigation, while other NETs that geologically store the sequestered CO_2 such as BECCS or direct air capture with carbon capture and storage (DACCS) do not qualify. (Minx *et al.* 2018, 3)

20. In 1983, the US NRC (1983) laid down a similar distinction between pre-emissions and post-emission CO_2 control (Keith 2000, 255).

21. Defined as the "sum of gross positive CO_2 emissions from energy and industrial sources, gross positive land-use CO_2 emissions, and gross negative CO_2 emissions" (Rogelj *et al.* 2019, 360).

22. The carbon lock-in "rises through a combination of systematic forces that perpetuate fossil fuel-based infrastructures in spite of their known environmental externalities and the apparent existence of cost-neutral, or even cost-effective, remedies" (Unruh 2000, 817). The lock-in results from the embodiment into and support of specific, fossil-based, technological infrastructures by private and public institutions. This entanglement renders the introduction of any technological alternative extremely difficult. *De facto,* alternatives are "locked out." For an analytical framework recapping the main factors, the reader can refer to Trencher *et al.* (2020).

23. The remaining volume of carbon to be emitted for securing a given probability (e.g., 30 percent, 50 percent, and 66 percent) of meeting a specific target (e.g., 2°C increase in global average temperatures).

24. IAMs combine climate projections with socioeconomic features.

Integrated models explore the interactions between multiple sectors of the economy or components of particular systems, such as the energy system. In the context of transformation pathways, they refer to models that, at a minimum, include full and disaggregated representations of the energy system and its linkage to the overall economy that will allow for consideration of interactions among different elements of that system. Integrated models may also include representations of the full economy, land use and land-use change (LUC) and the climate system. (IPCC 2015, 124)

25. The issue is more intricate than a simple trade-off between immediate and delayed economic growth/decarbonization investments.

26. To be sure, depending on how it is implemented, conventional mitigation can come with harms of its own, especially if it incurs increased poverty and related ills like diseases, hunger, and premature deaths.

27. Adaptation measures could be classified along a durability/resilience scale. Despite its importance, such an issue is not addressed in these pages.

28. In addition, SRM would reduce temperatures only during the daytime, leaving nighttime temperatures untouched (Michaelson 2013, 103).

Chapter 2

An Evaluative Framework

The denomination of climate engineering (CE) is an umbrella term that stands for a variety of methods, which often differ substantially from one another. It defines *anthropogenic interventions in the climate system that predominantly aim at altering radiative forcing locally or globally to slow down, stop, or reverse climate change and/or related adverse events.* These interventions range from reforestation and preservation of wetlands to stratospheric injections and orbital sunshades. The two main categories of interventions are *carbon dioxide removal* (CDR), which consists in capturing and sequestering carbon, and *solar radiation management* (SRM), which attempts at cooling the Earth by enhancing its albedo (reflectivity) (Shepherd *et al.* 2009).

HOW TO EVALUATE CLIMATE ENGINEERING?

Assessing CE methods is not straightforward, especially when it comes to the multiplicity of relevant aspects, such as technicity, scale of deployment, magnitude of the effects, risks, and uncertainties, and populations and ecosystems potentially affected. Several evaluative grids have been offered. In one of the first texts in moral philosophy on the topic, Dale Jamieson (1996, 326) proposes four criteria for assessing CE methods: technical feasibility, predictable consequences, relative superiority of CE outcomes compared to alternatives, and respect for fundamental "ethical principles or considerations." The Royal Society (Shepherd *et al.* 2009, 6–7) offers nine dimensions regrouped into two categories: *technical* (effectiveness, timeliness, safety, costs, and reversibility) and *nontechnical* (public attitudes, social acceptability, political feasibility, and legality). The US National Research Council (NRC) identifies four factors: efficacy, cost-effectiveness, risks and co-benefits, and governance.

How effective is the strategy at achieving predictable and desirable outcomes? How much does the strategy cost to implement at a scale that matters? What are the risks for unintended consequences and opportunities for co-benefits? What governance mechanisms are in place or are needed to ensure that safety, equity, and other ethical aspects are considered (e.g., intergenerational implications)? (NRC 2015a, 3)

This chapter does not dispense an in-depth analysis of those evaluative schemes, but nevertheless offers a grid that accounts for the main frameworks proposed in the literature (e.g., the Royal Society and the US NRC) (table 2.1). At the core of all those frameworks is the distinction between *feasibility* and *permissibility*.[1] For deciding whether an action is worth undertaking, it should be determined whether that action is feasible, that is, whether it *can* be done, and whether it is permissible, that is, whether it could be acceptable according to social, legal, political, and ethical standards. Although such a distinction is sometimes less clear-cut in practice, it remains useful for acknowledging the importance of both empirical and value judgments when it comes to assess technologies.

Along with feasibility and permissibility, a third dimension is preferability, that is, a "'comparative criterion,' which compares ideal policy with other politically feasible alternatives" (Morrow and Svoboda 2016, 86). Echoing Jamieson's fourth criterion, it highlights that some options may prove superior over, and therefore preferable to, alternatives. If feasibility and permissibility assessments usually focus on a given method typically apprehended in isolation from other climate tools, they are silent on how such a method could perform when weighed against other feasible and permissible avenues. The point of comparison commonly included in the permissibility assessment is the default position, which constitutes the status quo where no action is taken, often approximated by a business-as-usual scenario with slightly or not abated emissions (e.g., representation concentration pathway [RCP] 8.5,[2] which is a controversial choice [Hausfather and Peters 2020]).

Table 2.1   The Evaluative Dimensions

Feasibility	*Permissibility*	*Preferability*
• Technological maturity	• Intrinsic moral value	• Paretian rule
• Scalability	• Risks and uncertainty	• Gross benefits maximization
○ Technological	• Distributive justice	• Cost-effectiveness
○ Physical	• Procedural justice	• Net benefits maximization
○ Economic	• Social acceptability	• Overarching qualitative principles
• Efficiency	• Intergenerational justice	
	• Regulation and governance	

Preferability also reflects the fact that societies and individuals face dilemmas in which no option appears desirable per se, since all of the options imply harms, risks, and uncertainty,[3] but on closer inspection, one is preferable because it is "less worse" than alternatives on one or several morally relevant aspects. The preferability dimension highlights the nonideal circumstances[4] under which CE interventions are evaluated. In a perfect state of affairs, such interventions would not be needed. However, because countries are falling short of sufficiently mitigating climate change, CE *may* be justified everything else considered (Morrow and Svoboda 2016; Svoboda 2017). Finally, preferability judgments are necessary in a world of scarcity in which all strategies come with opportunity costs, defined by Gregory Mankiw (2021, 4) as "whatever must be given up to obtain some item," thus "[w]hen making any decision, decision makers should take into account the opportunity costs of each possible action."

FEASIBILITY

Material opportunities and constraints constitute the heart of feasibility. Can a technology be realistically researched, developed, and deployed within a given timeframe? This is the core issue dealt with in this first evaluative category. Feasibility assessments are muddled by the difficulty of anticipating methods' viability. Future possibilities and challenges are partly concealed at the time when the allocation of resources to research and development (R&D) is decided. Thus, the fact that a CE method appears impracticable at an instant *t* may not necessarily rule out any pertinence at *t+n,* and *vice versa*. Besides, investing in a specific technology may lock humanity in a given trajectory while impeding alternative routes, thus creating path dependency (Preston 2013, 28; Robock 2008, 17).[5] For instance, the development of carbon capture and sequestration/storage (CCS) devices to be installed on coal- and gas-powered plants might trap humanity with fossil fuels for many more decades.

Technological Maturity

The first dimension of feasibility is *technological maturity*, which reflects R&D advancement. Have CDR and SRM methods been tested in a laboratory or experiment? Have they led to successful pilot or demonstration projects? Are they ready to be deployed? Maturity is not binary, but a matter of degrees. Controlled settings in a laboratory may hint at opportunities, invalidated by subsequent *in situ* experiments. Various assessment tools exist, one of the most prominent being the nine technological readiness levels (TRLs) developed by the National Aeronautics and Space Administration (NASA 2016) for evaluating its spatial program.

- TRL 1: "Basic principles observed and reported."
- TRL 2: "Technology concept and/or application formulated."
- TRL 3: "Analytical and experimental critical function and/or characteristic proof-of-concept."
- TRL 4: "Component and/or breadboard validation in laboratory environment."
- TRL 5: "Component and/or breadboard validation in relevant environment."
- TRL 6: "System/subsystem model or prototype demonstration in a relevant environment (ground or space)."
- TRL 7: "System prototype demonstration in a space environment."
- TRL 8: "Action system completed and 'flight qualified' through test and demonstration."
- TRL 9: "Actual system "flight proven" through successful mission operations."

According to Olechowski *et al.* (2020, 3),[6] those levels could be bundled and mapped out into four phases: TRLs 1–4 correspond to "material solution analysis," TRLs 5–6 to "technological maturation and risks reduction," TRL 7 to "engineering and manufacturing development," and TRLs 8–9 to "production and deployment." The end of phases 1, 2, and 3 may host "gate reviews," that is, step evaluations during which the R&D direction is evaluated according to the initial goals and future prospects, which could serve as safeguards against slippery slopes (cf. chapter 5). Quite clearly, CE methods diverge in terms of TRL. Some remain nascent, such as space reflectors (somewhere between TRL 2 and TRL 3), while others are more mature, like bioenergy with carbon capture and sequestration/storage (BECCS) (between TRLs 5 and 7) (Low *et al.* 2022). In a CE context, the most advanced TRL is the very last stage before implementation at full scale, which brings the next criterion: scalability.

Scalability

The second criterion is the capacity of a system to expand in a sustained manner. Such capacity stems not only from the technology itself but also from such surrounding conditions as the institutional context (e.g., regulation, economic support, and industrial networks). In the case of CE, scalability designates the potential of being employed at a scope which offers a significant probability of reducing carbon emissions and/or concentrations or of enhancing albedo. For interventions characterized by high maturity like CCS, scalability is a strong concern (Rogelj *et al.* 2019, 362). Scalability can be broken down into three subdimensions: technological, physical, and economic.[7] The *technological* aspect refers to the mastery of core technical processes that allow the method to be deployed to an appropriate extent to be effective, that is, to deliver its expected benefits. Therefore, this dimension is a direct function of maturity, but not exclusively, since private and public institutions might favor or hinder specific developments through support to research, established standards, regulation, and so on.[8] *Physical* scalability is not directly bound by internal limits but by outside constraints, especially resource availability. In the case of CE, those constraints could be linked to the amount of required land for cultivating feedstock crops (BECCS) or planting trees, the existence of large and accessible geological reservoirs (in the case of methods involving carbon sequestration), or a sufficient supply of iron for fertilizing oceans or sulfur for stratospheric injection. In sum, physical scalability relates to whether the environment and material constraints do not impede CE deployment. *Economic* scalability denotes the possibility of garnering adequate financial resources to fund the intervention. It is usually assessed by looking at the potential revenues that could be produced by a given method or the funding and subsidies that could be levied (e.g., from governments and philanthropic initiatives such as the Bill Gates Foundation or Elon Musk's X Prize [Milman and Rushe 2021]). As a result, a CE intervention is economically scalable only if it can generate or attract enough financial resources to cover its costs, which introduces us to the importance of efficiency.

Efficiency

This is an umbrella category dedicated to evaluating the capacity of a given method to deliver valuable outcomes, given appropriate investments are available. Two approaches could be distinguished: *cost–benefit* and *cost-effectiveness* (Brohé 2016, 88–99; Heath 2020, 219). The former consists in maximizing the net benefits generated through a CE intervention by subtracting the gross costs from the gross benefits (Hausman 2021), whereas the latter aims at minimizing the costs for attaining a predefined goal (or set of benefits) like a decline in GHG emissions and concentrations, or a global/local cooling. The benefits dimension relates to the pertinence of the method, namely whether it tackles a serious issue. In the case of CE, such an issue is climate change, either its causes (GHG emissions and concentrations) or its effects (temperatures). For the sake of simplicity, both approaches are apprehended as the two facets of a broader optimization scheme. On the one hand, there is the efficiency of a given CE method at producing worthwhile results (e.g., radiative forcing reduction) and, on the other hand, the efficiency in the use of the required resources for attaining such a result.

Information about invested resources into researching, developing, and deploying a CE technology is worthwhile only if paired with positive and negative outcomes. Moreover, such outcomes are scenarios dependent in the sense that costs and benefits, in extent and nature, vary according to the forcing reduction strategy. Such dependency is particularly relevant for SRM. Halting, halving, or reversing global warming *qua* strategies performed within distinct emission scenarios (e.g., RCP 6.0 and RCP 8.5) do not gauge the same costs, harms, risks, and benefits (Keith and Irvine 2016). Hence, it is important to run efficiency assessments under different CE strategies and emission pathways. In addition, the evaluation should not stop at direct outcomes, but include indirect, spatially and temporally distant consequences, namely positive and negative externalities. An externality is "the effect of one person's actions on the well-being of a bystander" (Mankiw 2021, 516). More precisely, it is a benefit or a burden experienced by an agent who is not part of the interaction that generated the benefit/cost. The agent did/does not consent to shoulder the cost (negative externality) and did/does not contribute to generating the benefit (positive externality). Loud music from neighbors or air pollution are examples of the negative form. The adoption of healthy lifestyles in a country where health care is universal illustrates the positive form. Since the health costs pooled in the public insurance scheme are lowered, all policyholders benefit from the healthier habits.

Efficiency analysis raises epistemic and moral questions (Brohé 2016, 91), some being general whereas others are specific to CE. First, any assessment implies judgments about what can be counted as a cost/benefit, for whom, and under which form (e.g., monetary/nonmonetary estimations, climate variables, and rights). Value judgments must be passed about the appropriate community of affected entities, the kind of morally relevant impacts, and how such impacts are accounted for. Put differently, what should count as costs and benefits, approximated through which metric? Should efficiency assessments be restricted solely to dimensions that can be evaluated in monetary terms, or should they be broadened, and if so how (e.g., by including welfare or aesthetical criteria)? As an illustration, stratospheric aerosol injection (SAI) may cause so much disruption in isolated places like in the heart of Amazonia or Borneo, where whole ecosystems collapse, with rare species going extinct, but without any impact on humans. Should ecosystems and other living beings be integrated in the morally relevant community? Should the losses of such entities be included in the cost evaluation? If not under monetary form, then under which metric and with which weight?

No assumption here is made on the desirability or feasibility of broadening efficiency assessments. The previous remarks simply underscore that moral valuation cannot be eschewed when it comes to assessing CE efficiency. What to evaluate and how are not trivial issues. They emphasize that such an assessment is a moral enterprise. For initiatives as impactful as CE, especially on vulnerable groups, it could also be argued that choices about what to value and how ought to be collectively made, and not reserved to a narrow circle of engineers and decision-makers, which echo the Royal Society's advocacy for public participation (Shepherd 2009, xii). There is a risk of technocratic capture (cf. chapter 5). Decision-making raises procedural issues (cf. below). In sum, cost–benefit and cost-effectiveness analyses are not anodyne, purely technical, endeavors (Finkel 2018). Evaluative assumptions must be formulated and justified (Keith 2000, 278).

Second, models predict that some CE interventions may be prone to negative, potentially catastrophic, outcomes. For instance, marine cloud brightening (MCB), cirrus cloud thinning (CCT), space CE, and most notably SAI, may generate novel climates, with devastating consequences for developing countries (Flegal and Gupta 2018; McLaren 2018). Deployment interruption may cause a sudden surge of temperatures (termination shock) (NRC 2015b, 59–66). For these reasons, expected costs and benefits should account for risks and uncertainty, not only natural, but also social ones (Keith 2000, 274).

Economics' classic answer is to calculate the expected utility of a given CE intervention (CE_1), which is obtained by summing all foreseeable CE_1 outcomes (net benefits and losses) weighed by their relative probabilities, sometimes referred to as "magnitude" (Shue 2010).[9] The procedure is restricted to future events that are predictable, leaving uncertainty out (Svoboda 2017, 14). Following Frank Knight's distinction (Knight 1921), risks can be characterized as potential prospective states "where outcomes can be identified and probabilities assigned to various outcomes" (Sunstein 2005, 60), while uncertainty applies to situations "where outcomes can be identified but no probabilities can be assigned" (Sunstein 2005, 60).[10] Without delving further into detail, it is enough to acknowledge that risks (adverse events for which odds can be determined) and deep uncertainty (adverse events which are unknown or for which probabilities cannot be calculated) should be included in the assessment, which raises intricate epistemological, methodological, and moral questions about how to integrate risks and uncertainty and how to weigh them in public deliberation (cf. below).

Third, any efficiency evaluation of a given CE proposal carried in isolation of the remainder of climate strategy is of limited value. The appraisal is useful only insofar as it is replaced in the broader decision-making context (Keith and Irvine 2016), which includes: what kind of problem and with which potential damages, does the method try to tackle? Without a shadow of a doubt, the examination should be initiated by considering only a specific CE proposal (CE_1). However, CE_1 efficiency ought to be compared, first, to the default position (D) in which current climate policy trends are continued without drastic change, but also to alternatives: diverse degrees and combinations of traditional/conventional mitigation (TM_i), adaptation (A_i), and possibly CE interventions ($CE_2 \ldots CE_i \ldots CE_n$), a task that steps into preferability territory. *Ipso facto*, efficacy evaluation should be carried out at three levels: first, the *technology itself* (what are CE_1 costs, risks, and benefits?), second, weighed against *the default position* (How does CE_1 fare in comparison to D?), and third, compared with *other feasible, permissible, alternatives* (how does CE_1's cost-effectiveness measure up to CE_2 or CE_3?). For example, when determining whether SAI is efficient, it is necessary to calculate SAI's net benefits/costs for diverse injection strategies, or its cost-effectiveness for achieving various goals such as reversing or halving the impact of historic anthropogenic emissions. Then, SAI benefits/costs (cost-effectiveness) should be compared to the benefits/costs (cost-effectiveness) of D (insufficient mitigation or any continuing policy trend). Finally, SAI benefits/costs (cost-effectiveness) should be weighed against the benefits/costs (cost-effectiveness) of alternative courses of actions, like aggressive traditional mitigation, adaptation, BECCS, and so on.

Fourth, climate policy is not restricted to one dimension. CE will be a component of "policy portfolios" (Callies and Moellendorf 2021, 140), e.g. through the negative emissions technologies (NETs). The immediate implication is that the overall efficiency should be assessed for the complete set of measures, not for each component taken separately (Keith and Irvine 2016). A reason resides in the potential interactions among measures. For instance, by scattering incoming sunlight, SAI may disrupt photosynthesis (NRC 2015b, 70), which will affect forest management, the cultivation of energy crops used for BECCS, or solar power generation. On the contrary, SAI might, quite counterintuitively, increase the ocean and land carbon uptake (Wagner 2021, 52–53), although the possibility is only mentioned marginally in the literature. Thus, any robust efficiency assessment should account for the potential interactions between climate policy components.

PERMISSIBILITY

Permissibility reflects the force of the normative justifications for engaging or not into researching, developing, and deploying climate interventions. To simplify, if a given technology is feasible, do we have reasons to pursue its R&D and deployment, all things considered? The extent to which CE_1 is judged permissible stems from several criteria: CE_1's intrinsic moral characteristics (e.g., badness and wrongness); whether R&D and deployment can be assumed to remain within acceptable risk boundaries; the disparate distribution of costs and benefits; the procedures that preside over the decision to research, develop, and deploy; the public perceptions of CE_1; the impact on future generations; and the regulation and governance opportunities. In addition, if CE_1 permissibility derives from attributes like inherent badness, the magnitude of risk and uncertainty it creates, how it affects vulnerable populations, and so on, it also raises an institutional question. Can CE_1 be managed in a manner that produces less disparate impacts (or that opens the possibility of compensating the "losers"), diminishes risks/uncertainty, renders decision procedures more inclusive, is compatible with efficient and legitimate regulation, and so forth? Then, when invoking institutions, or more precisely the likelihood of designing mechanisms and instruments for addressing various issues bearing on permissibility, the topic of discussion shifts to whether CE_1 could be adequately governed under nonideal circumstances.

Intrinsic Moral Value

The first dimension is the *inner moral value* of the CE methods under scrutiny. Is there anything inherently bad, wrong, or evil (*malum in se*) about CE_1? An affirmative answer does not necessarily impose to reject CE_1, everything considered. CE_1 could be intrinsically bad (e.g., leading to a large number of innocent deaths), but still acceptable (e.g., because it may nonetheless help to minimize the overall number of innocent deaths or the loss of biodiversity). Such exemption is the essence of the lesser evil argument, sometimes branded the "principle of the lesser of two evils" (Spielthenner 2010).[11] Definitively ruling out an action under the motive of intrinsic badness requires demonstrating how such characteristic justifies an absolute prohibition to engage in CE_1 *under the current, nonideal, circumstances*.

Three common objections against CE represent variations of the intrinsic badness argument.[12] First, the *unnaturalness objection* posits that climate interventions are unacceptable because they oppose, violate, or disrupt nature per se, the natural order or evolution, which is morally bad (Morrow 2014, 133). The objection mirrors criticisms addressed to new technologies like genetic modification or nanotechnology (Corner *et al.* 2013),[13] and it implies to attribute an intrinsic value to naturalness as a normative reference. But naturalness needs to be defined (the heavy lifting of the moral evaluation lies here), and it is necessary to demonstrate why being unnatural is intrinsically bad or wrong.

A possible characterization of naturalness is the property of an entity or system of not being interfered with by humans (Taylor 1986, 3–4). It denotes the quality of remaining shielded from anthropogenic influence, the opposite state potentially being artificiality, a divide that seems to be at work in the public judgments of CE (Corner *et al.* 2013, 942). That interpretation may set the bar too high through the conception of pristine nature that should have been left untouched for qualifying as genuinely "natural." A good deal of the ecosystems regarded as such have, however, been shaped by humans who introduced and modified vegetal and animal species. For instance, many oak forests in Europe have been planted during the past centuries for the needs of the naval industry. Today they form "natural" biomes of their own. Agricultural and pastoral activities have also been impactful. It might be difficult to claim that those environments, which have been evolving along with and after anthropogenic interference, are artificial. Irrespective of the timing, they owe their existence, and occasionally their perpetuation, to human stewardship. It is nonetheless possible to situate those ecosystems somewhere between two extremities: pure naturalness and pure artificiality, some veering more toward one of the two ends.

Hence, the divide is less clear-cut and more continuous than sometimes supposed. Still a line could be drawn beyond which CE_1 would be deemed too unnatural, that is, too intrusive toward nature. On a case-by-case basis, and assuming that such a line could be drawn, some CE methods may be viewed as straying too far away from naturalness, which would constitute a bad thing. That being said, any blanket rejection of CE as unnatural appears to rest on shaky grounds since not all CE methods disrupt ecosystems to the same extent. A corollary is that critics who mobilize the unnaturalness objection in an indistinct manner need to justify in which sense interventions such as forest management, urban surface brightening (USB), CCS, or direct air capture with sequestration/storage (DACS) cross the threshold of unacceptable unnaturalness. This does not mean that *some* methods could not be convincingly judged as too unnatural, and thus morally objectionable, but it seems difficult to demonstrate that the *whole* CE project must be opposed in virtue of its intrinsic unnaturalness.

The second variation is the *hubristic objection* (Fleming 2010; Gardiner 2010; Hamilton 2013a, 181; Hulme 2014, 133; Jamieson 1996; Mann 2021, 162; Meyer and Uhle 2015; Preston 2012, 26; Robock 2016). Accordingly, CE would be motivated by a delusional belief that human beings could, through technological means, aptly control or influence the climate.[14] The objection is reinforced by the fact that some of the earliest and most vocal CE proponents were tied to the Lawrence Livermoore National Laboratory, infamous for proposing to use the atomic bomb as an engineering instrument (Kintisch 2010, 93–99; Hamilton 2014, 22).

What is wrong with hubris? Hubris *qua* arrogance is a vice, worsened in the case of CE by the "Promethean dream" of subjugating nothing less than the Earth's climate (Hamilton 2013a), which is opposed by "Gaians" (following James Lovelock's Gaia concept [Lovelock 1987]) (Thiele 2019).[15] Such a project may be judged as both consequentially and intrinsically bad, the former falling outside the scope of this section. On the one hand, CE would disregard fundamental natural mechanisms, principles, or harmony (left to be precisely defined), which would likely lead to unnecessary harms. Therefore, humans would be blinded by their delusive ambition of control, which would make them prone to reckless decisions. In that case, the problematic aspect of CE would less be the hubristic nature embodied in CE per se than the negative outcomes, for example, increased risk exposure. On the other hand, even if effective, CE "would still have a bad effect of reinforcing human arrogance and the view that the proper relationship to nature is one of domination" (Jamieson 1996, 332), which is a genuine intrinsic judgment. In that second sense, hubris would be bad *in itself*, namely as a vice corrupting humanity's moral character.

An iteration of the hubris objection is the *playing God objection* (Clingerman 2012; Clingerman and O'Brien 2014; Hamilton 2011 2013a, 177–183; Hartman 2017), which posits that by entertaining the grandiose project of altering the climate, humanity displays extreme arrogance, but also commits the further sin of posing as an omniscient and omnipotent force. The criticism is rooted on the assumption that mortals should refrain from claiming divine attributes. In the same vein as hubris, playing God is an attitude that can be challenged with intrinsic and consequentialist reasons. On the one hand, the badness may lie in the pretension itself of being like God or having God-like attributes such as controlling nature, the weather, and the climate. As for hubris, this attitude could have a corrupting effect on human moral character by nurturing a delusional sense of grandeur. It could also obfuscate the presence of the divine in the world, rendering it inaudible or invisible. On the other hand, humans may then be more likely to engage in damaging behaviors and take unacceptable risks, which is a consequentialist argument.

Whereas it could be arduous to show why CE *as a whole* would be bad per se, namely why *all* interventions would be inherently evil (e.g., capturing carbon from the atmosphere through reforestation, BECCS, or DACS), it might still be possible to demonstrate that some actions qualify as such. If intrinsic badness can claim any force, it will be on a case-by-case basis. Too often an intrinsic evilness judgment is imposed on all CE interventions without any nuance. To be robust, normative assessments must avoid committing the fallacy of composition, which consists in rejecting all CE methods based on a negative, demonstrated, judgment laid on some of them, for example, by extending the hubristic objection from SAI to all CE methods including forestry, biochar, USB, and CCS. Moreover, for claiming any exhaustivity, any moral assessment needs to go beyond intrinsic evilness and corruption. Although crucial, those dimensions do not exhaust the field of morality.

Risks and Uncertainty

The second criterion relates to foreseeable *and* unforeseeable adverse events potentially produced through the R&D and deployment stages. To some extent, risks are already tailored into the efficiency assessment, under the form of the probabilities that serve to calculated the expected outcomes/utility of a CE intervention. Considerations of risks and uncertainty also contain a comparative dimension when it comes to reducing deep uncertainty and potentially harmful consequences in the choice of the proper climate policy mix, for example, by deciding to pursue CE_1 instead of an alternative (CE_2). However, risks and uncertainty attached to a given CE method also pertain to permissibility. Specific methods may be deemed too hazardous or uncertain to be tried (Hartzell-Nichols 2012; McKinnon 2019; Shue 2010), not only because of their impact on the physical world, but also due to possible social outcomes (Baum *et al.* 2013; Morrow 2014; Tang and Kemp 2021).[16] Some CE interventions could constitute unacceptable gambles, especially when considering that most of the consequences will fall on future generations, that is, individuals who have no say in taking the initial "gamble."

Qualifying CE_1 as impermissible necessitates setting a threshold beyond which CE_1 should be rejected. A possibility is to refrain from researching or deploying methods that carry even infinitesimal, that is excessively tiny, catastrophic prospects (Hartzell-Nichols 2012). Imagine that a technology could immediately stop climate change, while presenting an extremely small probability of a global catastrophic risk,[17] for example, eradicating all life on Earth. Would the gamble be worthwhile? It could be argued that expected utility calculation could provide guidance, but not quite. No matter how low the probability of ending all life, the claim could be made that the mere prospect of such a horrific outcome should bar us from investigating the option further. Besides catastrophic events with extremely low probability, a second category consists of uncertainty, that is, events for which probabilities cannot be calculated or for which existence is unknown. While it may be impossible to identify in advance unknown events, it is still possible to *suspect* their existence in relation to specific actions. By their very nature, some CE interventions may be more prone than others to cause catastrophes that are undetectable for the time being. The claim is not that any "Black Swan" (Taleb 2007)[18] or suspicion of deep uncertainty justify discarding CE_1. Arguing so would set the threshold for impermissibility too low. Then, most if not all climate actions would have to be rejected, since starting from mitigation and adaptation, they contain uncertainty. Nonetheless, the core issue is to determine a threshold beyond which it would be immoral for humanity to venture further.

In the case of CE, time compounds the problem of risks with infinitesimal, extremely low, probabilities (Black Swans) and deep uncertainty. Since the kinds of interventions under discussion require a long and continuous deployment, from several decades for SAI up to millennia for CDR, with delayed effects, future generations are exposed to instances of risk transfer or externalization, in which present generations offload on the future generations part of the negative consequences of their decisions. In other words, the present generations, knowing well that they will not have to support the full costs of their decisions, may be tempted to engage in reckless and self-indulging climate policies, which constitute an occurrence of moral corruption (Gardiner 2006b 2011, 4548). Moreover, the fact that present generations may initiate CE, which will end up generating Black Swans or uncertainty several decades or centuries from now, raises issues of intergenerational justice (cf. below).

The importance of setting an acceptability threshold for catastrophic prospects explains the emphasis in the literature on the scalability and reversibility of CE methods, right from the R&D stage (Keith 2000, 275). To that respect, the distinction between "encapsulated and un-encapsulated technologies" (Hulme 2014, 10) takes on a particular moral significance. The former, in which interference in the environment is limited as is the case with DACS, would a priori be more permissible than the latter, in which effects cannot be contained such as in the case of SAI. Finally, such a threshold should apply also to ongoing climate change. A limit should be set for letting the Anthropocene unfold, beyond which the prospects of global catastrophic outcomes and uncertainty are so heightened that humanity is morally required to investigate and deploy CE.

Distributive Justice

The third permissibility criterion focuses on the consequences of CE methods, more specifically their disparate impacts. The possibility of CE_1 affecting various populations differently raises questions of *distributive justice* understood as "justice in the distribution of benefits and burdens to individuals, or (. . .) in the balancing of the competing claims persons make on the benefits that are up for distribution" (Olsaretti 2018, 2).[19] Climate models strongly suggest that particular CE interventions will create severe imbalances in the spatial and temporal repartitions of benefits and burdens (McLaren 2018; Horton and Keith 2016; Svoboda 2016; Svoboda *et al.* 2011; Tol 2016). This disparity problem is a distinct question from the overall generation of benefits and harms assessed by efficiency. It underscores that absolute global benefits, costs, harms, and risks are not the only pertinent dimensions that matter when judging of the acceptability of CE methods.

Evaluating disparate impacts triggers two preliminary issues. The first is the *spatial and temporal boundaries of the relevant moral community*. Distributive concerns apply to entities that have been deemed worthy of moral consideration. Put differently, when gauging CE outcomes, do distributive concerns apply to the inhabitants of the area in which the intervention is conducted, the adjacent regions, or the whole planet? Furthermore, should they extend to future human beings as well (Horton and Keith 2016)? If so, how should the interests of distant generations be balanced against those of the present ones? Should distributive concerns include animals and ecosystems? This last question highlights the issue of anthropocentrism in climate discussions. Simply put, anthropocentrism defines the ethical stance that ascribes priority or exclusivity to anthropogenic interests, welfare, rights, and so on by comparison with nonhuman ones (Moore 2017; Steiner 2005).

The previous interrogations underline the scope of the morally relevant entities, that is entities who are *prima facie* entitled to have their interests, rights, or well-being taken into account. The task of setting the boundaries of the moral community may have far-reaching implications. For instance, attributing substantial standing to marine life renders stratospheric sulfur injection (SSI)'s acceptability less straightforward. Since SSI does not tackle carbon concentrations, oceans will continue absorbing carbon and acidifying, which will be harmful for coral, shellfish, among other organisms. Thus, the moral frontier deeply shapes CE evaluation. Drawing such frontiers requires solid justifications, especially because a vast number of beings, entities, and ecosystems are and will be impacted.[20]

The second preliminary issue resides in the choice of a pertinent metric (or a set of metrics) for assessing the distributive characteristics of CE_1. To simplify, it is necessary to determine a currency of justice (Cohen 1989), which could offer guidance on which benefits, risks, and harms to count and how to count them. For instance, do climate parameters (average temperature and precipitation) constitute the (only) relevant distributive dimensions? Or should the consequences of the variation of such parameters on human (and nonhuman) rights, welfare, well-being, opportunities, and so on be included too (Carr and Preston 2017, 767; Flegal and Gupta 2018, 51)? If so, how can we measure aspects like welfare, opportunities, and rights? Normative thinkers have extensively analyzed those questions within the literature on justice. Those remarks should not be interpreted as pointing at dead ends, but as insisting on issues for which social sciences and humanities are of direct relevance. Indeed, political philosophers propose a plurality of conceptions of distributive justice that could be useful for thinking CE's disparate impacts (Svoboda 2017, 36–44).

Leaving such considerations aside, the permissibility of climate interventions depends, at a very general level, on how effects are distributed. Who is getting what in terms of benefits (e.g., temperature cooling and halt of sea-level rise), but also harms (e.g., precipitation changes, local warming, and alteration of ocean circulation) or risks/uncertainty (e.g., higher frequency of storms, floods, heatwaves, or hurricanes)? For instance, some models project that injecting aerosols over the tropical or Arctic regions could significantly interfere with the African and Asian summer monsoon, threatening food security and access to water, thereby hurting vulnerable populations (Incropera 2016, 144; NRC 2015b; Robock *et al.* 2008; Shepherd *et al.* 2009, xi).

In addition, CE methods may produce regressive results, that is disproportionally handicapping already disadvantaged groups. In the case of monsoon disruption, existing fragilities may be compounded by CE, deepening issues of justice, which is captured by the concept of "skewed vulnerabilities" (Carr and Preston 2017; Preston 2012; Gardiner 2011, 119120). Further, from an historical perspective, developing countries have had only a minor role in GHG emissions (low causal responsibility) while having benefited less from those than more industrialized nations. Thus, judgments about whether a given distribution of CE impacts is just, and how it could be made fairer, must be replaced in the whole picture of the responsibility for the Anthropocene. Overburdening populations who suffer from low socioeconomic development (stemming from past restricted access to fossil energies) and higher climate risk exposure, under the pretense of tackling harmful global changes for which those populations are marginally responsible, is morally problematic.

However, the fact that specific CE methods may further cripple already vulnerable and not-so-responsible populations is not necessarily unacceptable. Without endorsing crude consequentialism,[21] the possibility of compensation (offering something of equivalent value in exchange for the inflicted harms) or redress (correcting potential damages) could render such methods permissible, under stringent conditions.[22] Moreover, the duty to redress or compensate may not stop at the countries who decide to deploy CE, but could be extended to third-party nations, especially those who benefit from CE undertaken by others (Heyward 2014).[23]

A contrario, if redress or compensation mechanisms remain unattainable, then it becomes more arduous to justify engaging in climate interventions for which harms and costs are likely to be concentrated on vulnerable populations. Again, under nonideal circumstances, it is more difficult, but not impossible, to advocate for CE methods, which may disproportionally burden fragile groups, since extreme emergency, such as the prospect of a global catastrophe with dire and irreversible consequences, could vindicate sacrificing a minority. Another line of argumentation could be to demonstrate that disproportionally impacted, vulnerable populations would still be significantly better off than under the most plausible alternative scenario, or under all possible alternative states of the world.

The issue of disparate impacts underlines the importance of distributive requirements when evaluating the permissibility of CE methods: either initially, before adjustment, burdens and benefits should be equitably distributed, or it should be possible to re-equilibrate those effects ex post (i.e., after the intervention). The last option implies a degree of cooperation among states and international organizations that does not characterize current international relations. In any case, moral permissibility does not only depend on ex ante distribution, but on ex post compensation or redress, namely redistribution. And both distribution and redistribution presuppose the existence of a conception of fairness as well as of apt institutions, at the domestic and global levels.

In sum, if ex ante imbalance does not necessarily rule out specific CE methods, ex post unfairness, that is, disproportionate impacts that are left unattended, represents a more serious challenge for permissibility, especially if such impacts burden populations that are already vulnerable because of low socioeconomic development and/or due to the unfolding Anthropocene. Moreover, the distributive performance and redistributive prospects of various climate options or hybrid policies should be weighed against each other when deciding where to allocate resources for R&D or to permit deployment. Finally, part of the (re-)distributive prospects and performance attached to CE will depend on institutional arrangements distinct from the kind of CE involved. In other words, responding to the requirements of distributive justice presupposes the existence of well-functioning states as well as effective international treaties and organizations.

Procedural Justice

The requirements of justice transcend substantive, distributive, dimensions. They encompass how decisions which may generate disparate impacts as well as the rules for what is fair and right are made (Flegal and Gupta 2018; Hourdequin 2018; Meyerson and Mackenzie 2018; Preston and Carr 2018). "Unlike distributive justice, which pertains to how harms and benefits ought to be shared across persons, procedural justice pertains to processes whereby policy options are deliberated, evaluated, settled upon, and reviewed" (Svoboda 2017, 73). Procedural criteria bear on the characteristics of the decision-making process, the involved entities, and the form of their involvement. To be acceptable, decisions related to CE R&D and deployment should respect such criteria (Svoboda 2017, 73–94). Who should decide? About what? Under which constraints? Following which rules? With which safeguards? (Callies 2018; Gardiner 2014; McLaren 2018; Pamplany *et al.* 2020; Preston 2013). Moreover, the perceived fairness of the decision-making yields deeper trust in institutions carrying the process and, ultimately, higher public acceptance (L'Orange Seigo 2014, 853).

The foundation of procedural justice is encapsulated in the idea that, in order to be permissible, decisions and policies should not only generate outcomes that respect substantive principles (e.g., distributive fairness) but also abide by a specific decorum (Hourdequin 2018). On that matter, procedures can be subject to different rules and constraints. For example, Daniel Edward Callies (2019a, 115–116) offers two conditions in the case of solar CE: "fair terms of inclusion" (who should be part of the decision-making process?) and "fair terms of participation" (how those parties should be engaged?).[24]

Procedural concerns may expand into more exigent provisions, like *recognition* (Hourdequin 2019; Preston and Carr 2018; McLaren 2021; Whyte 2012) and *non-domination* (Smith 2018). Recognition constitutes a sphere of justice distinct but complementary to redistribution, centered on the standing of the different entities involved in a given social context, and their reciprocal attitudes (Fraser and Honneth 2003). To qualify as just, decisions not only need to meet specifications in terms of resources, benefits, harms, or risk distribution, they ought to respect the interests, views, rights, identities, and so on of the impacted parties, who should preferably be represented in the decision-making. For instance, participants should be placed on a rather equal footing, with their particular expectations acknowledged. "To enable people to engage 'full partners in social interaction' requires not only opportunities to participate in a formal sense in decisions that affect them; it also requires that people have similar status and standing as participants" (Hourdequin 2018, 273). In contrast, domination understood as the asymmetric capacity of an agent to arbitrarily interfere with less powerful entities (Pettit 1997) constitutes a significant risk that procedures ought to neutralize, especially in relation to CE (Smith 2018). Furthermore, domination in the context of CE research and deployment can carry a postcolonial or imperialistic tone when scientists from the Global North exchange with historically dominated groups (e.g., former colonies and indigenous peoples) (McLaren 2021; Whyte 2012, 2018).

Concretely, the necessity of involving a broad audience, arguably at the global scale (Jamieson 1996, 329), is stressed by official scientific bodies such as the Royal Society and the National Academies of Science (National Academies of Sciences, Engineering, and Medicine 2021, 9; NRC 2015b, 153; Shepherd 2009, xii). Such necessity motivates designing inclusion strategies, such as "upstream engagement" defined as "attempts [. . .] to involve members of the public in constructive dialogue about emerging and potentially controversial areas of science at the earliest possible stage" (Corner *et al.* 2012, 456). Beyond the specifics of what procedural justice requires, there are three justifications—*practical, moral,* and *political*—for including a large audience.

First, the practical justification is that CE R&D and deployment will presumably be facilitated by the affected parties' consent (Frumhoff and Stephens 2018; Visschers *et al.* 2017). A few research projects have already made efforts to integrate affected parties, such as indigenous peoples in an MCB experiment led by the University of Sydney and the Queensland University on the Great Barrier Reef (Oksanen 2023, 02). *A contrario,* the lack of consent could motivate citizens to obstruct projects that affected parties perceive negatively (Low *et al.* 2022), as it happened for ocean iron fertilization (OIF) (cf. chapter 3) and SAI (cf. chapter 4) research projects. More generally, insufficient participation may exacerbate the risk of litigation, nurture suspicion, and distrust vis-à-vis the scientific community and decision-makers (Raimi 2021). Substantial public participation may then reinforce trust in institutions overseeing CE R&D and deployment (L'Orange Seigo *et al.* 2014, 853). Nevertheless, any participatory initiative only motivated by the goal of securing acceptance to controversial initiatives at all costs will weaken permissibility. As a side effect, it could also strengthen research projects by exposing scientists to a broader spectrum of perspectives, feeding them with important information (e.g. about the Arctic local conditions when involving Inuits), or revealing hidden vulnerabilities.

Second, procedural concerns are indeed intrinsically important as manifesting the respect due to individuals as moral agents. Permissibility, however, demands more than incorporating individuals' interests and voices in the decision-making process under any form and shape. Participants ought to be involved in specific ways, in particular when it comes to collecting their consent. Their identities, cultures, and perspectives may have to be recognized as mentioned above, but more mundane constraints apply. A detour through medical ethics can help grasp the extent of such constraints. The consent of a patient, or of her representative if she is incapacitated, is a *prima facie* requirement for legitimizing any medical act. Not any form of consent is acceptable, though. It must be free, that is obtained without physical or psychological pressure, and informed, that is, based on adequate material about the nature, necessity, expected benefits, and potential risks of the medical procedure. The centrality of consent in medical contexts expresses the idea that when it comes to individuals' core interests, the existence of actual or prospective benefits is not enough for securing permissibility. Respect for individual autonomy is particularly critical (Beauchamp and Childress 2009, 101–149; Dworkin G. 1988).

Third, political legitimacy in democratic societies, whether of their institutions, policies, or actions, resides in the consent of the Sovereign, namely the People (Landes 2017, 87–91). For instance, the rulings of a dictator may be advantageous to his subjects, but those decisions will nonetheless lack basic democratic vetting and therefore legitimacy. A direct implication is that, whether explicit, implied, or hypothetical, consent is essential for legitimizing CE (Wong 2016).[25] Consent, however, constitutes only one form, probably the most minimal, of public participation. More ambitious and demanding mechanisms may involve substantial consultation or popular contribution to decisions about CE R&D and deployment.

In any case, the argument that collective decisions draw part of their legitimacy from the participation of those who are subject to them is captured by *the all-affected interests principle* or *all-affected principle* (Hourdequin 2018; Lenferna *et al.* 2017, 586). It stipulates that "everyone who is affected by the decisions of a government should have the right to participate in that government" (Dahl 1970, 49).[26] For being acceptable, a decision thus needs to fulfill a fundamental requirement, which is to accommodate for the participation of individuals and groups impacted by them *in some form or shape*. Although the principle has been mostly discussed within circumscribed political communities such as states, it does not preclude a duty to involve affected parties living beyond the frontiers of the country in which decisions related to CE are made.

However, CE's potential global scope complexifies the implementation of the all-affected interests principle. Since the effects of DACS, SAI, or space CE may be global and span across time, most existing humanity as well as future generations may have to be involved. This may turn out to be impossible, but practical limitations do not forcibly invalidate moral imperatives. As a comparison, the elusiveness of perfect equality or freedom does not render those principles obsolete. Thus, despite practical limitations, the all-affected interests principle retains its moral force, imposing to engage, *to a reasonably feasible extent,* the parties impacted by specific policies, particularly for such a primordial topic as climate manipulation.

For being realistically achievable, procedural requirements presuppose an ounce of cooperation among rather *legitimate*[27] states and international institutions with the intent of pursuing a global justice agenda. The concentration of the R&D capacities in the industrialized nations and the fact that vulnerable groups, often located in developing countries, may be disproportionally impacted by research and deployment may undermine CE permissibility (Carr and Preston 2017; McLaren 2018; Raimi 2021, 68). Permissibility is further undermined when such vulnerable groups are excluded from making the decisions that degrade their life conditions. A consequence is the impetus of broadening studies on CE perceptions to include the Global South (Carr and Yung 2018; Sugiyama *et al.* 2020; Visschers *et al.* 2017) and other groups as mentioned above such as indigenous peoples (Oksanen 2023).

The issue is not bounded to setting up ad hoc public consultation; it also reflects deeply ingrained, systemic, problems, a major one being the limited representation of developing countries among the researchers working on CE (Biermann and Möller 2019; Carr and Preston 2017, 768–769; McLaren and Corry 2021). A putative danger is the imposition by a "geoclique"[28] (Kintisch 2010, 8) of its favorable views on CE, more specifically on SAI, without conferring with potentially affected populations, especially in developing countries (cf. chapter 5). The current deficiencies regarding social engagement advocates for reinforcing participatory mechanisms, as early as the R&D stage (Winickoff *et al.* 2015). Such mechanisms could range from requirements of transparency and consultation (Carr and Yung 2018; Long *et al.* 2015, 31) to granting vulnerable groups in the Global South a preeminent role in decision-making processes (Rahman *et al.* 2018). To that respect, the unbalance may be progressively tilting as emerging economies like Brazil, China, India, and Indonesia are increasingly endorsing a prominent role in the research on CDR (Smith *et al.* 2023, 29) and developing their CE research capacities (Cao *et al.* 2015).

In sum, the procedural dimension is crucial for guaranteeing the permissibility of specific CE methods. Demonstrating that CE may be (overwhelmingly) beneficial does not exhaust what permissibility demands. Constraints apply to how decisions are made on the top of substantive conditions. Such constraints may mandate public involvement at different levels, under diverse forms (e.g., consultation, recognition, and direct participation), in various regions, especially where vulnerable groups, whose voice is usually inaudible, can potentially be affected (Hourdequin 2019). However, acknowledging the importance of such concerns, reinforced by recognition and non-domination requirements, offers limited practical guidance. Per se, it does not specify which kinds of procedures could strengthen permissibility. Thus, essential questions need to be answered, such as who should participate in the CE process? Under which conditions? At which stage? From the R&D phase (Frumhoff and Stephens 2018) or later? These interrogations raise deeper governance issues.

A last and important point: although it could be claimed that specific CE methods engender specific difficulties for respecting procedural principles, procedural justice mainly highlights the issue of constraints external to CE methods proper, constraints that depend on the institutional context. In other words, CE interventions may be permissible, or not, from a procedural point of view, not in virtue of their intrinsic characteristics, but in respect to how existing institutions could accommodate a broad spectrum of individual and collective interests and identities. Finally, the absence or ineffectiveness of dedicated institutions does not definitively rule out CE methods. Institutions can always be amended or created *ex nihilo*.

Social Acceptability

Social (or public) acceptability refers to the extent to which a given method could be viewed as permissible, and maybe endorsed, by society.[29] An array of factors determines acceptability (Raimi 2021), like perceptions of potential benefits and their distribution (Cummings *et al.* 2017, 259), scientific transparency and robustness, risk/uncertainty (Carr *et al.* 2012, 174–175; Carvalho and Riquito 2022, 908–911), controllability (Bellamy *et al.* 2017), naturalness (Burns *et al.* 2016; Sweet *et al.* 2021; Wolske *et al.* 2019),[30] as well as values, and beliefs (Clingerman and O'Brien 2014; Visschers *et al.* 2017). Less relevant factors may also shape social views, such as framing effects, which designate the influence of how the information is presented on people's beliefs and choices,[31] and spurious analogies (Burns *et al.* 2016, 537; Corner and Pidgeon 2015; Corner *et al.* 2013; Raimi *et al.* 2019), the influence of which stems from widespread unfamiliarity with CE (Corner *et al.* 2012; Corner and Pidgeon 2014; Cox *et al.* 2020; Cummings *et al.* 2017 253–254; Sugiyama *et al.* 2020; Wibeck *et al.* 2015).

In addition to low awareness, other findings regularly appear in various research studies on public perceptions. A prominent one is the belief that even the remote possibility of engineering the climate may lower the motivation to mitigate (Burns *et al.* 2016, 539; Corner and Pidgeon 2014; Raimi *et al.* 2019), independently of whether CE methods ultimately deliver on their promises to capture carbon or enhance albedo. That risk is frequently mislabeled as "moral hazard" (Hale 2012), more accurately referred to as "risk compensation" (Reynolds 2015).[32] De-incentivization could be even more prejudicial in that CE interventions may not tackle the root cause of climate change (fossil energies), just offering a fancy "band-aid" (Carr and Yung 2018; Carvalho and Riquito 2022; Cox *et al.* 2020; Wibeck *et al.* 2015). Other studies, however, show that learning about CE could galvanize people in support to more aggressive conventional mitigation (Corner and Pidgeon 2014). Respondents are usually also concerned with controllability, containment, and predictability of the consequences and make a distinction between research and deployment (Bellamy *et al.* 2017; Burns *et al.* 2016).

When considering the academic knowledge on public perceptions, the studies display a clear sample bias (Sugiyama *et al.* 2020). Surveys are heavily skewed toward populations located in Northern, industrialized, countries while largely disregarding the Global South and indigenous peoples (Carr and Yung 2018; Preston and Carr 2017; Whyte 2018; Winickoff *et al.* 2015). As per the procedural requirements seen above, the relative absence, not only in the decision-making process, but in the opinion surveys of groups who are prominently exposed to the Anthropocene while potentially being vulnerable to CE-induced harms, raise moral concerns (Suarez and van Aalst 2017; Raimi 2021, 68). Broadening the scope of studies of public perceptions may also help to better understand the diversity of the judgments toward CE and of the governance challenges across the globe (Visschers *et al.* 2017). For instance, the views on CE, notably SRM, appear more positive in the Global South than in many industrialized nations (Sugiyama *et al.* 2020).

Public perceptions, and ultimately acceptability, of specific CE interventions may be divorced from experts' assessments. Public opinion could reject a technology because of supposed risks despite scientific evidence pulling in the opposite direction. Such rejection may handicap R&D and deployment (Lin 2019, 543–545; Low *et al.* 2022). Thus, securing social acceptance appears crucial in particular for field experiments (Shrum *et al.* 2020, 2). In addition to rampant reticence, unfamiliarity may also undermine or distort acceptance (Burns *et al.* 2016; Carvalho and Riquito 2022, 906; Raimi 2021; Scheer and Renn 2014; Smith *et al.* 2023, 46), which constitutes a widespread problem since CE is invisible in the mainstream media in most countries (e.g., Fujiwara and Sugiyama 2016; Wibeck *et al.* 2015). However devising studies with self-reflective, deliberative, settings may not significantly change respondents' original perceptions of CE (Carlisle *et al.* 2022). In any case, other considerations, such as the trust in governing bodies or scientists could impact social acceptability (Carr *et al.* 2012, 181–182; Cummings *et al.* 2017; L'Orange Seigo *et al.* 2014; Raimi 2021).

Social acceptability could be interpreted as an external constraint that ought to be respected as such, even in the case of prevalent ignorance and misunderstanding. However, due to rampant unfamiliarity with the CE techniques, studies and communication on CE might influence social views through framing effects (Bellamy and Lezaun 2017; Corner *et al.* 2013, 945; Raimi *et al.* 2019). Although it may be impossible to avoid any frame when communicating scientific information (Scott 2012), such influence may sometimes be bordering on manipulation, which raises issues of research ethics and democratic governance. Upstream engagement could be used, but with extreme caution, for correcting the lack of familiarity and building legitimate acceptance. In other words, public engagement could serve to generate the social license (Cox *et al.* 2020; Gough and Mander 2019). However, nothing guarantees that strengthening stakeholders' involvement regarding specific CE methods, for instance by organizing public consultations and focus groups and collecting people's views and concerns, will necessarily improve acceptance (Frumhoff and Stephens 2018). Moreover, participatory tools could be used opportunistically for forcing acceptance, which would undermine legitimacy (McLaren and Corry 2021).

In conclusion, it could be argued that obtaining the consent of a substantial portion of the population, preferably the majority and/or the most vulner-able,[33] in respect of the all-affected interests principle, is required for secur-ing the permissibility of any CE intervention. Significant social opposition is not definitively crippling. R&D and deployment may proceed even in the presence of a strong public rebuttal, but under strict conditions like in cases where the rejection of CE is based on widespread misconceptions, or under the extreme emergency caused by looming catastrophes. As a practical mat-ter, too stark a rejection of CE, even if based on misconceptions, could none-theless impede feasibility by stirring violent reaction, which may undermine the legitimacy of political institutions. Lastly, studies on public acceptance possess another advantage. They provide priceless information about the dangers as well as the social, political, and ethical issues that public opinion thinks CE methods raise. In democratic regimes, those are precisely the topics on which scientists and decision-makers should take a stance.

Intergenerational Justice

As evoked above, for CE to be permissible the foreseeable impacts on future generations need to be accounted for. Present generations make all sorts of arrangements that will affect humans to come, for example, by creating path dependency (e.g., the carbon-based economy and result of past generations' research and investment decisions), building institutions that will endure (e.g., constitutions, laws, and state agencies), passing along substantial costs (e.g., nuclear waste), or creating risks and uncertainties. Future generations may also benefit from accumulated socioeconomic and technological growth, assuming that all the gains will not be wiped out by some catastrophes or major economic crises. However, the preferences of future generations are often unknown, which constitutes a problem when important decisions need to be made.

Intergenerational concerns underscore the fact that generations situated at different points in time influence each other, triggering obligations (Gosseries and Meyer 2009); the stream of influence is mainly going forward, from present to future generations. When assessing the permissibility of specific CE methods, it is important to consider, not only their potential impacts on current generations, but also how they may affect the interests of future generations. Intergenerational duties are not restricted to future fellow citizens, but conceivably extend to all humans.

Concretely, intergenerational justice demands us to answer questions like, for example, does SAI R&D or SAI deployment benefit future generations everything considered?[34] Is there a risk that they will be trapped into SAI (Keith 2013, 16–17) or face moral dilemmas that could have been avoided if previous generations had adopted other strategies such as aggressive mitigation (Ott 2012)? The response to such interrogations varies depending on the specific CE intervention under discussion (e.g., BECCS, MCB, and DACS). The general point is that researching some solar CE and CDR methods may make future generations better off under certain conditions. Substantial investments in R&D may improve the prospect of technologies maturing enough for reducing GHG concentrations or temperatures. In short, well-thought-out CE research may end up in "arming the future" (Gardiner 2010). Nevertheless, investment decisions have opportunity costs. They may detract resources from other valuable and possibly more pressing goals such as mitigation and adaptation.

Deciphering the interests of future generations represents a complex endeavor, in particular the further away from us those generations are placed. The difficulty is epistemic, that is related to knowledge. Although such interests might be arduous to identify for the time being, it does not lower the moral imperative of trying to approximate and include them in present decisions. The obligation is even more pressing because, by definition, future generations are vulnerable due to their exclusion from current decision-making processes. They face procedural injustices, which often result in "intergenerational buck-passing" (Gardiner 2011, 35) and cost externalization.[35] For instance, the fact that present generations are delaying conventional, aggressive, mitigation means that future generations will be experiencing a more hostile climate. Therefore, future generations will partly foot the bill of the inaction and self-serving behavior (maintaining the use of fossil fuels because it is cheap) of the present generations. As a result, the latter could be viewed as dumping onto the former part of the burdens, negative consequences, and risks of their choices.

Once the principle of including future interests is recognized, further issues emerge. To which extent, and how, such interests should be integrated into nowadays decision-making? How to balance future with present interests? How to weigh conflicting concerns, such as not remaining trapped into SAI versus alleviating harms due to heat waves?

A common answer in economics to the "to which extent" in the first question is the social time discount rate (STDR)[36] (Arrow *et al.* 2013). Discount rates are generally used for calculating the value of goods and services at various points of time. For example, what is the current value of US$100 received in twenty years from now? To calculate the present value, one needs to apply a discount rate, usually yearly, for every year, starting from the year 20 and then going backward. The higher the rate, the lower the present value of this future sum. STDRs are particularly important for figuring out if a current investment is worth its future benefits. To simplify, if the sum that is required to invest now is lower than the discounted value of the future benefits, the current investment is worth making. If the sum to be invested is higher than the discounted benefits, then the investment is not worth making. Applied to climate change, the STDR represents "the relative weight of the economic welfare of different households or generations over time" (Nordhaus 2007, 690). It allows comparing diverse generations' interests at a given moment, which explains its centrality in intergenerational justice. "Discounting is a factor in climate-change policy—indeed in all investment decisions—that involves a relative weight of future and present payoffs" (Nordhaus 2007, 689).

The higher the STDR, the lower the priority of the interests of or benefits to the future generations when weighed against the interests or benefits of the present generations. The explanation is that future interests or benefits are rapidly losing value as they move away from the present. Several reasons can justify this devaluation. Future generations may be speculated to end up better equipped than current ones for tackling climate change due to technological progress, accumulated wealth, better knowledge and research, more efficient institutions, and so forth. In that context, an STDR offers a "way of rendering commensurable costs and benefits that occur at different points in time" (Heath 2021, 159), and its rate expresses the underlying assumptions about how better off future generations will be.

Determining an STDR value is complex and controversial as illustrated by the debate between Nicholas Stern, author of the emblematic *Stern Review* (Stern 2007), and William Nordhaus, recipient of the Nobel Prize in economics in 2018. Stern defends a near-zero STDR, which captures the climate emergency by applying a low rate on future interests. His proposal is justified by the prospect of catastrophes, which would impose to aggressively cut carbon emissions without delay. In contrast, Nordhaus (2007, 2008) supports a higher STDR, based on the return on investments of financial markets, which leads to more modest mitigation at first, before progressively ramping up carbon abatement toward the end of the century.

Since it can be reformulated into distributive and procedural terms, the intergenerational dimension may appear to be redundant within this evaluative framework. The dimension nevertheless deserves to be singled out for not losing sight of the misaligned interests across generations (Gardiner 2011, 3244). It also constitutes a welcome reminder of the intertemporal requirements of justice when it comes to climate change and engineering. As for distributive and procedural concerns, it could be claimed that no CE method intrinsically carries intergenerational qualities. Such qualities would depend on the institutional framework, arrangements that make future generations bear the burdens of current CE implementation or not. Against this view, it could be argued that some CE methods may in themselves convey more serious future risks or be more prone to buck-passing and cost externalization. For instance, if deployed in a context of too modest mitigation, SAI represents a threat of path dependency; the longer the intervention, the stronger the lock-in. In consequence, SAI and similar techniques would be more difficult to justify.

Regulation and Governance

Defined as the "the sustained and focused control exercised by a public authority over activities valued by the community" (Selznick 1985, 363),[37] regulation plays a decisive, but not definitive, role for the permissibility of climate interventions, in the sense that some CE methods may be more difficult than others to administer. Regulation is understood in this section as mostly belonging to the legal and institutional realms and part of a broader field, governance, which is "the process of steering society and the economy through collective action and in accordance with common goals" (Ansell and Torfing 2022, 3). Governance tools' range is quite large, including codes of conduct, assessments and reviews, permits, patents, committees, and so forth (National Academies of Sciences, Engineering, and Medicine 2021, 11).

Along with morality, law is one of the dimensions of normativity, whose direct aim is to regulate human behaviors and interactions. Projects of climate interventions against which there is no serious intrinsic, distributive, procedural, or intergenerational objection and which do not cross some unacceptable risk/uncertainty threshold might nonetheless be hindered by domestic and international laws or the impossibility of governing them. A moratorium could be imposed on SAI deployment or even research. Constraints may apply to carbon storage in deep aquifers. Such limitations may not be solidly justified.[38] They could be arbitrary or motivated by various particular interests. In addition, they could be anachronistic, resulting from outdated tradition or constituting some imperfect transposition of legal provisions originally crafted for an adjacent topic such as biodiversity protection or marine pollution. But the law is not set in stone. It can be reformed, even with difficulties. Just because a CE intervention is illegal at a given moment, that does not mean it will always be, and *vice versa*. Legislation changes, for strong and weak reasons.

With those caveats in mind, it should be acknowledged that domestic and international regulations of CE research as well as deployment are fragmentary (McLaren and Corry 2021). At the time this book is written, no country possesses a dedicated regulatory and governance apparatus for CE. Further, no binding international instruments (e.g., treaties) directly bear on climate methods (Reynolds 2018; Talberg *et al.* 2018a). Tangential texts can be applied to CE, while being subject to interpretations, like the London Convention on the Prevention of Marine Pollution by Dumping of Wastes and Other Matter (1972) (and its afferent Protocol [1996]), the Convention on the Prohibition of Military or Any Other Hostile Use of Environmental Modification Techniques (EMTs) (ENMOD) (1977), and the Vienna Convention for the Protection of the Ozone Layer (1985) (and the Montreal Protocol [1987]). But those legal instruments did not initially target CE.

The Conference of the Parties (COP) under the Convention on Biological Diversity (CBD) is one of the few international institutions which tackles CE (Reynolds 2018, 96–100), with a focus on the ocean fertilization. The CBD assembles 195 States and the European Union, but with the notable exception of the United States. The COP enacted at least four decisions related to CE (IX/16 [2009]; X/33 [2010]; XI/20 [2012]; XIII/14 [2016]). The main points reasserted throughout those texts are the lacunar state of scientific knowledge on CE, the necessity of limiting CE research to "small scale scientific studies that would be conducted in a controlled setting" (decision X/33, [w]), the priority to be given to mitigation within climate policies, the importance of avoiding transboundary harms, and the precautionary principle, in particular concerning biodiversity. The CBD is calling for a moratorium on large-scale CE experiments (Tollefson 2008), especially if some impact on biodiversity is suspected, which is sometimes misinterpreted as an enforced ban. However, its decisions lack full legal force, being nonbinding, and their implementation is left to the signatory parties. In any case, the CBD represents one of the few international instances explicitly tackling CE, therefore constituting a potential regulatory body along with the London Protocol (Grisé *et al.* 2021, 10–11).

The fragmentary domestic and international legal framework is mirrored by a vacuum in the formal structures of governance, especially regarding research (Burger and Gundlach 2018; National Academies of Sciences, Engineering, and Medicine 2021), which has led to self-regulation initiatives, such as the Oxford Principles (Rayner *et al.* 2013) and the Tollgate Principles (Gardiner and Fragnière 2018) (cf. chapter 5). The lack of formal structures should not be equated with a total absence of governance. Those deficiencies have offered favorable conditions for what has been branded as "governance-by-default" (Talberg *et al.* 2018a), characterized by unresolved tensions between two international norms: precaution and harm minimization. In short, some "de facto governance" would be already in place, defined as "sources of governance that are unacknowledged and unrecognized as seeking to govern, even as they exercise governance effects," which will be "distinct both from formal, state-led, legally binding de jure forms of steering, as well as informal, non-state sources of steering, which share the characteristic of intentionally seeking to steer the behavior of certain actors or institutions" (Gupta and Möller 2019, 481). *De facto* governance would stem from small, quasi-seclusive, epistemic communities (McLaren and Corry 2021, 5–6), that is, experts contributing to assessments on CE, for example through the Royal Society and the National Academies of Sciences, Engineering, and Medicine reports.

A consequence of the fragmented framework is that a significant dose of interpretation is necessary when evaluating the promises and hurdles that the law or various institutional arrangements could present to CE R&D and deployment. When considering any CE method, it is important to conduct an ad hoc review of not only the texts and institutions which directly tackle this kind of intervention, but also peripheral material and structures. The next two chapters will offer more specific assessments of the legal and governance opportunities attached to different CE initiatives. But, so far, no solid ground exists for prohibiting the R&D and deployment of CE.

PREFERABILITY

For policy purposes, any analysis of a given CE method (CE_1) ought to combine feasibility (is CE_1 research or deployment possible?), permissibility (is CE_1 acceptable on moral, political, legal, and other normative grounds?), and preferability (how does CE_1 fare against the *status quo* as well as feasible and permissible alternatives? Or could CE_1 be the preferable strategy considering the range of available options, *status quo* included?). Preferability reflects the fact that R&D and deployment decisions about CE are taken under conditions of resource scarcity, therefore imposing to carefully evaluate the *relative* efficiency and permissibility of various strategies. It also captures the importance of moral trade-offs. As previously underlined, the partitioning between the three dimensions is not airtight. Elements of permissibility are present in feasibility assessments, and preferability plays an important part in permissibility judgments. For one thing, when contemplating a CE intervention that implies potential harms and catastrophes, permissibility should be grounded in how such a decision performs relatively to other feasible alternatives. In any case, those three dimensions are conceptually distinct, and for the sake of exposing rather clearly their specificities and significance, the tripartite, although a bit artificial, division retains some cogency. However, before reviewing some of the many rules for assessing preferability, it is important to acknowledge that issues muddy the ability to compare alternatives.

Complicating Factors

First, assessing preferability is complicated by *the nonideal circumstances characterizing climate politics*, namely the partial, or absence of, compliance with the duties of justice (Callies 2019a; Morrow and Svoboda 2016; Svoboda 2017). The fact that CE is increasingly considered as a legitimate component of climate policy illustrates such circumstances. Had humanity acted on its duties to vulnerable populations, future generations, and other living beings, there would be no need for CE. But a long streak of global inertia, leading to piling up GHG concentration, has contributed to a situation in which suboptimal initiatives such as CE have gained traction. Further, contextual elements reinforce the nonideal conditions. Material constraints and contingencies cripple climate policy efficacy, making first-best, ideal, perfect outcomes unattainable. No policy portfolio will maximize or even secure the welfare, rights, or any other morally relevant variable for all-affected parties. Strategies will underperform, trade-offs will have to be made, and unforeseen circumstances will arise. There will be net winners and losers, and those positions may shift through time as the Anthropocene and the consequences of policies unfold. CDR interventions like BECCS, which are included in most of the IPCC scenarios, will necessitate the use of vast areas, compromising food security (Anderson and Peters 2016) and biodiversity (Creutzig *et al.* 2014). SRM may prompt catastrophic outcomes and establish novel climates that will harm many, particularly among vulnerable groups, through droughts, floods, and alterations of monsoon patterns and ocean circulation (Flegal and Gupta 2018; McLaren 2018). Aggressive carbon abatement itself carries opportunity costs under the form of foregone socioeconomic growth and increased poverty, especially in developing countries (Callies and Moellendorf 2021). Thus, in the context of climate policies, preferability consists in balancing unsatisfactory alternatives in relation to a baseline that cannot be reasonably constituted by a state of affairs unaffected by the unfolding Anthropocene. The benchmark should be a future world which follows current, inadequate because too modest, mitigation. As a result, policymaking, *a fortiori* when including CE interventions, puts agents in front of dilemmas,[39] which explains the prominence of lesser evil arguments in the discussions surrounding climate alteration.

The comparative task is, however, entangled in deeper hurdles. A second obstacle resides in what moral philosophy refers to as *the fact of pluralism* (Rawls 2005, 36), namely that individuals usually endorse different, potentially conflicting, values, principles, conceptions of the good, and so on. Moral pluralism exacerbates well-known difficulties for establishing social preferences (Arrow 2012). A dimension so crucial as the value of human life is subject to dissents that cannot be settled through monetary estimation as economists often do (Broome 2012, 156168; Svoboda 2017, 22–27). Even when setting aside abhorrent beliefs (e.g., racial gradient of lives' value), there is still room for a fair dose of discontent among reasonable views when it comes to regulate the life in political communities, for instance about the place of religious symbols, the possibility to voluntarily interrupt life through death penalty or abortion, and the kind of priority (absolute or relative) to confer to mitigation. Individuals disagree to such an extent that the prospect of a unanimous accord on preferability evaluations is exceedingly low, especially on matters as contentious as CE. So, at minimum, any assessment and comparison framework need to pass a test of justifiability, that is, be potentially justifiable to reasonable agents.

A related, third, difficulty resides in the *diversity of goods to be integrated in the analysis*: the number of saved lives, improved or preserved well-being, upheld freedoms and liberties, disparate impacts and risk distribution on human and nonhuman beings, ecosystems, and so on. Comparisons need to be conducted through a broad spectrum of dimensions. It may be desirable to weigh goods against each other and aggregate them, or at least pool them in composite indicators. If the normative assessment has any ambition of being somehow exhaustive, other aspects than straightforward harms and benefits ought to be included. For this comparative exercise, economists typically use money, a choice that has limitations. A serious one is that climate change and engineering (will) affect human and nonhuman entities in ways that may not be monetarily evaluable in a reliable or satisfactory manner. For instance, how to convert in *meaningful* monetary terms the loss of biodiversity, which does not generate economic services, imputable to OIF and extensive feedstock cultivation, or the whitening of the sky due to SAI? In the absence of a market valuation, economics mobilizes the willingness-to-pay ("the maximum amount that a buyer will pay for a good" [Mankiw 2021, 132]), which assumes that individuals can produce robust and internally consistent monetary estimates not crippled by cognitive biases, an assumption that can be challenged (Diamond and Hausman 1994; Heath 2016 2020). This third difficulty is distinct from the previous one. Even if individuals shared identical, or at least mutually compatible values and principles, dilemmas would still exist. This does not imply that balancing various evaluative dimensions is an impossible endeavor. The social sciences and moral philosophy host multiple attempts at overcoming the issue. This simply means that the task should not be underestimated.

A fourth obstacle is that some CE methods such as SAI and MCB carry *significant probabilities of disparate impacts*. As a result, any preferability judgment ought to distinguish CE_1 (or the overarching hybrid policy) global aggregated outcomes from the set of locally and socially dispersed outcomes. Benefits and burdens of a varying nature and extent affecting diverse populations should be compared. Despite the expectations of net global benefits, CE_1 may still have to be ruled out due to impermissible strains it imposes on vulnerable groups or the impossibility of compensating the harmed populations in an acceptable manner (as per Kaldor–Hicks requirement). This point underlines distributive concerns.

A fifth difficulty resides in the fact that, to be worthwhile for policy-making, *preferability assessments need to go beyond simply contrasting separated climate initiatives against each other* (mitigation, adaptation, CDR, and SRM) or against the *status quo* (low abated climate change) (United Nations Environment Programme 2023, 20–22). Comparing SRM/ CDR interventions against each other (e.g., MCB versus DACS, SAI versus MCB, and BECCS versus DACS) is of limited informative value. No serious researcher advocates for a climate strategy restricted to a single CE method, for example SAI. CE methods will be combined with other options, including mitigation and adaptation. The dominant stance among experts is to ponder over the implementation of "hybrid" (Svoboda 2017) or "portfolio" (Keith 2013, xix; NRC 2015a, 2015b) strategies that incorporate CDR and/or SRM with mitigation and adaptation. Preferability assessments should apply to the characteristics and potential outcomes of specific CE initiatives *and* the interactions of such methods with different interventions. For instance, it is essential to evaluate SAI on its own merit, but it is equally crucial to consider how well it performs within a hybrid policy, by reinforcing or interfering with mitigation, adaptation, and other CE measures. For example, due to risk compensation (Keith 2013, 128–130), several CE methods, in particular SAI, may deter carbon abatement (Shepherd *et al.* 2009, 37). Or, to the contrary, the perspective of its research could incentivize agents to accelerate mitigation (Reynolds 2015, 178–179). Since SAI will most likely dim the sunlight and alter the amount of UV-B reaching the ground (NRC 2015b, 97), it may be detrimental to crop growth (Proctor *et al.* 2018). Furthermore, the impact on cultivating energy feedstock used for BECCS needs to be cautiously studied, especially considering the prominent role of BECCS in integrated assessment models (IAMs).

A sixth difficulty is the *uncertainty that characterizes climate change as well as the potential climate responses to CE* (Svoboda 2017, 10–13). The impossibility of foreseeing adverse events or calculating their probabilities not only complicates the evaluation of CE initiatives taken individually, but also muddies preferability assessments. Various methods may generate diverging levels of uncertainty, which could constitute a criterion of preferability in itself, for instance, by justifying the adoption of a decision rule that aims at reducing deep uncertainty. Uncertainty might be compounded by the combination of different climate initiatives within policy portfolios.

The next two chapters will introduce to the potential impacts of the main CE methods. They will therefore offer elements of comparability. However, to hammer the point once more, it is one thing to analyze a CE intervention in isolation and quite another thing to assess the effects and interactions between components of *a hybrid strategy*. Concerns raised during the evaluation of separate components may gain traction with adverse effects being reinforced when considered within a hybrid approach. Other concerns might be lessened or disappear.

A Glimpse into Comparative Principles

Assuming those caveats, how can we compare different CE options or hybrid policies? This section does not offer an exhaustive presentation of all the existing principles. Instead, the goal is to provide a rough sense of how comparison could be undertaken, as well as the respective advantages and disadvantages of diverse rules. Moreover, it aims at unveiling the rationale as well as the limits in the different dimensions, burdens and benefits, that could be subject to comparison.

A first, somewhat obvious, possibility needs to be evacuated. It will consist in adopting a *Paretian* rule. Pareto efficiency stipulates that "a situation is efficient if no change is possible that will help some people without harming others" (Frank *et al.* 2022, 132). Then, a Pareto superior CE method, or hybrid policy, represents an optimal situation in the sense that any departure from it will necessarily harm people. Thus, any departure should be opposed because it will create losers. However, as indicated above, and developed further in the two forthcoming chapters, CE will most likely spur disparate impacts on various populations, significantly harming some groups, while still generating overall benefits (e.g., reduced global average temperatures). This suggests that almost any choice implying CE would presuppose trade-offs that infringe Paretian efficiency. Even CDR methods, which may appear as relatively safe in that regard, may nonetheless cause damages. For instance, massive reforestation or energy crop cultivation could trigger land-use conflicts, rising land and food prices. There may be exceptions: if it becomes mature and scalable, DACS might come close to constituting a Pareto improvement since the method does not require extensive resources (e.g., land and water) while potentially benefiting everyone and making no one worse off at first sight. More generally, the tragedy at the heart of the Anthropocene is that any climate action, especially hybrid strategies including mitigation and adaptation, will produce losers. This does not mean that nothing can be done about it (e.g., redress, compensation, and redistribution), but it appears to limit the applicability of the Paretian rule.

A less demanding rule is to determine the climate strategy that *maximizes gross* (as opposed to net) *benefits*, while benefits are understood broadly, for instance under the concept of utility. An initial methodological issue, also pertaining to distributive justice as shown above, is to decide what could be counted as a benefit (e.g., temperature reduction, avoided deaths, increased well-being, and diminished suffering) and to whom (e.g., humans, nonhuman beings, and ecosystems). An inclusive, but challenging, choice would be to optimize gains across a multivariable function for all living beings and ecosystems. The task would be daunting, yet not impossible. Authors argue against limiting assessments to monetary payouts only, since some benefits may not be convertible (Svoboda 2017, 12). Other reasons could lead to questioning an overfocus on money, such as a critique of anthropocentrism, since using money as a metric equates attributing value only to entities and dimensions that are valuable for and by humans. In any case, any multifactorial evaluation requires a common metric (e.g., utility, welfare, and capabilities), or a basket of metrics, as well as *some* rather unified aggregative framework in which different goods for different moral subjects are pooled and, potentially, weighed against each other.

However, any policy generates benefits *and* costs. Therefore, appraising only the gross benefits leaves out of the picture a large part of the relevant information. An alternative is *cost-effectiveness*. As noted above, cost-effectiveness consists of comparing strategies on their efficiency at achieving a given objective or a set of given objectives (e.g., the reduction of radiative forcing by x percent). The assessment could be subject to different principles. An obvious one is cost or harm minimization. However, in the case of such emerging technologies as renewable energies or CE, models might be limited in their capacity to account for rapid innovation and the ensuing decline in costs (EASAC 2022).

A related rule is the *maximin* "according to which we ought to prefer that policy whose worst outcome is better than the worst outcomes of any other available policy" (Svoboda 2017, 15). Used for decisions under uncertainty, maximin could be interpreted as a formulation of the precautionary principle (Gardiner 2006a). In the context of CE, it forces to pick the method or hybrid strategy that is characterized by the worst outcome that is the best among all the (feasible and permissible) alternative methods or hybrid strategies under consideration. SAI research is often defended on maximin ground. For instance, hybrid strategies with SAI, at least under the form of an R&D program, are advocated for as not containing the worst consequences of climate change because they offer temperature peak shaving (The National Academies of Sciences, Engineering, and Medicine 2021, 113) or as maximizing the benefits/minimizing the damages to the poorest, most exposed to climate change groups (Horton and Keith 2016).

A more complete principle is *the maximization of net benefits,* which focuses on the differentials between benefits and costs. Compared to the previous rules which optimize only one variable at the time, either costs or benefits, *the maximization of net benefits* accentuates the complexity of the decision-making process. Akin to the cost–benefit analysis (CBA) discussed in the feasibility section, it is applied within a comparative framework. As mentioned above, the valuation and summation tasks, prior to the final preferability evaluation, remain tricky, considering that contrary to more "traditional" cost–benefit calculations, not all CE costs and benefits could be translated into monetary terms (e.g., endangered human cultures, damaged ecosystems, and reduced biodiversity). But there are more or less satisfactory manners of answering the challenge, such as using willingness-to-pay (Heath 2020, 187–253).

A possible objection against preferability assessments based on a CBA is to point out the pervasive risks and uncertainty tied to CE, which could be compounded if various initiatives are simultaneously implemented (mitigation, adaptation, CDR, and SRM) and interact in unpredictable ways. Such amplified risks and uncertainty may jeopardize optimization. A conventional solution for integrating risks is to rely on *expected utility theory.* Simply put, expected utility consists of summing the different outcomes of a course of action (e.g., a given policy portfolio) multiplied by the respective odds attached to each of those outcomes. Such a solution works well as long as all of the probabilities can be calculated or approximated. As uncertainty grows, probabilistic evaluations become hazier.

Finally, preferability assessments could eschew quantitative optimization and rest on *overarching qualitative principles* such as moral constraints. As an illustration, Callies and Moellendorf (2021) delineate a comparative framework that, without relying on the straightforward maximization of quantitative variables, is based on two "moral considerations": the "avoidance of catastrophic climate change" and the "right to sustainable development." Those principles draw their legitimacy from the fact that they constitute the foundation of international climate politics. Most States acknowledge the normative force of these two principles as they figure in the 1992 United Nations Framework Convention on Climate Change (UNFCCC, article 2) and the 2015 Paris Agreement (article 2). Thus, they can be safely assumed to represent, to a significant degree, the will of the peoples whose governments signed and ratified the UNFCCC. In other words, preferable CE interventions or hybrid policies would be those which protect the best against climate dangers while securing the capacity of societies to grow economically and socially.

TECHNOLOGICAL DIFFERENCE AND MORAL NONEQUIVALENCE

In conclusion, CE should not be apprehended as an indiscriminate bloc of interventions (Heyward 2013). It constitutes an umbrella denomination that lumps together distinct methods. The use of a common label is cogent insofar as it is apt at characterizing the project of altering radiative forcing, globally or locally. However, each CE initiative is more or less specific in terms of benefits, risks, political and ethical challenges (e.g., regarding governance), and so forth (Schäfer *et al.* 2015, 21). At the fundamental level, SRM and CDR technologies work in radically different ways (Keith 2017, 75), which is recognized by the US NRC when it published two separate reports (NRC 2015a, 2015b).

Outside climate sciences and engineering, CE is frequently discussed and rejected *en bloc* without nuance and attention to those details. The propensity for apprehending CE as an indistinct category is reflected in public perceptions (Cummings *et al.* 2017, 257). And when the diversity of CE methods is acknowledged, it is typically not taken seriously. Another prejudicial tendency is to assume that a specific intervention, usually *SAI*, is representative of the challenges posed by SAI, SRM, or even CE in general. It remains nonetheless crucial to appreciate the various methods for what they are, first by distinguishing between SRM and CDR and then by deepening the distinctions within each category (Pamplany *et al.* 2020, 3,099). The urgency of climate change commands to exercise caution when evaluating the promises and pitfalls of CE.

The necessity of adopting a finer-grained view is reinforced by the magnetic pull on researchers and policymakers of a handful of methods, among which are prominently CCS, BECCS, forest management, and SAI. Such focus, in particular on SAI, tends to distort the overall discussion on CE. However, CE methods are not technologically equivalent. They do not raise identical issues in the ethical and political realms. As an illustration, the interference with "natural" cycles differs across CDR interventions. DACS requires a reduced land footprint, as well as limited meddling with ecosystems in comparison to BECCS or reforestation. So, those methods cannot be treated as being interchangeable regarding their projected disruption of environmental cycles or the uncertainty they could trigger. The diversity of technical specifications should be reflected in the normative (i.e., legal, political, and ethical) assessments. Nothing would be more damaging for climate policy and future generations than to hastily reject potentially beneficial initiatives. The stakes for humanity are too high. To repeat the point, CE does not represent a uniform set of substitutable methods which would display similar qualities and could be subject to overarching, sweeping, and judgments (natural/unnatural, good/bad, just/unjust, fair/unfair, and promising/unpromising).

In sum, all CE methods are often presented as *materially, and then morally, equivalent*. Against this shortcoming, it is crucial to recognize a fundamental *nonequivalence*. SAI, MCB, CCS, BECCS, surface brightening, forest management, and OIF do not carry identical promises and risks across space and time. Thus, any evaluation of CE needs to start by a rather complete exposé of the main methods as *per* the scientific literature. Consequently, the next chapter will be devoted to introducing CDR, while the subsequent will be dedicated to SRM.

NOTES

1. Some authors use variations of this distinction. For instance, Toby Svoboda (2016) advances that climate policy proposals should be evaluated based on their effectiveness, political feasibility, and moral permissibility.

2. Scenarios that include time series of emissions and concentrations of the full suite of greenhouse gases (GHGs) and aerosols and chemically active gases, as well as land use/land cover (. . .). The word representative signifies that each RCP provides only one of many possible scenarios that would lead to the specific radiative forcing characteristics.

The term pathway emphasizes that not only the long-term concentration levels are of interest, but also the trajectory taken over time to reach that outcome. (IPCC 2015, 126)

Basically, four main RCPs, or emission scenarios, are used: RCP 2.6, RCP 4.5, RCP 6.0, and RCP 8.5. The number indicates the increased radiative forcing in W/m^2 reached by 2100 due to GHGs' accumulation in the atmosphere.

3. (Deep) uncertainty characterizes circumstances under which adverse events are unknown or their probabilities cannot be calculated (cf. chapter 1).

4. In moral philosophy, nonideal circumstances describe situations where compliance to justice requirements (e.g., John Rawls' two principles) can only be partial, which characterizes the world in which discussions on climate change take place (Heyward and Roser 2016).

5. [P]ath-dependent processes emerge initially from contingent (chance, random) circumstances that confer an initial advantage on a particular technology, followed by self-reinforcing processes or positive feedback, such as cumulative cost reductions and learning effects linked to increasing returns to adoption. (Cairns 2014, 650)

The next chapter offers concrete examples as well as a discussion of path dependency in the case of CDR technologies.

6. Olechowski *et al.* (2020) also offer an overview of the main challenges, perceived from practitioners' perspective, that is, those who are using TRL in their professional activities.

7. A fourth dimension, institutional, could have been added since the potential of a CE method for deployment at scale technologically, materially, and financially depends on the existence and support of private and public institutions. But since this dimension is intimately intertwined with the three others, it has been included in the three other categories.

8. The literature on path dependency and lock-in discusses the influence of social contexts, institutional structures, economic interests, and the like (the "techno-institutional complexes" [Unruh 2000; Unruh and Carrillo-Hermosilla 2006]) on technological developments. The reader can refer to Trencher *et al.* (2020), who offer an encompassing analytical grid for understanding path dependence and lock-in phenomena.

9. The expected utility of an act is a weighted average of the *utilities* of each of its possible outcomes, where the utility of an outcome measures the extent to which that outcome is preferred, or preferable, to the alternatives. The utility of each outcome is weighted according to the probability that the act will lead to that outcome. (Briggs 2019)

10. This presentation is debatable. Knight can be interpreted as separating *objective* and *subjective* probabilities (LeRoy and Singell 1987). Uncertainty would define situations where, in the absence of deductive distribution of events ("objective probabilities as those everyone would agree to" [LeRoy and Singell 1987, 398]), individuals can still infer "estimates," building *subjective* statistical distributions of adverse events. Leaving aside Knightian exegesis, uncertainty can be used to qualify known events for which odds are unknown as *per* Sunstein's reading, known events so rare that they are fully discounted in risk assessments, such as Black Swans (Taleb 2007),

or unknown events for which neither probability nor estimate can apply (sometimes characterized as deep uncertainty or ignorance).

11. For a critique of the lesser evil in the context of CE, cf. Gardiner (2010).

12. There are other reasons why CE could be considered as intrinsically bad, wrong, or evil. Particular methods could be judged as inherently dominating, disrespectful, or antidemocratic (Horton *et al.* 2018). However, such criticisms are less common than unnaturalness, hubris, and playing God.

13. The perceived (un-)naturalness of CE methods may have a determining influence on public acceptance (Sweet *et al.* 2021; Wolske *et al.* 2018) although such influence is still debated.

14. Persons (or groups of persons) show hubris, if they act with a reprehensible overestimation of their abilities. In the case of GE these persons overestimate their technical abilities and the epistemic abilities connected to them. They have inadequate beliefs about the probabilities that certain technologies will be successfully implemented. (Meyer and Uhle 2015, 5)

15. The Gaian position originates in Rachel Carson's *Silent Spring* and Bill McKibben's *The End of Nature* (Thiele 2019).

16. In addition to the divide between climatic and social risks, another distinction is between direct and indirect/compound risks. The first category is constituted of negative outcomes that directly stem from CE, for instance droughts due to SAI. The second category is made of harmful consequences that are distantly related to CE, for example, the establishment of authoritarian regimes following the political turmoil following CE-induced droughts.

17. Global catastrophic risks are risks of events that would significantly harm or even destroy humanity at the global scale. (Baum *et al.* 2013, 169)

18. According to Taleb (2007, xvixvii), a Black Swan possesses three attributes.

First, it is an *outlier*, as it lies outside the realm of regular expectations, because nothing in the past can convincingly point to its possibility. Second, it carries an extreme impact. Third, in spite of its outlier status, human nature makes us concoct explanations for its occurrence after the fact, making it explainable and predictable.

19. Distributive justice is a central topic in political theory, especially since the publication of John Rawls' *A Theory of Justice* (Rawls 1971). The philosophical debates on questions such as the currency of justice and the conception of equality have been particularly rich (e.g., Cohen 2001; Dworkin 2000; Nozick 1974; Sen 1980; Walzer 1983). For readers interested in those questions, many overviews are available (e.g., Olsaretti 2018).

20. The question of the moral standing ascribed to nonhuman entities is not addressed in this book. It is nonetheless worthwhile to note that the central question is not whether attributing a moral value or worth to different entities, but the kind of value or worth. Despite the importance of the issue, it exceeds the scope of an introduction on CE.

21. Consequentialism (Sinnott-Armstrong 2022) is one of the three major ethical theories, besides deontologism (Alexander and Moorer 2021) and virtue ethics (Hursthouse and Pettigrove 2022). Put simply, it stipulates that the acceptability of an

action or a rule should be determined by evaluating the consequences of such action or rule and their properties (goodness/badness, happiness/unhappiness, pleasure/pain, etc.) and not based on other considerations such as intentions or moral character.

22. The Kaldor–Hicks criterion ventures further by deeming efficient any action that harms individuals as long as the overall benefits exceed the overall harms and there is the possibility to compensate the losers (Heath 2021, 187–188).

23. Such spillover underscores the "risk triangle" (Hermansson and Hansson 2007) at work between three categories of agents in CE: the decision-makers, the burden-bearers, and the beneficiaries (Wolff 2019, 573).

24. Other procedural standards are possible for CE research and deployment (e.g., Hourdequin 2019).

25. Explicit consent characterizes situations in which "individuals *directly* and *actually* endorse the decisions (or, the institutions that make those decisions) through explicit consent, and thus, the decisions that individuals have consented to can be seen as made by them" (Wong 2016, 178). Implied consent refers to circumstances in which "consent is *inferred* from individuals' actions (or their lack of actions) in another situation which is sufficiently similar to the situation requiring consent" (Wong 2016, 179). Contrary to the two previous forms, hypothetical consent strays away from actual consent by considering what "they [individuals] *would* consent to under certain background conditions" (Wong 2016, 180).

26. The all-affected interests principle predates modern democracy. Its origin can be found in the Justinian Code (Warren 2017, 1): "what touches all must be approved by all (*Quod omnes tangit debet ab omnibus approbari*)." The interpretation of its content, scope, and implications are, however, debated (Goodin 2007).

27. Another hurdle to the all-affected interests principle is that it requires that the political representatives of the affected populations (governments, parliaments, and public institutions) are, to some extent, responsive to their constituency and their interests (Lenferna *et al.* 2017, 586). In other words, it demands a modicum of responsiveness from political elites to popular expectations, which is not present in a large part of the world.

28. The derogative denomination of "geoclique" loosely refers to a variable group of individuals and institutions usually including researchers identified as "proponents" of CE such as David Keith, Ken Caldeira, or Bjorn Lomborg as well as more controversial political figures or institutions such as Newt Gingrich, Heartland Institute, or the American Enterprise Institute. Sometimes, the oil industry and military sector are associated with the geoclique.

29. Acceptability figures more as a property of the intervention, while acceptance designates the fact of agreeing (L'Orange Seigo *et al.* 2014). This section does not maintain a watertight distinction between the two concepts.

30. The relation between naturalness and public acceptability may be more complex than the simple dichotomy that would identify natural with permissible and unnatural with impermissible (Corner *et al.* 2013).

31. "(. . .) [P]eople's choices—sometimes choices about very important things—can be altered by irrelevant changes in how the alternatives are presented to them. How the problem is framed affects what they choose" (Stanovich 2010, 23). For early

investigations of framing effects, the reader can refer to Daniel Kahneman and Amos Tversky (2000).

32. Both concepts describe the adoption of riskier behavior by agents who benefit from risk protection or compensation. The nature and reality of the phenomenon are debated, as illustrated by chapter 5. For now, it is enough to note that

> moral hazard is a socially inefficient increase in risk-taking by one party once another party absorbs some of the potential negative consequences of the first party's actions, typically through an insurance-like agreement between the parties and typically without the latter party's full knowledge of this increase. (Reynolds 2015, 176)

In contrast, "risk compensation is an increase or decrease in risk-taking once an individual perceives that risk to be lower or higher, respectively" (Reynolds 2015, 177). Although both concepts describe seemingly similar adjustments to changes in the expected utility of adverse events an agent faces, risk compensation does not imply the transfer of part or all risk-induced damages to another party as for moral hazard (Reynolds 2015).

33. It should be noted that many principles of legitimizing consent are possible. Among the different rules of legitimacy, it may be required that a decision or policy garner the consent of all-affected parties (unanimity/consensus), most of them (majority), or the most exposed/vulnerable to the consequences of the decision or policy.

34. It is to be noted that such a formulation simplifies the issue by disregarding potential disparate impacts.

35. Gardiner (2011, 143–184) considers that future generations are vulnerable to the "tyranny of the contemporary." Present generations may decide not to tackle pressing issues such as climate change or pollution because the costs of inaction will be passed on. Intergenerational buck-passing constitutes a second tragedy of the commons, adding its effects to the first tragedy characterizing international climate politics (cf. introduction).

36. The STR should be distinguished from pure time preference, which is the preference for receiving a benefit or a good as close as possible to the present instead of some point in the future. For a general introduction to the discounting method, see Broome (1999, 44–67). For a discussion on social discounting and climate change, see Heath (2021, 230–268).

37. It is worthwhile to note that the concept of regulation is subject to important debates (Baldwin *et al.* 2010; Koop and Lodge 2017).

38. Legality and morality do not always overlap. An illegal action is not necessarily immoral (e.g., the criminalization of same-sex relationships or opposing racial segregation) as a legal one is not necessarily moral (e.g., adultery and petty lies).

39. Moral dilemmas are situations in which agents face conflicting ethical duties or commitments (e.g., incompatible promises made to two different persons), implying that the final decision is "tainted," unsatisfactory, or harmful.

Chapter 3

Climate Engineering Methods I

Carbon Dioxide Removal (CDR)

According to the Royal Society, carbon dioxide removal (CDR) regroups "techniques [that] address the root cause of climate change by removing greenhouse gases from the atmosphere" (Shepherd *et al.* 2009, ix). CDR is the first of the two categories of climate engineering (CE) methods, the second being solar radiation management (SRM). In this book, CE is defined as *anthropogenic interventions in the climate system that predominantly aim at altering radiative forcing locally or globally to slow down, stop, or reverse climate change and/or related adverse events.*

As mentioned in the previous chapters, CDR and SRM interventions substantially differ, not only on technological ground but also regarding the scale and scope of risks/benefits as well as challenges for public policy and governance (Keith 2017, 75). This sheer variety explains why sweeping judgments should be avoided, especially since some CE methods might ultimately support climate policy in "prevent[ing] dangerous anthropogenic interference with the climate system" (as per United Nations Framework Convention on Climate Change [UNFCCC] formulation, article 2). However, hasty generalizations are fed by the fragmentary knowledge among the public at large about CDR methods, and more generally CE (Cox *et al.* 2020; Pidgeon and Spence 2017; Raimi 2021; Smith *et al.* 2023, 11). Informative discussions stay confined to a handful of experts and educated policymakers (Shrum *et al.* 2020). To grasp the full extent of CE's opportunities and shortcomings, getting acquainted with its main forms is crucial. Thus, this chapter and the next one elaborate on CDR and SRM methods, respectively. The content remains broad enough for being accessible to people without any engineering or climate science background. The focus is on the key elements necessary to initiate a general assessment along the evaluative criteria spelled out in the previous chapter.

WHAT IS CDR?

Often referred to as negative emissions technologies (NETs)[1] (EASAC 2018; Minx *et al.* 2018), CDR methods rely on extracting carbon from the atmosphere to store it biologically or geologically (Smith *et al.* 2023; Vergragt *et al.* 2010, 288). Biological sequestration in the biomass and in the upper soil, for example, is less durable than geological sequestration like in empty oil reservoirs and aquifers on land or at sea, and, presumably, in the oceanic depths. CDR efficacy is determined by the capture rate (gigatons [Gt] of carbon, CO_2, or CO_{2eq} captured per year), long-term storage potential (total Gt C, CO_2, or CO_{2eq}), and overall costs (National Research Council [NRC] 2015a; Shepherd *et al.* 2009). However, evaluating efficiency faces accounting issues that fall in the domain of monitoring, reporting, and verification (MRV) (Brander *et al.* 2021; Mace *et al.* 2021, 70–71), defined by the World Bank (2022) as

> the multi-step process to measure the amount of greenhouse gas (GHG) emissions reduced by a specific mitigation activity, such as reducing emissions from deforestation and forest degradation, over a period of time and report these findings to an accredited third party. The third party then verifies the report so that the results can be certified, and carbon credits can be issued.

TEXTBOX 3.1 CARBON, CO_2, AND CO_{2E}

Three indicators are widely employed for representing anthropogenic emissions and concentrations: CO_2, carbon, and CO_2 equivalent (CO_{2e} or CO_{2eq}). CO_2 refers to carbon dioxide. Adopting this metric leads to underestimating the seriousness of the Anthropocene by not accounting for other, more potent GHGs like methane and nitrous oxide (N_2O). Carbon is often used. In a loose manner, it could designate the CO_2 molecule, or, in a more precise manner, it could refer to a specific kind of atoms, which is practical for aggregating all GHGs made of carbon (e.g., methane [CH_4]). However, some powerful GHGs do not contain carbon, such as N_2O or sulfur hexafluoride (SF6). The third accounting option is to mobilize CO_2 equivalent (noted CO_{2e} or CO_{2eq}), in which CO_2 constitutes the currency in which the global warming potential (GWP) of all GHGs is converted (cf. textbox 0.1 in the introduction).

For analytical purposes, different taxonomies of CDR methods are possible. In this book, the classification is built around two categories.[2] The first category regroups methods that stimulate natural processes (e.g., photosynthesis and rock weathering), thus strengthening natural carbon sinks (cf. textbox 1.1). Those methods could be further subdivided according to the location of the sinks, namely in soils (and land-based biomass) and in oceans (and maritime biomass). The second category bundles solutions, which heavily rely on engineered systems that could qualify as "artificial" sinks (e.g., direct air capture with sequestration/storage [DACS]) and which use geological reservoirs.[3] Table 3.1 details the content of each class, all of which, except for grassland and wetland restoration and part of soil carbon sequestration/ storage (SCS), will be covered in this chapter.

At first sight, CDR methods seem less controversial than SRM for several reasons. They can be viewed as more natural owing to more limited intrusion into "nature," and as tackling the origins, not the symptoms, of climate change (Cummings *et al.* 2017; Shepherd *et al.* 2009, ix). However, a recent research survey conducted in the UK shows that the public considers that CDR does not confront the root cause of climate change (Cox *et al.* 2020). Perceptions of naturalness vary across the different CDR interventions (Raimi 2021, 67; Wolske *et al.* 2019). Moreover, international texts and organizations such as the European Commission[4] as well as states acknowledge the necessity of capturing and storing carbon, often under the denomination of NETs, turning CDR into a key tool for respecting the 2°C, preferably 1.5°C, threshold set

Table 3.1   CDR Taxonomy

Natural Sinks Enhancement	*Fully Engineered Sinks*

• Land-based carbon sinks enhancement
  ○ Land management and ecosystem
    restoration
    ■ Forest management/tree sequestration
    ■ Grassland and wetland restoration (not
     covered)
  ○ Land sequestration
    ■ No-/low-tillage
    ■ Biochar
  ○ Land-enhanced weathering
• Ocean-based carbon sinks enhancement
  ○ Ocean-enhanced weathering
  ○ Ocean pumps
    ■ Physical pump
    ■ Biological pump/ocean iron fertilization
     (OIF)
  ○ Macroalgae cultivation
	• Carbon capture and sequestration/
	  storage (CCS)
	• Bioenergy with carbon capture
	  and sequestration/storage (BECCS)
	• Direct air capture and storage/
	  sequestration (DACS)

by the Paris Agreement in article 2 (Buylova *et al.* 2021; Fuss *et al.* 2018; Smith *et al.* 2023).

Indeed, the foundation of international climate politics—the UNFCCC (1992) explicitly mentions in its article 4 the responsibilities of the parties to

promote sustainable management, and promote and cooperate in the *conservation and enhancement, as appropriate, of sinks and reservoirs of all greenhouse gases* not controlled by the Montreal Protocol, including biomass, forests and oceans as well as other terrestrial, coastal and marine ecosystems. (emphasis added)

Moreover, states which are part of the UNFCCC have a duty to keep a national inventory of carbon sources, sinks, and reservoirs on their territory. Such a monitoring system constitutes a prerequisite for any climate efficient policy. The Paris Agreement (2015) reaffirms in its preamble "the importance of the conservation and enhancement, as appropriate, of sinks and reservoirs of the greenhouse gases referred to in the Convention." While article 4 mentions the necessity to reduce emissions for balancing the sources and the sinks, article 5 lays down a national responsibility for enhancing carbon sinks and reservoirs, which includes forests. This last article directly points at CDR technologies, which have already been *de facto* embedded in climate models for several years, for example, since 2006 for BECCS (McLaren 2021, 80).

CDR has indeed made its way into scientific assessments and policy documents. Since the Intergovernmental Panel on Climate Change (IPCC)'s Fourth Assessment Report (AR4) in 2007, CDR has been embodied in most of the IPCC projections respecting the 2°C threshold (de Coninck *et al.* 2018; EASAC 2018; Minx *et al.* 2018; OECD/IEA 2016; Smith *et al.* 2023). In its Fifth Assessment Report (2014), out of the 116 scenarios that maintain CO_{2eq} concentrations within 430–480 ppm (therefore limiting the risk of temperature overshoot), 101 include NETs deployment starting from 2050 (Fuss *et al.* 2014, 850; Smith *et al.* 2016, 43). As put by Peters and Geden (2017, 619), "the governments that signed and ratified the Paris Agreement accept the IPCC consensus that CDR cannot be avoided if ambitious climate targets like 1.5°C or 2°C are to be met." In 2018, land carbon sinks enhancement (branded as Agriculture, Forestry and Other Land Use [AFOLU]) and BECCS were presented as crucial for reaching negative net emissions by the mid-twenty-first century, maximizing the chance of remaining within the 1.5–2°C boundaries (IPCC 2018). Net-zero targets are therefore attracting more attention from policymakers (Schenuit *et al.* 2021). As of 2023, in addition to aggressive conventional mitigation, all pathways respecting the Paris Agreement include CDR (Smith *et al.* 2023, 9), which renders its relative absence from public debates and policy discussions rather worrying (Lenzi *et al.* 2018; Lin 2019).

The concern is reinforced by the fact that CDR appears as a necessary component of overshoot scenarios in which the bulk of the abatement efforts is delayed, leading to exceeding the carbon budget as well as possibly the 1.5–2°C threshold before returning under the targets, preferably before 2100. Overshoot pathways could be deliberate or accidental. Deliberate overshoot could be justified by the potential gain to be extracted from cheap energy like natural gas for a longer period while massively investing in fossil-free technologies (Nordhaus T. 2018). Within that perspective, CDR may provide some headroom for pursuing important goals such as socioeconomic growth, poverty alleviation, and innovation at lower costs, resulting in future generations who will be better prepared for decarbonization, robust mitigation/adaptation, and CE. Accidental overshoot refers to emissions and concentrations surging above the threshold owing to insufficient carbon abatement caused by inept international coordination, lack of development of fossil-free energies, and so on. In those scenarios, CDR interventions could nonetheless offer some control by leaving open the prospect of re-establishing a less dangerous climate.[5] The strategy becomes uncontrolled when CDR cannot offer to return below the carbon budget. Then, the temperatures will soar, and catastrophes will multiply.

CDR is currently researched and discussed in scientific studies, and integrated, usually without further critical evaluation, into policy documents (Lin 2019; Mace *et al.* 2021; Rockström *et al.* 2017; Smith *et al.* 2023; Tollefson 2018). Such prominence is anything but a surprise. In 2018, the existing and planned energy infrastructure was judged as already committing humanity to overshoot the carbon budget left for respecting the 1.5°C target with a 50–66 percent chance, while potentially consuming two-thirds of the budget for staying under the 2°C boundary (Tong *et al.* 2019). In other words, CDR offers to bridge the mitigation gap, that is, the insufficient carbon abatement for securing fair chances of respecting the Paris Agreement (Tamme and Beck 2021). However, CDR development is lagging behind what is required by the Paris Agreement "to prevent dangerous anthropogenic interference with the climate system." To give an order of magnitude, annual CO_2 emissions represented ca. 36 $GtCO_2$ in 2022 (IEA 2022a), while yearly extraction was quantified to be ca. 2 Gt (Smith *et al.* 2023).

This chapter is divided into four sections. The first presents interventions that aim at improving land-based carbon sinks, while the second focuses on ocean-based sinks. Those two parts reflect the main forms of natural sinks enhancement. The third section offers a succinct overview of the main fully engineered solutions, that is, techniques that rely on artificial processes of sequestration. The chapter ends on a general assessment of the feasibility, permissibility, and preferability of the CDR methods.[6]

This chapter does not present an exhaustive technical account of CDR. The objective is more modest: to provide the readers with an accessible summary of the main interventions as well as to highlight some promises, drawbacks, and risks based on the framework elaborated in chapter 2. The readers who are interested in more substantial assessments can consult the scientific literature (e.g., EASAC 2018; IPCC 2005, 2019; NRC 2015a, 2019; OECD/IEA 2016; Shepherd *et al.* 2009; Royal Society and Royal Academy of Engineering 2018). In addition, the meta-study in three parts published by Fuss *et al.* (2018), Minx *et al.* (2018), and Nemet *et al.* (2018) provides a robust evaluation of CDR nature, mechanisms, and challenges, which is solidly supplemented by the Oxford University's report *The State of Carbon Dioxide Removal* (Smith *et al.* 2023).

LAND-BASED CARBON SINKS ENHANCEMENT

The first type of CDR interventions aims to enhance land carbon sinks, which is by far the dominant mode of carbon extraction currently in use (Smith *et al.* 2023, 8). Through better land stewardship, the goal is to support natural processes, for example, biological capture by trees, wetlands, or upper soils. Usually, those interventions are deemed useful, at least in the short run, while waiting for more ambitious mitigation or CDR to kick in. This method is also perceived as more natural (along with, to a lesser extent, ocean carbon sinks enhancement) than alternatives such as BECCS or DACS (Raimi 2021, 67; Wolske *et al.* 2019). Since land carbon sink enhancement implies preserving, strengthening, or restoring ecosystems, it is classified as part of the "natural climate solutions" (Griscom *et al.* 2017), but it arguably constitutes a CE tool (Keith 2000, 266). The method also overlaps with what the IPCC (2018) designates as AFOLU. Mirroring the interventions dominating the literature, this section is divided into three parts: land management and ecosystem restoration, with a focus on forest management, land sequestration, and land-enhanced weathering.

Land Management and Ecosystem Restoration

As forest management (tree sequestration) is the most discussed type of interventions within this category (Bastin *et al.* 2019; IPCC 2018), it will be the focus of this subsection. It subsumes deforestation halting, afforestation, and reforestation. Those three types of interventions are already subject to corporate (e.g., Microsoft, Shell, and Mastercard) and international initiatives (e.g., Bonn Challenge, Trillion Tree Campaign, and UNFCCC's Reducing Emissions from Deforestation and Forest Degradation in Developing Countries [REDD+]) (Seymour 2020). Land management and ecosystem restoration also include initiatives for preserving, restoring, or expanding grasslands and wetlands (Griscom *et al.* 2017). However, those are less present in the literature (with exceptions, e.g., Macreadie *et al.* 2017).[7]

Deforestation is a significant source of GHG, contributing to approximately 10 percent of all anthropogenic emissions (NRC 2015a, 39). Forest clearing releases the carbon stored in the vegetation. Therefore, curbing deforestation, everything else being equal, reduces GHG emissions. Moreover, irreversible damages to boreal and tropical forests constitute two major tipping points (Lenton *et al.* 2019), beyond which the probability of positive feedback skyrockets, with the risk of embarking upon a more unpredictable and hazardous climate. Halting deforestation comes with significant co-benefits, such as preserved biodiversity, increased soil moisture and reduced erosion, and less expensive rehabilitation programs when former woodlands need to be reforested.

Afforestation consists in planting trees in areas where forests have been absent for a prolonged time (usually over fifty years [NRC 2015a, 39]). *Reforestation* applies to restoring forests that have been more recently deteriorated or cleared. Both raise difficulties for choosing adapted species ahead of a changing climate, especially in boreal regions where the tree life cycle is longer than in the subtropical regions.[8] Afforestation and reforestation require predicting which species will be fit to the local climatic conditions after several decades or centuries. The sequestration potential greatly differs according to variables like the extent of the lands dedicated to the intervention, the latitude, the kinds of species, the type of soils, and how forests are taken care of.

As a climate policy option, forest management carries strengths and side benefits. First, forestry is technologically mature, and humans have been planting trees for millennia and in a more scientifically cogent manner for decades. Thus, there is no specific difficulty in terms of technological feasibility, although physical scalability appears more challenging. Smith *et al.* (2023, 19) evaluate its technological readiness level (TRL) at 9 (cf. previous chapter for the meaning of the TRL).

Second, forest management *prima facie* seems more "natural," or less intrusive into "nature,"[9] than engineered interventions. The perceived naturalness, or lower intrusion, may incline toward judging the whole category as being intrinsically good, or not objectionable per se, and thus more acceptable than "artificial" methods (Shrum *et al.* 2020, 5; Sweet *et al.* 2021; Wolske *et al.* 2019).[10] In sum, forestry may appear less risky, more familiar, and less dreadful than CDR/CE alternatives (Shrum *et al.* 2020, 6), generating less public opposition and making it easier to implement.

Third, forest management per se does not a priori raise any extraordinary political and legal hurdle. Furthermore, the method can be initiated by a state (e.g., Iceland's afforestation efforts since 1945 and Ethiopia's National Green Development programme) or large private actors without requiring international cooperation, even though such initiatives need to be globally pursued for being effective at capturing carbon on a large scale. However, any ambitious intervention will probably generate land-use conflicts (with agriculture or energy crops dedicated to BECCS for instance) and biodiversity conservation issues depending on the modalities of deployment (Dooley *et al.* 2021).

Finally, forest management may produce important co-benefits in the areas where it is undertaken. If it is mindful of the existing biome, forest management could help restore biodiversity, reduce soil erosion, and support biomass yield. Again, many of those potential positive side effects will be contingent on the way afforestation/reforestation activities are carried out, how much they are adapted to the local ecosystems, how stable they are throughout time (for keeping the carbon stored), and how adapted they are to the unfolding Anthropocene.

Forest management faces serious shortcomings too, the major ones residing in its lack of effectiveness and physical scalability. Alone it will not be sufficient for averting dangerous climate change. Moreover, it may represent a harmful distraction from substantial carbon abatement and fossil-free energy development (Ellis *et al.* 2020). It is also prone to be exploited for greenwashing (Khadka 2022; Neimark 2018). In addition, planting trees in a magnitude that could mitigate the Anthropocene would necessitate vast areas. Such demand would become problematic, in particular because of population growth (Searchinger *et al.* 2019), which could reach approximately 11 billion people by 2100 (United Nations 2019).[11] Massive reforestation/afforestation would generate land-use conflicts with agricultural utilization (Shepherd *et al.* 2009, 10), potentially undermining poverty alleviation and food security in developing countries (IPPC 2019, 27). Moreover, such potential pressure on the land underscores the related issue of sustainable diet. According to the United Nations Food and Agriculture Organization (FAO), agriculture mobilizes 38 percent of the total soil. Out of which, crops use one-third, while the rest serves for cattle grazing. Thus, any ambitious tree sequestration may require diminishing or abandoning meat production (National Academies of Sciences, Engineering, and Medicine 2019, 13), along with taming population growth.

Another issue is decreased albedo, that is, the decline in Earth's capacity to reflect incoming solar radiation (Veldman *et al.* 2019). Massive reforestation/afforestation will substantially reduce the albedo in boreal areas compared to different vegetal surfaces such as grasslands, especially during winters where forests are darker than treeless, snow-covered, landscapes. It is important to point out, however, that the interaction between forest coverture and albedo is complex and varies regionally (NRC 2015a, 42). Depending on the latitude, afforestation/reforestation could lead to significant localized or global warming, the tropical areas being the most thermic neutral (Anderson *et al.* 2011; Wang *et al.* 2014).[12]

The sustainability of afforestation/reforestation poses another difficulty. In comparison to methods that offer soil (e.g., biochar) or lithospheric sequestration (e.g., CCS, BECCS, and DACS), forest management keeps carbon away from the atmosphere only for a few decades or centuries, at best (Ellis *et al.* 2020). When the vegetation dies, it decomposes and releases carbon back into the atmosphere.

Ultimately, the perceived naturalness and low disruption of nature are partly misleading. At the global scale, this intervention requires to plant trees in areas with predating ecosystems, especially in the case of afforestation (Dooley *et al.* 2021, 36–37). Therefore, humans will act less as a restorative than a perturbing force for the indigenous vegetal and animal species. Some will adapt, while others will recede and disappear. If naturalness is to be assessed based on interference with natural cycles and evolution, then afforestation/reforestation may be less natural than is often assumed. Halting deforestation represents the most natural intervention since it refrains from any further interference. A less intrusive method is natural regeneration, which consists in leaving former wooded zones to regrow at their own pace, the colonizing species resulting from natural selection (Chazdon 2014).

Fuss *et al.* (2018, 16) project a capture potential for tropical afforestation/reforestation around 3.6 $GtCO_2$/year by 2050,[13] followed by a decline to 0 in 2100 at a cost of \$5–50/tCO_2 with a land use of 500 Mha. The assessment is limited to tropical regions due to the negative albedo effect implied by a deployment in temperate and boreal areas. The low potential supports the efficiency concerns mentioned above. A removal rate of 3.6 $GtCO_2$/year barely represents 10 percent of the 2021 global emissions, which were estimated to be 36.3 $GtCO_2$ (IEA 2022a). Smith *et al.* (2023, 19) provide a broader evaluation at the 2050 horizon with a capture potential situated in a large interval between 0.5 and 10 G_tCO_2/year at a cost of \$0–240/tCO_2.

Land Sequestration

Land sequestration characterizes interventions supporting the carbon storage capacity of soils. Two main initiatives dominate the literature: no-/low-till agriculture, sometimes included in SCS, and biochar. Other methods feature practices that increase the carbon intake of farmlands, for example by selecting specific crop species (Shrum *et al.* 2020, 3).

Since "tillage disrupts the soil, opening it to decomposer organisms and generating aerobic conditions that stimulate respiration and release of CO_2" (Shepherd *et al.* 2009, 10), *reduced or no-tillage* aims at preserving the soil microbial biomass, which is pivotal for capturing and keeping carbon underground (NRC 2015a, 43–44). In low-tillage, soil disruption is limited to the exact location in which seeds are introduced. In addition to sequestration, the initiatives could generate co-benefits such as decreased erosion and better retention of nutrients. The intervention is already field-tested through projects like the *"4 per 1,000" Initiative* launched by France during the COP21 in Paris. The ambition is to increase the carbon intake of the top thirty centimeters of the soil. Although meta-studies show that low or no-tillage may contribute to lower atmospheric carbon (Haddaway *et al.* 2017), the global efficacy remains controversial (Ogle *et al.* 2019). Moreover, as briefly mentioned in the next chapter, no-tillage could enhance soil albedo (Davin *et al.* 2014).

Biochar[14] results from low-temperature pyrolysis, that is, biomass combustion at 300–600°C in an anoxic (oxygen-deprived) environment. The stability of the product (charcoal) reduces *ex post* release of carbon, which represents an opportunity for long-term storage in the soils. Biochar may generate co-benefits, such as stimulated plant growth leading to increased carbon capture (National Academies of Sciences, Engineering, and Medicine 2019, 105). There are a few uncertainties. One is the possible disruption of the microbial biomass through the tillage incurred by charcoal burial. Another is the energy demand for generating enough biochar to lower global carbon concentrations (Hamilton 2013a, 45).

Both land sequestration and land management are moderately to highly mature techniques. Biochar TRL is evaluated at 6–7, and 7–9 for SCS (Low *et al.* 2018; Smith *et al.* 2023, 19). No significant technological challenge impedes land sequestration. No massive investment is needed, a shortcoming that obstructs other CDR technologies like BECCS and DACS. Moreover, land sequestration seems more natural than the rest of CDR methods, excluding forest management, since the engineered apparatus is limited and the interference with nature appears to be minimal, a fact which is confirmed by few studies about SCS (Sweet *et al.* 2021). In comparison to other initiatives, land sequestration does not require large areas, so it drastically reduces the risk of land-use conflict. A last potential impact is the co-benefit in terms of increased soil productivity.

However, one of the main obstacles encountered by zero or low-till agriculture is to coordinate farmers' efforts globally and maintain those efforts throughout time. This illustrates the deeper problem of sustainability. Tillage interventions do not guarantee resilient sequestration. Although promising more enduring storage than in biomass like trees, biochar faces similar issues of durability, depending on the surrounding conditions (Fuss *et al.* 2018, 26). Moreover, the pyrolysis, transportation, and burying necessitate energy. Overall, the life cycle might be carbon positive, that is, net emitting, or at least the efficacy might be reduced. Finally, the impact of placing biochar in the soil organic life needs to be assessed (Fuss *et al.* 2018, 25).

For SCS, which includes more interventions than solely no-/low-tillage, "the best estimate (with range) of realistic technical potential is considered to be close to the median of the minimums of the ranges provided, which for SCS is 3.8 (2.3–5.3) $GtCO_2yr^{-1}$" (Fuss *et al.* 2018, 28) with costs falling between US\$0 and US\$100/tCO_2. For biochar, Fuss *et al.* (2018, 26) evaluate capture for 2050 around 0.3–2 $GtCO_2$/year at US\$90–120/tCO_2. Smith *et al.* (2023, 19) offer higher approximations with 0.3–6.6 $GtCO_2$/year for US\$10–345/tCO_2 for biochar and 0.6–9.3 $GtCO_2$/year at US\$-45–100/tCO_2 for SCS (the negative lowest bound reflects the projected increase in land productivity).

Land-Enhanced Weathering

Also labeled land accelerated weathering, its core mechanism is to support capture through chemical reactions with water, carbonic acid, or other acids. In the absence of human intervention, the weathering of calcium and magnesium silicates takes millennia during which rocks breakdown due to the physical and chemical influences of natural elements (wind, water, and thermic amplitude). The cycle could be hastened in dedicated facilities by crushing and then spreading minerals like silicates (e.g., basalt), which are known for causing a chemical weathering reaction when in contact with atmospheric CO_2, usually under the form of carbonic acid (NRC 2015a, 46–47). The Urey reaction is the interaction happening between silicates and carbonic acid that leads to the durable capture of CO_2 (Kellogg *et al.* 2019). Potentially, the carbon could be trapped indefinitely (Keith 2000, 267).

Natural weathering is limited by the surface of silicate minerals in contact with natural elements at a given time. Enhanced processes tackle this issue by grinding silicate rocks and then exposing them to the ambient air in facilities or dispersing them on croplands. Soil amendment with minerals has long been practiced by farmers, for example, with crushed limestone, due to its positive impact on nutrients (Van Straaten 2007). Enhanced weathering can generate co-benefits like reduced soil erosion and acidity (Swoboda *et al.* 2022), improved crop yields, and preserved nutritional value (Beerling *et al.* 2020), thereby reinforcing food security.

Nevertheless, enhanced weathering on croplands is perceived as being less natural, or tampering more with nature, than interventions like SCS, forest management, or biochar (Shrum *et al.* 2022, 5). Surveys conducted in

Australia, New Zealand, and the UK have nonetheless shown that the method prompts more positive than negative judgments (Pidgeon and Spence 2017). Land-enhanced weathering does not seem prone to grave negative externalities, with the impact being circumscribed to the amended soils with little or no possible adverse effects. Nevertheless, negative external costs may stem from the fact that the potential areas for intervention are located in subtropical regions, in which disadvantaged groups may be exposed to harms, like toxic residues of silicate weathering (Lawford-Smith and Currie 2017, 3).[15]

However, only a few research projects implying modest deployments have been taking place, a major one being conducted by the University of California Davis' Working Lands Innovation Center (Low *et al.* 2022, 7). More field experiments, performed under strict conditions, are required before venturing into substantial conclusions (NRC 2015a, 55), which is reflected in public expectations (Pidgeon and Spence 2017). This explains why technological maturity remains moderate, ranging from TRLs 3 to 5 (Low *et al.* 2022, 2; Smith *et al.* 2023, 18).

In any event, scalability might constitute an obstacle to efficiency. Although arable land is broadly available, and the existing supply chain and mining capacity appear to be adequate (Houlton 2020), doubts subsist about the prospect of ramping up the land-based weathering (NRC 2015a, 60). The main worries bear on the costs and energy penalty of mining, crushing, transporting, and spreading the minerals (NRC 2015a, 49–56, Shepherd *et al.* 2009, 14).[16] Silicate residues from the industrial and mining sectors could be used in an acceptable manner if measures are taken for averting soil pollution (Beerling *et al.* 2020). Efficiency heavily depends on the energy requirements for the mining and transportation phases, but more importantly on the use of nonfossil energies. Such a constraint could be partly lifted by relying on waste rock from prior mining activities instead of virgin material (Cox and Edwards 2019).

The potential and costs of enhanced weathering applied on croplands rest on the location of intervention. A study shows variations between a pool of fast-growing nations (Brazil, China, India, Indonesia, and Mexico) and the most mature economies (Canada, France, Germany, Italy, Poland, Spain, and the United States). Under a pathway of a yearly extraction of 0.5–2 $GtCO_2$ by 2050, the first group of countries offers lower costs (approx. US\$55–124/t) than the second (approx. US\$157–220/t) (Beerling *et al.* 2020, 243–245). The higher efficiency flows from the broader amount of available farmland. Other studies situate the global potential for carbon capture to 2–4 $GtCO_2$/year at US\$50–200/t by 2050 (Fuss *et al.* 2018, 23; Hepburn *et al.* 2019, 92). In comparison, according to the Environmental Protection Agency, the United States emitted 6.5 $GtCO_2$ in 2019.

OCEAN-BASED CARBON SINKS ENHANCEMENT

This category aims at stimulating the ocean-based sinks. It includes three main types of intervention: enhanced weathering, improved pumps, and macroalgae cultivation. As a preliminary remark, it should be noted that the TRL of storing carbon in the ocean depths is evaluated as being rather low, well below geological sequestration (Low *et al.* 2022, 2).

Ocean-Enhanced Weathering

Similar to its land counterpart and also described as "liming the seas" (Hamilton 2013a, 36) or ocean alkalinization (Minx *et al.* 2018), the method consists in hastening the cycle at the conclusion of which weathered rocks end up in the oceans, resulting in their alkalinization. The minerals still need to be mined,[17] grinded, transported, and scattered. Offshore dispersion would most likely rely on loading the grinded rocks on vessels, for instance in the ballasts of commercial ships (Kölher *et al.* 2013).[18] The fine powder could be silicates, like olivine (which could weather faster than other silicates [Velbel 2009]), or carbonates, like limestone. The dispersed minerals would help to capture the carbon dissolved in the water with the additional benefit of decreasing ocean acidity.

This method suffers from similar drawbacks as the land intervention, namely costs and energy use during grinding, transportation, and disposal (Gonzales *et al.* 2018, 7122). On average, the energy use promises to be more consequent since the materials need to be brought to sea in very specific locations where the dispersion would be the most efficient, therefore extending transportation distance. Maritime weathering has been subject to fewer studies than its land counterpart, which makes it more difficult to appraise its efficiency (Fuss *et al.* 2018, 23).[19] The capture potential could range from 100 $MtCO_2$ to 10 $GtCO_2$/year with costs between US\$14 and US\$500/tCO_2 (Fuss *et al.* 2018, 22). The overall technological maturity is assessed to be lower than that for land-based interventions, at TRLs 1–3 (Low *et al.* 2022, 2; Smith *et al.* 2023, 18). An additional source of concern is that a sudden interruption may generate a termination shock[20] similar to some SRM methods (MCB, CCT, and SAI), thereby leading to a surge of ocean acidification and surface temperatures (González *et al.* 2018).

Besides OIF, ocean-enhanced weathering is one of the few CE interventions that are tangentially covered and constrained by international law. Under the UN International Maritime Organization, the Convention on the Prevention of Marine Pollution by Dumping of Wastes and Other Matter—so-called London Convention (1972)—and the London Protocol (1996) prohibit the illegal dumping of waste and pollutants at sea, a priori applying to OIF from the research and development (R&D) stage (Reynolds 2018, 88–93). Indeed, the Conference of the Parties (COP) of the UN Convention on Biological Diversity (CBD) released its decision IX/16 in 2008, based on the London Convention/Protocol as well as the precautionary principle, calling for a moratorium on large-scale commercial projects, making exceptions only for limited coastal experiments (Tollefson 2008). On October 18, 2013, the London Convention/Protocol were amended by the Parties for proscribing the "placement of matter into the sea from vessels, aircraft, platforms, or other manmade structures at sea for marine geoengineering activities." Despite having only fifty-three state parties, the London Protocol could serve as a basis for future governance and regulation of ocean-enhanced weathering and OIF (Grisé *et al.* 2021, 10–11).

Ocean Pumps

The sink capacity of the oceans could be stimulated by enhancing the "physical (solubility) pump" or the "biological pump" (Lampitt *et al.* 2008; Schäfer *et al.* 2015, 126). The former consists of carbon dissolution into boreal cold waters, which due to their high density and circulation patterns plunge into great depths, taking some of the carbon from the surface. The physical pump can theoretically be boosted by generating an artificial upwelling/downwelling exchange through vertical pipes for instance. Downwelling implies injecting the surface water, which is warmer and more acidic, into the lower layers. The upwelling process is about lifting profound seawater, which is cooler, richer in nutrients, and less acidic and offers better capture capacity (Tyka *et al.* 2022). Being complex and expensive, the method requires the offshore installation and maintenance of millions of tubes (Oschlies *et al.* 2010). Moreover, it could disturb the thermocline (buffer zone between the surface and the deep water) and trigger significant additional global warming (Kwiatkowski *et al.* 2015). Finally, the efficacy of the physical pump is likely to collapse with unabated climate change due to the potential saturation of the ocean sink and the decline of carbon solubility in warmer waters (increased vertical stratification).

The second, biological, pump consists in active "fertilizing" the oceans with macronutrients (e.g., phosphorus or nitrogen) or micronutrients (e.g., iron) to stimulate capture by marine microorganisms (mostly phytoplankton) through photosynthesis (De Coninck *et al.* 2018, 346). Iron is the nutrient which has attracted most of the attention. The limited studies available focused on areas characterized by iron deficiency like the Antarctic Ocean (Martin *et al.* 1990). OIF mimics the offshore deposits of large iron dust storms, which boost algae blooms (Goodell 2010, 139).[21] Those blooms of "planktonic algae and other microscopic plants take up CO_2 at the ocean surface and convert it to particulate organic matter" (NRC 2015a, 56). In the end, part of this organic matter sinks to the seafloor, (hopefully) durably trapping the carbon.[22]

The technology lacks maturity since no ambitious field experiment has been undertaken to date. Only a dozen limited controversial initiatives, such as in the Galápagos islands in 2007 (Tollefson 2017), the LOHAFEX project in 2009 (Kintisch 2010, 154–164), and British Columbia in 2012 (cf. textbox 3.2), have been conducted so far (Low *et al.* 2022, 3). As for ocean-enhanced weathering, R&D activities are further impaired by treaties like the London Convention and the London Protocol. In addition, the Parties to the UN CBD have reached a series of decisions (e.g., IX/16 [2009] and X/33 [2010]) that constitute reiterated calls to the precautionary principle in order to impose a moratorium on large-scale experimentations in international waters, thereby restricting *de facto* tests to modest coastal ones (cf. previous chapter). The core of the legal and ethical issues is to (a) figure out whether OIF is a polluting activity and (b) whether such pollution is morally permissible (Hale and Dilling 2011).

A hypothetical deployment gauges a strong concern. The costs and energy use of the mining, transporting, and scattering operations will most likely be significant. Uncertainties exist about the nature and extent of the afferent ecological impacts of massive dispersion, most notably on marine life. Toxic algae blooms, hypoxia/anoxia (lack/depletion of oxygen), and acidification of the deep ocean layers are potential adverse effects. There could be "'regime shifts,' with associated large-scale changes in regional biogeochemistry and the structure of the food web" (Burns 2016, 15). OIF is often judged as one of the riskiest CDR interventions (Smith *et al.* 2023, 47).

Since the publication of the Royal Society report (Shepherd *et al.* 2009, 17–18), evidence of OIF's efficacy at capturing and durably storing carbon has remained elusive at best (Burns 2016; Lauderdale *et al.* 2020). Smith *et al.* (2023, 18) place the TRL at 1–2, the least advanced of all CDR methods reviewed, on par with ocean alkalinization. Those uncertainties are reflected in paradoxical assessments. On the one hand, Fuss *et al.* (2018) judge OIF potential as being extremely low, impairing its viability as a NET. On the other hand, Smith *et al.* (2023) confirm such a low capacity (1-3 $GtCO_2$/year for US\$50-500/tCO_2). Moreover, as mentioned above, for the physical pump, climate change may impede the efficacy of the sequestration process.

TEXTBOX 3.2 THE HAIDA SALMON RESTORATION CORPORATION (HSRC)'S EXPERIMENT

In 2012, an infamous experiment took place in British Columbia. An entrepreneur, Russ George, convinced the Haida Old Massett Band to finance a company (Haida Salmon Restoration Corp.) that would dump 100 tons of iron sulfate at sea, in the Haida Gwaii's waters. The intention was to stimulate phytoplankton bloom on which salmons feed. Since the band was operating a salmon hatchery on the Yakoun River, and the runs were declining, the proposal was welcomed by the indigenous community. The sequestration ensuing from the unconsumed phytoplankton sinking on the ocean floor would represent an additional benefit. If scientifically verified, it could generate emission reduction credits (carbon credits). However, in addition to inconclusive results, the method stirred public outcry and was immediately associated with hubristic climate manipulation (Tollefson 2012). The story is, however, more complex (Buck 2019, 163–169). It implies a destitute indigenous first nation, using its meager natural and financial resources for a scheme conceived as a component of a broader plan of environmental restoration. In addition, the intervention was perceived as in accord with the duty of natural stewardship embodied in the Haida culture. Many scientists who were not part of the project were nonetheless interested in accessing the results. The experimentation led to investigations from the Canadian authorities, most notably because HSRC received funding from the Canadian Research Council, without official authorization for this kind of trial, as well as some equipment from the US National Oceanic and Atmospheric Administration.

Macroalgae Cultivation[23]

Another, marginal, proposal is to farm kelp and additional seaweed (macroalgae) (Buck 2019, 82–87; Chung *et al.* 2017; Energy Future Initiatives 2020). Historically, algae have been used for various purposes: human food, animal feed, cosmetics, and briefly considered in the 1970s–1980s as a potential biofuel. As a CDR method, the idea is to grow seaweeds for either sinking them, so the carbon remains trapped in the sediments (Krause-Jensen and Duarte 2016), or burning them as biomass in power plants, which could be counted as BECCS (cf. below) (Energy Future Initiatives 2020, 20).[24] An advantage is that the algaculture infrastructure, knowledge, and practices have reached some maturity, most notably in Asia (FAO 2018). The TRL of algal BECCS and coastal blue carbon is evaluated at 6 (Low *et al.* 2022, 2).[25] The main obstacle, as for other ocean-based initiatives, is the uncertainty surrounding the reality and durability of marine sequestration.

Active farming is only one possibility for marine biomass capture; an alternative is to restore kelp and other macroalgae (e.g., *Posidonia*) forests, for example, in California, England, or Tasmania, which have been receding due to global warming and the multiplication of sea urchins. An advantage is that intervention eschews water and land-use conflicts that cripple forestry and BECCS. In addition, the surface occupied by oceans (70 percent of the planet) boasts a significant potential for deployment. A side benefit is that kelp forests could serve as a refuge to marine life (Shelamoff *et al.* 2020).

A few shortcomings emerge from the literature. First, scalability and efficacy remain unclear. Can macroalgae culture or restoration have a noticeable impact on carbon concentrations? Various factors influence efficiency such as the seagrass species, the location, the nature of the sediments, and so forth. Some assessments situate the capture potential ca. 1–5 GtCO$_2$/year (Energy Future Initiatives 2020, 15). However, doubts have been voiced about the possibility of sequestrating large carbon volumes as well as the methodology used for calculating the potential (Johannessen and Macdonald 2016). This concern highlights the necessity for solid MRV for CDR in general (Fuss *et al.* 2014; Mace *et al.* 2021, 70–71; the Royal Society 2018), and kelp farming in particular for gauging the efficacy of the sinking and sedimentation phases. The magnitude and duration of the removal ought to be evaluated as well as the impact of carbon on marine life. Such focus on MRV is not particular to macroalgae but spans throughout CDR. Second, if the seaweeds are not sunk but convert to biofuels (carbon capture and utilization), the carbon footprint of transportation and transformation should be taken into consideration. The use of fossil energy will negatively affect the carbon balance of the method. Moreover, used as biofuels, transformed macroalgae will release carbon in the atmosphere. Third, even though seaweed cultivation is not heavily regulated, international texts (e.g., London Convention and Protocol) could apply depending on the intervention's location (territorial sea, exclusive economic zone, or the high sea) (Energy Future Initiatives 2020, 31–33). Since macroalgae cultivation is not extensively discussed in the literature, the rest of this book will not insist on it.

FULLY ENGINEERED CARBON SINKS

Instead of supporting natural processes, the second category relies on specifically designed industrial facilities. The interventions discussed in this section include CCS (2.3.1), BECCS (2.3.2), and DACS (2.3.3). All are viewed as less natural than methods like forestry or SCS (Shrum *et al.* 2020; Sweet *et al.* 2021; Wolske *et al.* 2019). Moreover, geological sequestration generates safety concerns while being perceived as the part that is the most intrusive into nature (L'Orange Seigo *et al.* 2014).

Carbon Capture and Storage/Sequestration (CCS)

While CCS *qua* a denomination can apply to any CDR method, CCS in the specialized CE jargon designates only "post-combustion capture" (Hulme

2014, 9), namely trapping devices installed on industrial equipment. CCS describes the extraction of flue gas emitted by a power plant or a factory and its injection into un-mineable coal pockets, deep aquifers, former oil/gas reservoirs, or marine sediments (Schrag 2009). The carbon can be stored on-site or transported to reservoirs through pipelines or by vehicles. If properly hermetic, geological sequestration could last up to 1 million years. CCS can be implemented in large stationary facilities like gas/coal-fired plants, diverse factories (e.g., ethanol, steel, and fertilizers), refineries, or gas fields. The potential for CCS is significant when one considers that "electricity and heat generation correspond to over 40% of global CO_2 emissions from fuel combustion, with coal plants emitting over 70% of the associated emissions" (IEA 2022b).

A core argument in favor of CCS is that the currently operational and planned fossil energy infrastructure is incompatible with the Paris targets (Fofrich *et al.* 2020; Tong *et al.* 2019). Whereas the first best option would be to substitute fossil-free plants for existing ones, the staggering scale of the required investment and infrastructure implies that full replacement is out of reach in the short run. Decarbonizing sectors like cement and energy will be extremely challenging. Considering the lifetime of coal power plants (forty-six years on average, but with the possibility to extend operations up to fifty to sixty years [Cui *et al.* 2019]), retrofitting them with CCS devices may be necessary for keeping temperatures within the 2°C limit without retiring plants.

Finally, the technology has reached *some* maturity, as illustrated by enhanced oil recovery (EOR) activities (OECD/IEA 2016, 17–18),[26] the TRL of which is 9 (Low *et al.* 2022, 2). The first EOR projects, like the Val Verde Natural Gas Plant and the Enid Fertilizer and Shute Creek Gas Processing Plant were initiated half a century ago in the 1970s and 1980s. In 2021, twenty-seven commercial CCS facilities were in operation, four were under construction, fifty-eight were in advanced development, and forty-four were in early development. All the projects in development were planned to be completed during the 2020s (Global CCS Institute 2021, 62–66).

CCS faces shortcomings. A significant issue is unclear cost-effectiveness and financial scalability. CCS suffers from an "efficiency penalty," which designates the amount of energy dedicated to the capture and storage phases of the process (Supekar and Skerlos 2015). The penalty underscores the difficulty of ramping up CCS. Despite being operational in twenty-seven industrial sites worldwide and in development in many more, critics doubt the technical and economic feasibility of retrofitting existing facilities and building new ones (Anderson and Peters 2016, 183). To become viable, CCS, as well as other NETs such as BECCS or DACS, would require substantial funding, in particular from public institutions (Bednar *et al.* 2019). The issue of unaccounted for costs highlights the importance of properly assessing the CCS life cycle through MRV. The energy necessary for transporting and storing carbon ought to be properly recorded, especially when it is sourced from fossil-fuel combustion (Fuss *et al.* 2018, 3). If confined to the sole costs of carbon extraction, CCS evaluation conveys low informational value. Financial scalability is further undermined by the immaturity of carbon markets and inadequate pricing, both of which obstruct channeling resources to the development of CCS projects. Any deployment required for a net-zero strategy is conditional on prices high enough for stimulating private investments, that is, prices which would include the social cost of carbon, that is, the damages and harms caused by anthropogenic climate change.

A second problem is the stability of geological sequestration (Smith *et al.* 2023, 47). Leakages and tremors may ensue from lithospheric storage (Mace *et al.* 2021, 71), which appears to be a central public concern (L'Orange Seigo 2014, 854). The issue is not limited to efficiency but extends to safety if carbon is stored under a gaseous form (CO_2). Since highly concentrated CO_2 is lethal, major leakages could endanger humans, animals, and ecosystems. In 1987, a natural, sudden, CO_2 release on the shores of Lake Nyos in Cameroon killed 1,700 people and 3,500 livestock. Although studies suggest that sequestration is relatively secure in the long run, over a period of 10,000 years (Miocic *et al.* 2019), uncertainties still remain (Alcade *et al.* 2018).

A third concern is that CCS represents an end-of-pipe intervention. Although it may contribute to curb emissions, more or less significantly, it leaves concentrations untouched. Thus, classifying CCS (and, to a lesser extent, BECSS) as CDR constitutes a misnomer, which explains why several overviews exclude CCS from the CDR taxonomy (Smith *et al.* 2023, 11). Additionally, it may serve for greenwashing purposes (Mann 2021, 153) through the "clean coal" rhetoric as seen in Australia (Schenuit *et al.* 2021). Worse, CCS promotion could jeopardize mitigation by softening the incentives to invest in fossil-free energies. Whereas it could represent a bridging technology for building a carbon-free economy (L'Orange Seigo *et al.* 2014), CCS can also reinforce the existing "carbon lock-in" (Unruh 2000), that is the institutional, technological, and behavioral factors trapping societies into fossil fuels (Asayama 2021; Vergragt *et al.* 2011).

The potential for capture depends on future investments in CCS facilities. For securing at least a 50 percent chance of remaining under the 2°C threshold, the Organisation for Economic Co-operation and Development (OECD) and the International Energy Agency (IEA) estimate that emissions should be cut by 60 percent compared to 2013, which will necessitate a cumulative capture of 94 $GtCO_2$ by 2050 (OECD/IEA 2016, 51–53). Such a projection appears out of reach. In 2021, according to the IEA (2021c), capture costs ranged from US$15–25/t in the case of high CO_2 concentration such as at coal power plants or in natural gas to US$60–120/t for more diluted flue gas like exhaust gas from cement factories. The transportation and storage costs span from negative for EOR to a bit below US$60/t. More importantly, due to technological progress, accumulated experience, economies of scale, and so forth, the IEA projects a steady decrease in costs.

Bioenergy with Carbon Capture and Storage/Sequestration (BECCS)

BECCS partly addresses the lock-in concern by redirecting carbon from biological sources to geological reservoirs. BECCS couples CCS with biomass energy production from crops (e.g., canola, sweet sorghum, sunflower, and hybrid poplar), forestry residues, and organic waste. The principle is to burn ("convert") biomass and then capture and sequestrate the emitted carbon according to similar processes as for CCS (NRC 2015a, 63). At first sight, BECCS presents two advantages: (1) to achieve carbon neutrality in power generation and (2) to durably store carbon that would otherwise be released by decaying or burnt biomass. These advantages partially explain why BECCS is the CDR technology *par excellence* for integrated assessment models (IAMs),[27] climate scenarios (Anderson and Peters 2016, 183; Fajardy *et al.* 2019a; Fridahl 2017, 90; Fuss *et al.* 2014; Geden *et al.* 2018; IPCC 2018), and scientific reports (NRC 2015a, 63–67; National Academies of Sciences, Engineering, and Medicine 2019, 137–188). Political pressures also explain the inclusion of BECCS into scenarios and models (Haikola *et al.* 2018). Despite such prominence, only eighteen large-scale facilities were in operation worldwide in 2019 (Consoli 2019), which is mirrored in an average TRL (5–6) (Smith *et al.* 2023, 19).

Ambitious BECCS deployment raises five significant challenges. The first has to do with environmental sustainability defined as the capacity of an activity to preserve natural resources to an extent that allows the self-perpetuation of ecosystems. As for forest management, BECCS may necessitate immense areas for energy crops. The surface could represent between one and two times the size of India (Anderson and Peters 2016, 183), which may undermine biodiversity by degrading vast natural habitats (Creutzig *et al.* 2014). In addition, cultivation will require tremendous amounts of water, nutrients (e.g., potassium, phosphorus, and nitrogen), fertilizers, and so forth (Dooley *et al.* 2021). Such massive monocultures may durably impoverish and pollute soils. However, if the criticism has some teeth against the first generation of biofuels (energy crops), the second generation (forestry residues) seem less prone to endanger ecosystems (Lin 2019, 544–545), although doubt remains about scalability (Burns 2016, 16). In any case, BECCS presents serious enough difficulties to environmental sustainability for calling into question its characterization as a green method that would necessarily help restore ecosystems (Heck *et al.* 2016).

The second challenge is related to social sustainability, which is the capacity of an organization to preserve human resources while taking into consideration the interests and rights of the affected individuals. In a context of population growth, declining access to water, and losses of arable land (due to erosion, salt contamination, and pollution), BECCS requirements may exacerbate those vulnerabilities and endanger food security. Severe land-use conflicts will most likely emerge or worsen, jeopardizing, in many regions, the right to sustainable development as stipulated in the article 3.4 of the UNFCCC (Dooley *et al.* 2018; Mace *et al.* 2021; Shue 2017). The issue is particularly troubling since the Global South is more exposed to land and water issues than the Global North. An aggravating factor is that scientists from the Global South are underrepresented in research projects and institutions such as the IPCC (Biermann and Möller 2019).

Third, BECSS would require massive infrastructure investments (e.g., biomass plants, sequestration sites, wells, and dams), with direct consequences for carbon emissions. To minimize the risk of dangerous climate change, a staggering number of facilities should be rapidly put into operation. For limiting global warming to 2°C, 16,000 plants should be put into service before 2050 (Lenzi *et al.* 2018, 304). Such fast-paced deployment will demand extensive logistics, from feedstock transportation up to sequestration (Anderson and Peters 2016, 183). BECCS plants have to be built close to reservoirs (e.g., deep aquifers) or the carbon will have to be carried to the storage sites through pipelines or dedicated vehicles, which will worsen the carbon footprint. The BECCS supply chain may be split, with the three segments (crops, biomass facility, and repository) located in different countries (Peters and Geden 2017, 621). In addition to international cooperation, BECCS raises the need for robust accounting backed by independent MRV (Fajardy *et al.* 2019b; Fuss *et al.* 2014).

Fourth, concerns have been voiced in relation to the efficiency of energy production as well as the carbon neutrality of the whole process. BECCS capacity is estimated as ranging from 40 to 100 exajoules (EJ) per year, which represents only a small proportion of the world energy production, evaluated to be 617 EJ in 2019 by the IEA (EASAC 2022, 2). This implies, *ceteris paribus*, that BECCS cannot cover all needs, and a diverse mix of CE solutions ought to be implemented, potentially including fossil fuels. Moreover, the final carbon balance will depend on the kind of land use the energy crops replace (Geden *et al.* 2018; Mace *et al.* 2021, 76).

Ultimately, it is not clear whether the technology can accomplish net carbon removal, especially within the limits set by the Paris Agreement (EASAC 2022). Foremost, BECCS needs to reach carbon neutrality, the point where the amount of trapped carbon compensates for the accumulated emissions resulting from the construction of the BECCS plant, as well as transportation and sequestration infrastructure and the feedstock farming. Then, negative emissions (net carbon extraction) should be achieved at a rate that makes a noticeable impact on atmospheric concentration. Moreover, the massive land cultivation implied by ambitious BECCS will generate difficult-to-predict effects on the carbon cycle (Heck *et al.* 2016).

BECCS capture capacity by 2050 has been estimated to be 3.5–5.2 $GtCO_2$/year, when using only biomass residues, for example from forestry, and 10–15 $GtCO_2$/year, when additionally relying on energy crops (National Academies of Sciences, Engineering, and Medicine 2019, 355). Smith *et al.* (2023, 19) provide a semblable calculation (0.5–11 $GtCO_2$/year) with costs varying from US\$15 to \$400/tCO_2. Those estimates are challenged as being too optimistic (EASAC 2022). For instance Fuss *et al.* (2018, 14) quantify a significantly lower potential: 0.5–5 $GtCO_2$/year for US\$100–200/tCO_2.

Direct Air Capture and Storage/Sequestration (DACS)

Also labeled direct air ambient capture, the third and last category of fully engineered interventions consists in trapping CO_2 from the surrounding atmosphere (NRC 2015a, 67–75; National Academies of Sciences, Engineering, and Medicine 2019, 189–246). While sharing the CCS apparatus with CCS and BECCS, DACS differs in that CO_2 is extracted from more dilute sources than flue gas, for instance. In 2021, nineteen facilities were in activity (IEA 2021a). In 2023, there were 27 commissioned plants and 130 in development according to the IEA..

The major advantage of DACS in comparison to CCS and BECCS is its geographical flexibility (Lin 2019, 540). The technology is not spatially constrained as BECCS (proximity with biomass sources) or CCS (combined to the existing industrial sites). DACS plants can easily be built in the vicinity of geological reservoirs, lessening the costs of carbon transportation. In addition, the method has a much lower land footprint and water use than BECCS, which results in no significant threat to water access and food security. Due to minimal natural resources requirements, DACS seems to pose less danger to environmental sustainability and biodiversity than BECCS.

However, DACS suffers from shortcomings. First, the development of the technology represents a serious challenge. Although its TRL is being evaluated ca. 7–8 (Low *et al.* 2022, 2; Smith *et al.* 2023, 18), in comparison to CCS and BECCS, DACS captures carbon from less concentrated sources (Bellamy and Healey 2018, 6). DACS is at the demonstration phase with a few facilities in service, the Orca's plant in Iceland being the largest (up to 4,000 tCO_2/year). Strong doubts remain in the literature about its physical and financial scalabilities (Fuss *et al.* 2018, 17).

Second, it requires a dedicated source of energy for its operations, contrary to BECCS that incorporates biomass combustion, or CCS that is either attached to fossil-powered plants or tap into the external energy alimentation of the plants (e.g., for cement factories). To qualify as a NET, the type of energy devoted to the DACS facility is crucial as it should be nonfossil or the ratio *trapped/emitted carbon* ought to be, preferably largely, superior to 1. In any case, the method necessitates considerable amounts of energy, which is reflected in the cost estimates for the capture (and sequestration), usually the highest in all the CDR interventions (e.g., NRC 2015a, 37–38).

Third, since DACS does not provide any co-benefit, such as energy cogeneration (as for BECCS) and biodiversity preservation (as for forestry), its deployment will be fully contingent on carbon pricing (Cox and Edwards 2019) under a regime of trading carbon credits and more generally on public subsidies. An alternative, direct air capture with utilization where the trapped carbon is repurposed (e.g., building materials and carbonated drinks), offers financing opportunities, although the overall carbon balance may be positive (net emissions) depending on the type of energy used for the capture. The problem is compounded if the carbon rapidly returns to the atmosphere, as in the case of beverages.

The capture potential is evaluated by Fuss *et al.* (2018, 20) to be 0.5–5 $GtCO_2$/year by 2050. However, they mention that actual capture may turn out to be significantly greater, up to 40 $GtCO_2$ if the constraints bearing on the storage are less draconian than envisaged and if DACS does not prove harmful for the environment. Other studies corroborate the ambitious estimates (Smith *et al.* 2023, 18). The costs are currently estimated as being between US\$600 and 1,000/tCO_2, although they may decrease to US\$100–300/tCO_2 due to factors such as innovation and better mastery of the technology and economies of scale (Fuss *et al.* 2018, 20). Smith *et al.* (2023, 18) support the low-cost projections.

TEXTBOX 3.3 IS CDR CLIMATE ENGINEERING?

Classifying CDR as CE is questionable. For instance, Wagner and Weitzman (2015, 32) claim that CDR fundamentally differs from CE. The latter would constitute a technofix circumscribed to tackling the consequences of climate change, whereas the former would genuinely engage the origins of the Anthropocene by promising to reduce carbon emissions and concentrations. CE is widely characterized as the "deliberate large-scale manipulation of the planetary environment to counteract anthropogenic climate change." Critics such as Wagner and Weitzman dispute the classification of CDR as CE by arguing that CE does not *substantially* address climate change, contrary to CDR. By doing so, they challenge the third component of CE definition. The underlying assumption is that counteracting does not equate to tackling the root of the Anthropocene.

Thus, demonstrating that CDR is not a CE intervention mandates one of the two things: either to show how CDR does not fit the Royal Society's characterization or to propose a more restrictive definition that will more aptly capture the true nature of CE. The former is tricky since CDR ticks all the boxes: intentional, potentially large scale (if we consider both the effects and the required deployment), and dedicated to counteracting climate change by removing carbon. According to the *Oxford Learner's Dictionary*, counteract is "to do something to reduce or prevent the bad or harmful effects of something." It is then difficult to see in which sense CDR, if effective, would not reduce or prevent climate dangers. In the UNFCCC's parlance, CDR could serve the "stabilization of greenhouse gas concentrations in the atmosphere at a level that would prevent dangerous anthropogenic interference with the climate system" (article 2). The second path—redefining CE in a more restrictive fashion—requires justifying what is gained in terms of conceptual clarity. To our knowledge, no tentative redefinition along those lines has been attempted. It does not mean that it could not be done, simply that the Royal Society's formulation remains the widely accepted standard.

A last-resort option would be to stop using the CE qualifier. It could be claimed that CDR and SRM are so different from the perspective of the technology, intentions, means, potential risks, political/ethical challenges, and so on that bundling them under the CE umbrella makes little sense. Thus, it would warrant distinguishing them while dropping any reference to an overarching category. To some extent, this is the methodological choice operated by the US NRC when it produced two, separate, reports. Diverse reasons could justify separating CDR from SRM. Besides descriptive accuracy, a convincing rationale is of a strategic nature: the concern is widespread that CDR's image could be tarnished by its association with SRM downsides such as the termination shock or the threat of militarization.

Nonetheless, the CE label has some cogency. It situates CDR and SRM in the category of policy instruments that differ from mitigation and adaptation while sharing the trait of attempting to alter the radiative forcing through more or less engineered interventions.

CDR: A GENERAL ASSESSMENT

As mentioned before, any attempt to offer general conclusions on CDR runs the risk of overgeneralization and oversimplification, both of which characterize many discussions on the CE topic. Beyond shared characteristics, CDR methods like afforestation, no-/low-tillage, macroalgae cultivation, ocean-enhanced weathering, or DACS significantly diverge in terms of technology, risks, legal challenges, and so forth. Despite such diversity, it is nonetheless possible to draw valid conclusions, point at key issues, and underline CDR's ethical and political challenges. This assessment presents a condensed review of feasibility, permissibility, and preferability issues without providing a detailed account of each evaluative category or a definitive assessment. A sketchy summary could be that feasibility at a scale that significantly impacts carbon concentrations remains unassured for all the interventions considered in isolation (But there is still the possibility of combining them within an hybrid policy). Moreover, permissibility may be impeded by issues of justice, especially related to the use of resources like water, land, and minerals for some of the most prominent methods (forestry and BECCS). The possibility of carbon lock-ins raises strong concerns that underpin the necessity of a strong commitment to conventional mitigation independently of any stance in relation to CDR. Securing the participation of the impacted populations from the R&D stage constitutes a priority as well as developing appropriate regulative and governance frameworks.[28]

Feasibility

The current and prospective feasibility of the various CDR methods varies to a great extent. Doubts remain about the efficiency of all the methods when it comes to offering significant and durable CO_2 capture. Many interventions are accessible, like forestry, no-/low-tillage, macroalgae cultivation, CCS, or BECCS, at least in a context of limited deployment, but none, under the current technological conditions, can guarantee a decent probability of counteracting climate change at the required scale, that is staying within the 1.5–2°C threshold set by the Paris Agreement. This is true especially if conventional mitigation is not aggressively ramped up.

The *technological maturity* across methods varies from high (TRLs 8–9) for forest management to low (TRLs 1–3) for ocean-enhanced weathering and ocean sequestration, passing through average maturity (TRLs 4–7) for biochar, blue carbon, BECCS, DACS, and SCS (Anderson and Peters 2016; Low *et al.* 2022, 2). The consensus is that the most promising CDR interventions, especially BECCS, which prominently figures in IAMs and various scenarios, are not ready to be deployed at the scale that is required by the Paris targets.

This gap between the reliance of climate scenarios on CDR methods and their immaturity is stirring concerns (Lin 2019; Smith *et al.* 2023).

Thus, to date, all the CDR interventions display *questionable scalability owing* to varying reasons: a tremendous need for natural resources (minerals, water, biomass, and nutrients), energy (for transporting of material, operating facilities, mining, etc.), land, access to geological reservoirs, the required global coordination for maintaining the carbon stored (forestry and land sequestration), and so forth (Mace *et al.* 2021). Even forest management, the most mature CDR intervention, necessitates monumental stretches of land. Moreover, besides forestry and tillage techniques, other interventions still must be tested at a large scale. CCS, BECCS, and DACS are only implemented in a handful of facilities, and in more of an experimentation setting for DACS. Only tiny ocean patches have been fertilized for a brief period, which cannot ground any conclusion. Land-enhanced weathering experiments were conducted on limited land areas, while not having been tested at scale at sea. But the fact that a given technology is not mature or scalable at an instant t does not imply that it will still be untested at $t+n$. This explains the importance of R&D and the necessity of distinguishing arguments in favor of R&D from those advocating for full-scale deployment (cf. chapter 5).

Assuming that some interventions are technologically and physically scalable, feasibility will depend on proper carbon pricing and regulation (Carton *et al.* 2020, 14), as well as public subsidies (Bednar *et al.* 2020), especially for methods with few co-benefits outside capture like DACS. Based on the previous experience with CCS, an essential aspect of CDR from the R&D stage to deployment is that carbon pricing will not be sufficient (Lin 2019, 571). Therefore, carbon markets ought to expand, and the full social cost of carbon (and other GHGs) should be tailored in. The central issue is whether markets constitute the proper institutions for including the whole cost of fossil energies, including the negative externalities. While environmental economists advocate for setting the global price of carbon at a much higher level, over US$100/t, as soon as possible (Bhat 2021), 2023 prices are set at between a few dollars per ton in markets like China or South Korea up to US$60–100/t for the European Union carbon permits market, and over US$130/t for countries which implemented a carbon tax like Liechtenstein, Sweden, Switzerland, and Uruguay.[29] Carbon prices should drastically increase throughout the century. A minimum price of US$400/t by 2050 would be desirable for reinforcing mitigation and CDR (Rockström *et al.* 2017, 1271). Forms of support other than carbon pricing, from both private and public entities, appear as indispensable complements (Lin 2019, 571), for instance under the form of taxes and direct regulation. Limited market efficiency advocates for more ambitious public intervention (Cox and Edwards 2019). However, studies on the political conditions for implementing CDR methods such as BECCS (Fridahl *et al.* 2017) are lacking, while public policy, regulation, and governance do not support CDR at the level assumed by IAMs for respecting the Paris targets (Geden *et al.* 2018; Lin 2019; Schenuit *et al.* 2021).

The distance between the advancement of CDR projects and the assumptions embodied in policy documents is described as the "commercialization gap" (Tamme and Beck 2021). In order to bridge the gap, investments are required in CDR methods as well as the whole surrounding infrastructure (e.g., transportation and storage). The "accounting gap" constitutes a second obstacle (Tamme and Beck 2021). To efficiently implement CDR, states and international organizations should establish a standardized system of accounting. Moreover, there is the necessity to undertake a "full carbon accounting" (Lin 2019, 542), which raises a recurrent issue: the fact that CDR lifecycle assessments are often incomplete (e.g. excluding the energy or costs required by carbon transportation and storage; or not adequately accounting for carbon leakages). This task is usually undertaken by various MRV apparatus developed around the world, for example in the European Union Emissions Trading System (EU-ETS).

MRV taps into *efficiency* evaluation, which is pivotal for the allocating of resources to research, and, later, to deploying various CDR methods within a climate portfolio. Such evaluation ought to be performed as soon as possible to avoid technological dead ends, which illustrates MRV's importance for assessing various interventions on carbon sinks and reservoirs (EASAC 2022). MRV schemes are already embedded in climate policy, for example in the intended nationally determined contributions, which constitute a country's pledges to reduce carbon emissions under the UNFCCC (Wiener 2015). With varying degrees of robustness, countries report their carbon sources and sinks. Recordkeeping needs to be strengthened, for example when it comes to evaluating the performance of reforestation or for including potential geological reservoirs. To determine whether specific interventions are worth pursuing, reliable measures of the amount and duration of the capture are crucial (Fry 2007; Lin 2019; Lövbrand 2009; Mace *et al.* 2021; Weiner 2015). MRV is also essential for preferability assessments when comparing CDR methods (Fajardy *et al.* 2019b, 3). In addition to the elaboration and adoption of international standards, MRV necessitates auditing carried by independent parties (Wiener 2015, 192–193).

CCS accounting is further justified by factors that potentially aggravate the risk of wasteful investments in sterile techniques. One factor resides in the fact that CDR interventions, primarily forestry and BECCS, are already integrated in most of climate scenarios and models such as IAMs, and by organizations such as the IPCC. As a result, various mechanisms may trap policymakers in investing in or placing their hopes in methods that cannot deliver significant climate benefits during the twenty-first century. One such mechanism is sunk costs, which designates irrecoverable costs dedicated to an activity (e.g., R&D of a specific CDR). Sunk costs act as psychological barriers that prevent decision-makers from abandoning inefficient or undesirable courses of action. This could lead to mounting inefficient investments under the form of "escalating commitments" (Staw 1976). The risk of remaining trapped into dead-end technologies due to prior committed resources is serious, especially if past investments have been substantial as they have been for CCS (Gardiner 2010, 289; Rayner *et al.* 2013, 502). In a world of scarcity, such resources would be lost for more efficient usage like mitigation, adaptation, or different CE interventions.

The ex ante identification of promising and feasible interventions is a daunting task due to the urgency of developing an internationally coordinated climate response and the uncertainty attached to many policy options, first and foremost to CE. Moreover, any early regulation of emerging technology faces the "Collingridge dilemma" (Collingridge 1980), namely that during the initial steps of the development of a technology, not enough is known for crafting efficient regulation, and later, it may be too late to back down from the technology due to various lock-ins and sunk costs. However, the necessity of identifying a threshold of reasonable doubt remains, beyond which it becomes obvious that further R&D will lead nowhere or that governing the intervention will be next to impossible.

In that respect, the leading role of the fossil industry for CCS development and promotion, among other CE methods, could aggravate sunk costs, thereby escalating commitments to inefficient technologies and creating carbon lock-in due to vested interests. However, a lock-in is not necessarily bad. It simply results from social commitments to technological standards. For instance, the EU is "locked in" to the 230 V/50Hz electric standard or to the right-hand roadway traffic. Those are conventions that render plenty of activities more efficient and less hazardous. Furthermore, lock-ins come in degrees. Shackley and Thompson (2012) differentiate two levels when it comes to CCS, which is the CDR method the most prone to carbon lock-in. First, "high-carbon fossil fuel lock-in" take place when "new unabated coal or gas power plants [are] being constructed on the basis that CCS will be applied to them at some point in the future, but then CCS never actually being implemented" (Shackley and Thompson 2012, 104). This constitutes a serious concern since CCS would then deter conventional mitigation, contributing to perpetuate a carbon-based economy. The second type is "low-carbon fossil fuel lock-in," which happens when "new coal and gas power plants are constructed and CCS installed (and possibly existing plants retrofitted), but that this perpetuates the use of fossil fuel such that development of alternative energy sources is inhibited" (Shackley and Thompson 2012, 104). The latter form is less discomforting than the former since, even though CCS still contributes to maintain the status quo, it nonetheless provides a solution for carbon removal. In addition, the problem resides less in the existence of lock-ins, which simply express path dependency, than the difficulty to get out of them. The implications for CE assessment are twofold: to determine which kind of lock-in different methods create and to evaluate how deep humanity could be trapped in each of them.

Finally, when it comes to evaluating CDR general efficiency at lowering GHG emissions and concentrations, a serious limitation resides in the fact that most, if not all, methods presented in this chapter focus on carbon dioxide, while mitigation of non-CO_2 GHGs constitutes a pressing challenge (IEA 2023, 10–12; Shepherd *et al.* 2009, 52). CDR methods disregard very potent gases. Due to its GWP (cf. textbox 0.1), higher than that for CO_2, and abundance, in particular in the permafrost, methane represents a serious climate threat. Despite such risks, too few research initiatives have been dedicated to developing methane capture devices (de Coninck *et al.* 2018, 347), which impedes the efficacy of GHG removal.

Permissibility

Like feasibility, permissibility varies according to the criteria exposed in the second chapter: the intrinsic moral value of the intervention, the prospect of unacceptable risks/uncertainty, distributive, procedural, and intergenerational consequences, social acceptability, as well as the opportunities for regulation and governance. This section does not pretend to offer a full-fledged normative account of the different CDR methods. Rather, it offers some indications about prominent questions as well as general trends that emerge from the literature.

Intrinsic moral value. The fact of removing carbon for storing it away from the atmosphere does not sound bad or wrong per se. On the surface, it is difficult to see what could be intrinsically objectionable in cultivating seaweeds, planting trees, or barely tilling farmlands. Those activities can disrupt the surrounding ecosystems if not carefully conducted (e.g., if the trees varieties used for afforestation are toxic or incompatible with the local flora and fauna). In such a case, the problem lies in the manner that the CE method has been implemented, not in its inner characteristics. It could be objected that the magnitude of the intervention required for significantly lowering carbon concentrations manifests arrogance, the will of dominating nature or of playing God. Then, the objection has less to do with the activity per se than its extent. A possible response is that forestry or agricultural interventions are objectionable because they constitute, by definition, instances of interference into the natural world, but in that case, any human activity could fall under a similar objection.

The other CDR methods aiming at enhancing natural processes such as the physical pump, OIF, or enhanced weathering could be perceived as shielded against, at least, the unnaturalness objection because they support existing biochemical mechanisms. However, they could be opposed to on the ground of representing human interference per se. The objection would be untenable since it would lead to a principled rejection of any anthropogenic interference, including some that are a priori tolerable. Those methods could also be criticized as representing instances of *improper* interference, for instance because they are potentially polluting (OIF and enhanced weathering) or extremely disruptive (e.g., of the thermocline with the physical pump). Without expanding further, a clear-cut line should be drawn between proper and improper interference. In addition, as for forestry and agricultural practices, the extent of the intervention could be claimed to be the issue. In that case, the moral objection is focused on how a method is deployed, not the method itself.

At first sight, no serious intrinsic moral issue appears to cripple engineered methods (CCS, BECCS, and DACS), except to raise a principled opposition to any engineered apparatus per se as being unnatural, hubristic, or playing God. BECCS, however, represents a specific case. The extent of energy crops cultivation could be seen as undermining existing ecosystems. But, again, the problem is located in the magnitude, not the intervention itself. The very idea of trying to artificially mimic parts of the carbon cycle could also be judged as expressing some unacceptable arrogance. In that case, an argument would have to be carefully crafted about the conditions under which reproducing natural processes are bad, evil, wrong, and so on.

In sum, the activity of removing and storing carbon or other GHGs away from the atmosphere does not appear particularly problematic at first glance. This does not mean that an argument cannot convincingly demonstrate that a specific method could be inherently objectionable. Furthermore, this does not imply that CDR methods are free of moral concerns, for example, regarding potential harmful consequences or violation of rights (Burns 2016). But intrinsically, no specific aspect of CDR methods stands out as impermissible per se.

A closer inspection of the consequences draws a different, more nuanced, picture. First, interventions may generate *risks and uncertainty* that exceed the boundary of permissibility even though the perceived risks appear generally low for CDR (Shrum *et al.* 2020). From a bird's eye view, the concerns seem to be twofold. On the one hand, a sudden large-scale carbon leakage could cause a catastrophe comparable to the Lake Nyos incident, but on a much broader scale. However, as indicated above, despite uncertainties, such a risk appears infinitesimally low (Miocic *et al.* 2019). Sequestrated carbon could and will leak for sure, but not at a rate and in quantity that could cause serious harms to humans and the environment. Moreover, carbon reservoirs could be situated far away from the main population centers. On the other hand, as indicated above, excessive optimism and overconfidence in CDR "fantasy" (Shue 2017) could deter mitigation and reinforce carbon lock-in, which appears to be particularly salient for CCS (Asayama 2021). In the advent of insufficient mitigation (and adaptation), the result could be to place humanity on an uncontrolled overshoot pathway that could ultimately lead to catastrophic consequences, especially in the future. Climate change could then unfold in very harmful ways with soaring temperatures well above the Paris target, and unfolding catastrophes (severe droughts, storms, heatwaves, etc.). Those delayed impacts carry intergenerational consequences by weakening future generations' resilience and condemning them to live in a more dangerous world.

Besides catastrophes *per se*, CDR could raise significant risks of harmful consequences. One of those consequences is undermined environmental sustainability and biodiversity (Shue 2017). The damage could happen, for instance, through changes in land uses and soil composition (e.g., forestry, biochar, and BECCS), chemical and nutrient pollution (e.g., enhanced weathering, ocean interventions, and OIF), the pressures on natural resources such as water (e.g., BECCS and enhanced weathering), introduction of invasive or toxic species (e.g., OIF), accelerated acidification (e.g., artificial upwelling), and so forth (Dooley *et al.* 2021). While these dangers exist, they have been insufficiently studied. More resources should be invested in mapping out different negative and positive CDR-induced externalities. Moreover, harmful spillovers are less likely to affect "encapsulated" technologies, that is, interventions that are "modular and contained" such as DACS (Hulme 2014, 11).[30]

CDR methods, especially those that are land-based, will likely produce disparate impacts, especially for risks. Due to their high-land requirements, forest management and BECCS will put a strain on available lands, driving up crop and food prices and undermining alimentary security for local populations, especially in developing countries (Burns 2016, 15–17; Nemet *et al.* 2018). Sustainable development may be undermined (Mace *et al.* 2021). It should be noted that some CDR interventions like enhanced weathering avoid endangering food security (Beerling *et al.* 2020). More generally, the problem here is redistributive in the sense that some CDR interventions would produce global benefits (reduction of carbon emissions or concentrations), the burdens of which will be unequally distributed, potentially impacting vulnerable populations (e.g., farmers in tropical areas) the hardest.

R&D and deployment also carry indirect issues of distributive justice in the sense that the increasing mitigation gap as well as past responsibility for/benefit from GHG emissions argue in favor of a special responsibility falling on the most developed countries. More precisely, it could be argued that they should shoulder most of the investments implied by researching, developing, and potentially deploying CDR (Buylova *et al.* 2021; Mace *et al.* 2021). Such responsibility could be supported by the UNFCCC principle of "common but differentiated responsibilities" confirmed under the Paris Agreement, especially if CDR is assimilated under mitigation (Honegger *et al.* 2021), which remains controversial (cf. chapter 1).

CDR also raises *procedural concerns,* especially for those interventions that will significantly impact or generate risks for vulnerable populations that are traditionally excluded from CE research and from the policy decision-making, such as indigenous peoples (Whyte 2018) and developing nations (Preston 2013, 28–29). By itself, the lack of participation and consultation from the R&D stage creates vulnerabilities that could concur to domination schemes, in which affluent decision-makers (have the capacity to) impose decisions onto more modest populations and countries. For instance, as already mentioned, the demand for energy crops might create land-use conflicts, which may be extremely prejudicial to vulnerable populations located in the regions where such cultivation is undertaken (e.g., tropics). Another example is OIF and its potential negative impact on marine ecosystems. In the case of alteration of the biodiversity, the intervention could threaten the food security of local fishing communities, especially with experiments undertaken in coastal waters, *as per* the CBD recommendations (cf. chapter 2).

Two moderating factors ought to be highlighted. First, populations in developing countries might reluctantly support CDR as well as SRM interventions, even though they will not fully control R&D or not initiate deployment (Carr and Yung 2018). Such a possibility strongly advocates for broadening the scope of participation for methods that carry significant negative transboundary externalities (e.g., OIF, ocean-enhanced weathering, BECCS, and forestry in the case of extensive land use in those countries). Geological sequestration presents safety issues that could be considered, if not global, at least regional, depending on the location and size of the reservoirs. Second, although CE research has gained the reputation of being dominated by the "Global North," the actual picture is a bit different (Cao *et al.* 2015; Smith *et al.* 2023, 29). Emerging markets such as Brazil, China, India, and Indonesia are among the leaders in CDR research and those places are often the focus of targeted research programs.

As noted in the previous sections, *public acceptability* and acceptance fluctuate across the methods. Moreover, the level of awareness and knowledge of the different interventions is rather low (Cox *et al.* 2020; Smith *et al.* 2023, 41–53), which explains the importance of framing effects (Corner and Pidgeon 2015). As previously noted, one of the major criteria for social acceptance is perceived naturalness (Shrum *et al.* 2020; Sweet *et al.* 2020; Wolske *et al.* 2019). Other criteria include how familiar the respondents are with the CE method. The higher the unfamiliarity, the lower the acceptance, and the higher risk perception. Trust in the institutions engaged in CE activities (R&D, deployment, and regulation) also constitutes a determinant factor. The public in various studies tends to favor forestry as the most natural, less risky, and least uncertain intervention, followed by no-/low-tillage. DACS, BECCS, and SCS appear as less natural (Wolske *et al.* 2019) and riskier, while biochar is perceived as carrying more uncertainty, ocean fertilization being riskier, and enhanced weathering being riskier and still more uncertain (Shrum *et al.* 2020, 6). In a recent study conducted in the United States, the methods supported by the majority of respondents were limited to forestry and SCS (Sweet *et al.* 2021). The latest data extracted from a sentiment study on Twitter from 2010 to 2021 (Smith *et al.* 2023, 52) show that the most positively perceived CDR interventions are forest management and enhanced weathering followed by biochar, ecosystem restoration, and DACS. The most negatively method, by far, is OIF followed by BECCS and DACS (with the positive sentiment for the latter overcoming the negative one). Public sentiment could have preferability implications.

At the end of the day, this chapter shows that *governance and regulation* are still in their infancy when it comes to CDR, except for OIF.[31] The COP under the CBD call for a moratorium on large-scale experiments, but without its decisions being legally binding. Such a lacuna creates a vacuum that could be exploited for rogue or uncoordinated deployment, but in the case of technologies that necessitate significant public regulation and support it undermines permissibility or at least the feasibility. In addition to the development of special regulatory instruments and institutions, there is a strong need for international cooperation, sanctioned by the UNFCCC. The Paris Agreement stipulates that its parties should "strive to formulate and communicate long-term low greenhouse gas emission development strategies [LT-LEDS]" (Buylova *et al.* 2021, 3). However, cooperation will have to expand beyond R&D and MRV for including the maintenance of the CDR intervention through time, which could be subject to a tragedy of the commons (Lawford-Smith and Currie 2016, 3), in which states are reluctant to adequately contribute financially or technologically to keep carbon capture at the appropriate levels. Since sudden interruption of the given interventions carries the risk of runaway dangerous climate change, CDR permissibility depends on the existence of redundant mechanisms that guarantee the persistence of carbon extraction and storage (Wong 2017).

Preferability

A full-blown comparison of the relative benefits of all CDR interventions falls outside the scope of this book owing to the variety of dimensions and technicity of the task. Nonetheless, brief examples of how to conduct this part of the assessment could be offered. As discussed at the end of chapter 2, various decision rules could be used for comparing CE methods. The most difficult to implement is the Paretian approach. From a bird's eye view, as already mentioned, under the conditions of not generating significant harms, the implementation DACS could constitute a Pareto improvement, even when included in a hybrid policy. Large-scale CCS might also allow to drastically improve a given hybrid strategy without potentially making any group worse off. As presented, other rules could be considered. For instance, policymakers could consider the projected net benefits. In that case, methods involving geological sequestration may be preferable to biological storage because the former promises higher efficiency. The latter, for example, forestry or OIF, can only promise a relatively short sequestration (Carton *et al.* 2020, 5), potentially a couple of centuries, most likely a few decades, except if the carbon captured by the biomass is converted into more stable nonorganic carbon like biochar. However, geological sequestration carries extremely low but still possible risks of massive leakages, with catastrophic consequences for human beings and ecosystems. Therefore, a maximin approach might insist on discarding such options or heavily constraining their usage. The preferability dimension is not limited to weighting one intervention against another. In the current context, in which the best estimates suggest that isolated CDR interventions can only capture a small portion of annual anthropogenic emissions, a portfolio or hybrid approach is increasingly favored by many commentators (e.g., Keith 2013, xix; NRC 2015a, 2015b; Svoboda 2017). Consequently, preferability judgments need to be rooted in comparisons between different policy bundles including mitigation, adaptation measures, and some mix of CDR (and potentially SRM) interventions. The necessity of optimizing climate policy portfolios is intensified by the fact that humanity will soon need to actively extract atmospheric carbon, that is reach net negative emissions, for avoiding dangerous climate change.

NOTES

1. A NET is a technology that achieves a net removal of greenhouse gases (GHGs) from the atmosphere, i.e. the removals achieved are greater than any positive emissions caused by the deployment of the technology. (Brander *et al.* 2021, 700)

2. A similar typology as the one adopted in this book rests on separating nature-based interventions from technological/geochemical-based ones (Buylova *et al.* 2021; Schenuit *et al.* 2021). Other classifications are possible (e.g., Dooley *et al.* 2021; Minx *et al.* 2018).

3. It could be objected that the "fully engineered" qualifier does not capture significant differences between, for instance, bioenergy with carbon capture and sequestration (BECCS) and ocean-enhanced weathering, since the former relies on crop cultivation, which is not heavily engineered, while the latter rests upon ambitious industrial, therefore engineered, processes (mining, transportation, and dispersion). Despite those and other objections, at the end of the day natural sinks enhancement stimulate natural capture and sequestration processes, with the support of engineered devices, while fully engineered sinks mobilize anthropogenic capture and sequestration (usually directly into geological reservoirs).

4. COM/2022/672 final available on: https://eur-lex.europa.eu/legal-content/EN/TXT/?uri=COM:2022:672:FIN and also https://climate.ec.europa.eu/eu-action/carbon-capture-use-and-storage_en

5. Although the delay before implementing CDR as well as its slow impact on GHG concentrations will keep the climate in a dangerous zone for an extended period (EASAC 2018, 13), a solution is then to turn to SRM methods such as SAI for smoothing out the excessive temperatures implied by the overshoot (peak shaving).

6. The readers only interested in a summary evaluation of CDR can directly jump to the last section.

7. Recently, the US NRC devoted a portion of its report on NETs to coastal blue carbon, which includes projects to capture carbon in coastal wetlands (mangroves, salt marshes, swamps, river deltas, etc.) (National Academies of Sciences, Engineering, and Medicine 2019). Restoring those areas could generate important co-benefits such as protection against storms or sea-level rise, restoration of biodiversity.

8. For instance, the Icelandic Forest Service is experimenting with different species, such as the Siberian larch, but the larch is not well adapted to mild summers, which will become even milder in Iceland due to climate change. Forestry research therefore is focusing on producing a hybrid between Siberian and European larches, as well as other species (Eysteinsson 2017).

9. Nature is a debated and fluid concept and is perceived differently especially in the case of debates surrounding CE (Corner *et al.* 2013).

10. The impact of naturalness judgments on the public acceptability of technologies, especially novel ones, is noticeable in other topics, for example, genetically modified organisms (Palmgren *et al.* 2015, 158–160).

11. Population growth trends constitute a debated topic. If the United Nations has been traditionally authoritative for long-term projections, they are now challenged by other models which provide lower estimates. For example, the International Institute for Applied Systems Analysis is forecasting a peak at 9.7 billion around 2080, followed by a decline to 9 billion by 2100 (Adam 2021, 464).

12. Other aspects influence the impact of forestry on carbon capture and cooling, for instance the availability of water in the soils, the trees species, the existing regional climate, the use of fertilizers, and the interactions between the soils and the surrounding atmosphere (Anderson *et al.* 2011).

13. The estimate of 3.6 $GtCO_2$/year appears to be the highest bound, but estimate goes as low as 0.5 $GtCO_2$/year (Hepburn *et al.* 2019, 92).

14. Along EOR, or the production of carbon-based building materials, urea, and polymers, biochar is sometimes classified as being part of carbon capture and utilization (CCU) (Hepburn *et al.* 2019).

15. Lawford-Smith and Currie (2017) offer a comprehensive overview of how different ethical concerns (e.g., moral hazard, dirty hands, and cost externalization) apply to enhanced weathering.

16. The inclusion of the whole life cycle explains why cost evaluations per sequestered ton of CO_2 vary a great deal. For instance, Renforth (2012) estimates the costs ranging from US\$24 and US\$538/tCO_2, some evaluations going as high as US\$1,000/tCO_2 (NRC 2015a, 37).

17. Industrial waste products could be used, like the cement kiln dust (NRC 2015a, 51), which raises concerns about chemical pollution.

18. Another, much less discussed, dispersion method could be to use major tropical rivers such as Amazonia and Congo, at the price of a rising river pH (Kölher *et al.* 2010).

19. However a few studies have seen the light over the recent years such as project VESTA, OceanNETS, and the Greenhouse Gas Removal by Enhanced Weathering project (Low *et al.* 2022, 7).

20.   The consequences of a sudden halt or failure of the geoengineering system. For SRM approaches, which aim to offset increases in greenhouse gases by reductions in absorbed solar radiation, failure could lead to a relatively rapid warming which would be more difficult to adapt to than the climate change that would have occurred in the absence of geoengineering. SRM methods that produce the largest negative forcings, and which rely on advanced technology, are considered higher risks in this respect. (Shepherd *et al.* 2009, 35)

21. Massive storms would have contributed to the last major ice age by decreasing atmospheric CO_2 (Martin 1990).

22. In some deep layers of the North Pacific, carbon could remain trapped up to 1,000 years before being released through the ocean currents (Siegel *et al.* 2021).

23. A portion of macroalgae cultivation falls under the blue carbon label (Dooley *et al.* 2021, 39), although the definition of blue carbon oscillates between any kind of carbon capture using marine biomass or the oceans and the capture taking place only into coastal biomass. For instance, the US National Oceanic and Atmospheric Administration defines it as "the term for carbon captured by the world's ocean and coastal ecosystems," while Johannesen and Macdonald (2016, 1) characterize it as "the carbon fixed by vegetated coastal ecosystems including seagrasses."

24. Other potential usages such as bioplastics are not mentioned because no sequestration is intended.

25. Smith *et al.* (2023, 18–19) provide lower estimate for coastal blue carbon TRL (2–3).

26. EOR consists in several methods, one being to inject CO_2 in a hydrocarbon reservoir for improving the collection of crude oil.

27. IAMs designate a family of climate models of various degrees of complexity that integrate natural components, but also dimensions such as human development

or socioeconomic conditions. William Nordhaus' Dynamic Integrated Climate-Economy model is an example of an IAM (Nordhaus 2008, 6–9).

28. For a typology of the early forms of CDR policy, the reader can refer to Schenuit *et al.* (2021).

29. Those prices are constantly fluctuating. Furthermore, the varying carbon prices result from various mechanisms: ETS like in the European Union or carbon taxes like in Uruguay.

30.   Some CDR proposals, such as mechanical air capture of carbon (sometimes called artificial trees), are well-bounded, closed systems. Others involve releasing chemical agents into the ambient environment. Any geoengineering method involving the introduction of such deliberate 'pollutants' into the oceans, the air, or on land is likely to prove controversial regardless of whether it is designed to reflect sunlight or remove carbon. (Rayner 2014, 5)

31. For a more detailed legal assessment, the reader can consult Gerrard and Hester (2018).

Chapter 4

Climate Engineering Methods II
Solar Radiation Management (SRM)

As seen above, climate engineering (CE) characterizes *anthropogenic inter-ventions in the climate system that predominantly aim at altering radiative forcing locally or globally to slow down, stop, or reverse climate change and/or related adverse events.* Besides carbon dioxide removal (CDR), the second category mainly consists of methods that enhance the albedo, namely the capacity of surfaces to bounce back incoming radiation. There is one exception, though, cirrus cloud thinning (CCT), which aims at increasing the amount of outgoing longwave radiation emitted by the Earth's surface. Those methods are typically grouped under the concept of solar radiation manage-ment (SRM), a denomination crafted by Ken Caldeira (Wagner 2021, 77) and defined by the Royal Society as "techniques attempt to offset effects of increased greenhouse gas concentrations by causing the Earth to absorb less solar radiation" (Shepherd *et al.* 2009, ix).

WHAT IS SRM?

Despite SRM being pooled with CDR under the CE label, they differ in terms of technology, but also in risks and promises (Keith 2017, 75). Echoing those differences, the lay public usually perceives SRM more negatively than CDR interventions (Carlisle *et al.* 2022; Pidgeon *et al.* 2013), although respondents in the Global South seem to be less opposed to SRM than in the North (Sugi-yama *et al.* 2020; Visschers *et al.* 2017). While CDR efficiency is evaluated by the mass (tons, megatons, gigatons, etc.) of carbon captured and ultimately sequestrated, SRM is assessed by the variation of net radiative forcing, which is measured by watts per square meter (watts/m^2 or W/m^2). As a reference, since 1750, the increase in net forcing has been estimated to be 2.4 W/m^2 (Bauer *et al.* 2018, 473; National Research Council [NRC] 2015b, 133), which translates into average global temperatures higher by 1.09°C in 2011–2020 than over the 1850–1900 period (IPCC 2021, 5). In contrast to CDR though, SRM does not stop the course of the Anthropocene. It leaves greenhouse gas (GHG) emissions and concentrations unabated. That explains why it is considered as a second choice to mitigation, only intended to supplement it. While CDR may be deployed for a long period, SRM in general is seen as a set of transient solutions, worth either for buying time before full decarbonization or for emergency deployment in case of runaway climate change.

Although widely used, the denomination of *SRM* is also contested. For instance, the US NRC utilizes albedo modification because it is a more straightforward and neutral description of the physical process involved, and it is free of the connotations of a precise and orderly process carried by the term "management" (NRC 2015b, 32). The Intergovernmental Panel on Climate Change (IPCC) prefers "solar radiation modification" (IPCC 2018). This book sticks to SRM, used in alternance with solar CE. The reason lies in the ambition of the methods, that is, to regulate, locally or globally, the amount of energy reaching Earth from the ground level up to the exosphere (the utmost layer of the atmosphere). Whether or not such ambition turns out being delusive, the managerial intent remains.

Albedo alteration is advanced as a temporary response to the Anthropocene. Stratospheric aerosol injection (SAI) is the most researched (de Coninck 2018, 350; Irvine *et al.* 2017; United Nations Environment Programme [UNEP] 2023) and controversial (Hulme 2014) of those methods. A prominent advocate is Paul Crutzen, recipient of the Nobel Prize in chemistry for his works on the impact of the nitrogen oxide on the destruction of the ozone layer. Drawing inspiration from the 1991 Pinatubo eruption, Crutzen (2006) offers one of the most famous defenses of SRM, among others (Wigley 2006), as a complement to mitigation. Before Crutzen and his colleagues, the idea of enhancing Earth albedo was entertained by scientists like Mikhail Budyko (1974). It also featured in various reports such as the one addressed to US President Johnson in 1965 (*Restoring the Quality of Our Environment*) and in documents published by the US NRC since 1977 (Keith 2017).

Different classifications are possible. The taxonomy followed in these pages mirrors the one crafted by the Royal Society (Shepherd 2009, 70). The methods are sorted by the altitude of the intervention: surface, troposphere (lowest atmospheric layer, 0–10 km on average depending on the latitudes and seasons), stratosphere (10–50 km on average), and space (beyond the thermosphere and outside the Earth's atmosphere) (Barry and Chorley 2010, 32–36). As this chapter will show, two of these methods have attracted most

Table 4.1   SRM Taxonomy

Surface	Troposphere	Stratosphere	Space
• Urban surface brightening (USB) • Crop albedo	• Marine cloud brightening (MCB) • Cirrus cloud thinning (CCT)	• Stratospheric aerosol injection (SAI)	• Space reflectors or scatterers

of the attention: marine cloud brightening (MCB) and stratospheric aerosol injection (SAI) (Irvine *et al.* 2017; NRC 2015b). As a result, the debate surrounding the whole SRM category is skewed toward a discussion of global issues and potentially catastrophic risks mostly carried by SAI, which is reductive considering the diversity of SRM methods.

The chapter's subdivision mirrors that taxonomy: (1) surface albedo modification, (2) tropospheric interventions, (3) stratospheric ones, and (4) space-based methods (table 4.1).

SURFACE ALBEDO MODIFICATION

Also called "regional land radiative management" (Seneviratne *et al.* 2018), surface albedo modification aims at enhancing the reflectivity of elements from the ground level (e.g., streets) up to several meters (e.g., canopy and rooftops). The main targets for intervention discussed in the literature are cityscapes and crops. If studies underscore both methods' potential for lowering local and possibly regional temperatures, with co-benefits such as increased moisture retention, their global impact remains uncertain and most likely minimal (NRC 2015b, 128–129). Although those interventions diverge from the ambitious "large-scale manipulation of the planetary environment," they can be classified as a local, or targeted, CE (cf. chapter 1).

Urban Surface Brightening (USB)

The Anthropocene is yielding drastic increases in temperatures that are especially disastrous in tropical and subtropical regions. Urban environments are particularly exposed owing to the extent of low-reflectance surfaces (e.g., asphalt, concrete, and dark roofs), which absorb more incoming radiation than rural areas which are often brighter owing to fields and grasslands. This thermic accumulation causes the urban heat island (UHI) effect, where cities are on average warmer than the surrounding area/countryside. Higher average temperatures and heatwaves frequency put urban residents at risk, which is a serious issue when considering that they represent 56 percent of the world's population in 2020 according to the World Bank. The most vulnerable groups include newborns, seniors, individuals with preexisting health issues (heart conditions, respiratory illnesses, allergies, etc.), and low-income households who cannot afford adaptation measures such as air conditioning.

USB constitutes a direct response to UHI by ramping up cityscapes' albedo. Urban surfaces such as rooftops, streets, and sidewalks are made more reflective by covering them in brighter colors (white and light gray). The expectation is a cooling outcome of potentially several degrees Celsius. USB initiatives are already implemented in metropoles like Chicago, Los Angeles, or New York (Lomborg 2020).

TEXTBOX 4.1 THE URBAN HEAT ISLAND EFFECT (UHI)

The UHI effect describes a thermic phenomenon in which urbanization leads to increased average temperatures, even in the absence of GHG-induced radiative forcing (Georgescu et al. 2014). Through the substitution of darker cover made of concrete, bricks, tar, and so on for vegetal cover such as woods and grassland, human-made buildings dampen the albedo. In addition, urban construction reduces the green canopy, which is essential for offering protective shadows, retaining moisture, and ultimately cooling the surrounding air. Despite its limited scope, UHI may have a broader impact than usually assumed, accounting for 2 to 4 percent of global warming (Jacobson and Ten Hoeve 2012). The UHI effect is particularly harmful in summertime, while being more benign or even positive in wintertime by lessening the risk of deaths by cold (Lowe 2016). Therefore, enhancing cities' albedo may help to alleviate a tiny portion of global warming, while shielding vulnerable urban populations from the most drastic consequences of the Anthropocene. Potential interventions include USB, but also green roof retrofits (Wilkinson and Dixon 2016), which consist of covering rooftops with vegetation after some engineering work (e.g., waterproof membranes and drainage systems).

Even if installing reflective material on rooftops and pavements could affect the microclimate and thus contribute to a moderate UHI (Oleson *et al.* 2010), it is very unlikely that even when conducted at a vast scale, it could generate a significant and durable global cooling effect (Akbari *et al.* 2009). The impact will remain local, or at best regional (Irvine *et al.* 2011). Moreover, for addressing UHI, nature-based solutions such as an expanded urban canopy, and sometimes rooftop greening, could constitute effective complementary measures.

The technological readiness level (TRL) of whitening rooftops has been evaluated at 9 (cf. chapter 2), the highest level (Low *et al.* 2022, 2). Besides uncertain global impacts, the efficacy of USB as well as greening initiatives (e.g., urban canopy and roof vegetation) fluctuates depending on the features like the already existing canopy, building density and height, paving, and so on (Makido *et al.* 2019). The feasibility appears robust without any significant scalability issue. The cost-effectiveness for large implementation needs to be further assessed by research. Finally, USB does not pose any specific threat for populations and ecosystems, while not being publicly opposed. The only limitation could be that cooling will also take place during winters, potentially increasing the energy needs for heating.

Crop Albedo

Crop albedo CE, or bio-geoengineering (Ridgwell 2009), "would involve growing crop plant varieties with a higher albedo than currently grown as a means to produce a cooling of the planet" (Irvine *et al.* 2011, 2). The category could be stretched to include initiatives undertaken for brightening farmlands during the fallow period such as low tilling, leaving harvest residues on the field (Davin *et al.* 2014), or using crop covers cultivated outside the growth season (Carrer *et al.* 2018). This intervention takes advantage of the fact that crops have greater reflectance on average than noncultivated vegetation, for example, barley and wheat possess a higher albedo (0.18–0.25) than coniferous (0.09–0.15) and deciduous (0.15–0.18) woodlands (Barry and Chorley 2010, 49).

However, albedo properties vary from species to species and, within the same species, diverge across varieties depending on characteristics such as leaves' waxiness or canopy morphology (Singarayer *et al.* 2009, 3). Thus, comparative studies are needed for picking high-performing specimens. The advantages of specific plants could be augmented by reinforcing albedo-enhancing traits (such as leaf waxiness) through crossbreeding or genetic manipulation (Singarayer *et al.* 2009, 8). The latter could shorten the whole selection process but may undermine public acceptance due to genetically modified organisms' perceived unnaturalness, potential threats to human health, and risks of contamination of the surrounding ecosystems (Palmgren *et al.* 2015).

The feasibility of crop albedo appears to be high. The maturity is evaluated at TRL 7 (Low *et al.* 2022, 2) while material and economic scalability seem robust. Despite potentially significant local effects (up to 1°C according to Ridgwell *et al.* [2009, 148]), the global impact remains unclear. More importantly, maximizing negative radiative forcing requires that most farmers worldwide pursue albedo enhancement. Nonetheless, the agricultural infrastructure and equipment are already in place, thus no drastic change in the farming structure would be necessary for deployment. However, uncertainties persist regarding the potential destabilization of the surrounding ecosystems induced by vast and sudden shifts in the crop cultures (Shepherd 2009; NRC 2015b, 129).

TROPOSPHERIC INTERVENTIONS

The second category includes interventions in the troposphere, the boundaries of which vary with latitude. In polar regions, the troposphere covers the zone between the ground level and somewhere around six kilometers, while under the tropics, the upper limit reaches as high as approximately eighteen to twenty kilometers. Also labeled as "cloud albedo" (Shepherd 2009), the two main methods are MCB and CCT. The former is the most studied and discussed of the two and second in popularity only to SAI within SRM (NRC 2015b).

Marine Cloud Brightening (MCB)

The US National Academies of Sciences, Engineering, and Medicine (2021, 2) defines MCB as "a strategy for adding particles to the lower atmosphere (near the surface) in order to increase the reflectivity of low-lying clouds over particular regions of the oceans." It aims at accentuating the Twomey effect (Salter *et al.* 2008, 3,990) "where increases in cloud condensation nuclei produce higher cloud droplet concentrations and cloud albedo" (Mitchell and Finnegan 2009, 2). The idea is to spray aerosols in the ambient air close to the ocean surface from vessels equipped with dedicated rotors or turbines (Latham *et al.* 2008; NRC 2015b, 102–103) or commercial ships. Although seawater droplets constitute the main candidates, dimethyl sulfide, dust, engine exhaust smoke, and soot might also be used as particles (Hamilton 2013a, 52–53; NRC 2015b, 122). The intended effect is to initiate or expand stratocumulus, which will bounce back to space more incoming shortwave radiation (Goodell 2010, 163–189; Hamilton 2013a, 52–55; NRC 2015b, 101–127).

The impact could be significant considering the poor reflectance of the water (0.06–0.10) compared to stratocumulus (0.6) (Barry and Chorley 2010, 49). However, the potential for albedo enhancement varies per region and season. Models show that some areas are more propitious to marine clouds than others, for instance the Southern Ocean during the North hemisphere winters or the North Pacific, South Atlantic, and offshore West African coasts during the North hemisphere summers (NRC 2015b, 104). In addition to laboratory research, MCB has only been subject to a few direct observations and small-scale field experiments (NRC 2015b, 107–109) on cloud formation such as the Eastern Pacific Emitted Aerosol Cloud Experiment (E-PEACE) in 2011 (Russell *et al.* 2013), or the MCB for the Great Barrier Reef in 2020 (Low *et al.* 2022, 5).

Keeping those limitations in mind, the average global potential for diminished radiative forcing has been estimated to be 0.97 W/m^2, with significant regional variance. Interventions in the most suitable areas could lower local radiative forcing by more than 30 W/m^2 (NRC 2015b, 111). However, because of MCB deployment in some areas, other regions may suffer from a decreased albedo (increased forcing), such as North Africa, Saudi peninsula, parts of India, the United States, and Canada.

MCB displays prospective merits (NRC 2015b, 143–147). First, it could substantially contribute to "mask some consequences of greenhouse gas warming" (NRC 2015b, 143). Moreover, MCB may help to lessen coral bleaching (Latham *et al.* 2013) or hurricanes' intensity (Latham *et al.* 2012), although side effects need to be investigated, particularly for potential adverse impacts, for example, reduced photosynthesis. Second, scalability does not seem to be seriously constrained. The level of technicity required for deployment and monitoring is moderate, suggesting full scalability once the technology has matured (cf. below). The material and financial scalabilities do not present any insurmountable obstacle, the estimated overall cost being low (NRC 2015b, 144). Furthermore, existing commercial fleets could be used for spraying the particles. Third, potential risks and uncertainty appear limited. No hazardous chemical or seeding substance seems necessary (Hamilton 2013a, 54; Latham *et al.* 2008), with seawater constituting the most discussed option. MCB favorably compares with SAI as being less intrusive. The absence of damage to the ozone represents an additional advantage over SAI. Finally, once mature, MCB could be scaled up within a few years (NRC 2015b, 143) or shut down in a matter of days if adverse outcomes should manifest, with the cooling impact dissipating immediately while SAI effects will be more enduring. It could be crafted as a targeted intervention in regions particularly exposed to climate change like the poles (Latham *et al.* 2014), for instance, for limiting Arctic Sea ice loss during summers.

Nonetheless, in addition to not tackling carbon emissions and concentrations, MCB suffers from shortcomings. First, the aerosol-cloud interactions and large-scale climate responses are not fully understood, therefore necessitating better models and monitoring systems (Diamond *et al.* 2022; NRC 2015b, 113). Clarity is needed about the optimal density of nuclei for cloud formation (Diamond *et al.* 2022, 3–4; NRC 2015b, 103), which carry engineering implications. What is the optimal dimension for injected particles, that is the most effective in terms of reflectance? How to deliver the particles? Any efficacy assessment will remain crude as long as "cloud models and global model parametrization of marine stratocumulus" are not improved (NRC 2015b, 117). The technical challenge is not restricted to unrefined models. It extends to monitoring systems, especially during field experiments. Adequate *in situ* tools (measurements from vessels or aircraft) and also satellite imagery are required for evaluating shifts in cloud albedo.

Second, such limitations impede risk analysis. MCB might pose significant threats that cannot be properly identified and gauged under the current circumstances (Diamond *et al.* 2022; Hamilton 2013a, 54–55; NRC 2015b, 120–121, 127). For instance, under brightened clouds, marine and land patches will cool down, interfering with local ecosystems through changes in precipitations and energy fluxes. Photosynthesis could be drastically diminished, with a noticeable impact on the carbon capture capacity of the oceans. Globally, the cloud cover and precipitations in distant areas may be altered. Those areas could experience reduced albedo, possibly between 5 and 10 W/m^2 (NRC 2015b, 111). Ocean circulation and thermic exchanges with the atmosphere could be disrupted, throwing off balance periodic climate variation patterns such as the El Niño-Southern Oscillation (ENSO). Those risks support the necessity of devising "exit ramps," that is criteria for interrupting research or a deployment program, and "exit strategies" (Diamond *et al.* 2022; Schäfer *et al.* 2015, 100).

Third, MCB efficacy will be constrained by the local weather conditions at the time of the deployment (Diamond *et al.* 2022, 4) and the ephemeral nature of the intervention. On the one hand, if the dispersion route is already thickly covered with large cumuli, additional seeding may do little for enhancing the albedo (NRC 2015b, 105, 114). Worse, the intervention might reduce the reflectivity of the existing cover. On the other hand, while being advantageous for risk management, the brief lifetime of the particles in the troposphere (theoretically up to 5–7 days, more realistically between 24 and 72 hours) seriously constrains efficiency (NRC 2015b, 109). It would impose constant vessel rotations, with no interruption lasting more than a few days for avoiding a termination shock (Hamilton 2013a, 55).[1] Since

> SRM only masks the warming effects of GHGs and is not designed to reduce their concentrations in the atmosphere, [. . .] if SRM were ever used to mask a high level of warming and its deployment were terminated suddenly, the temperature would rebound toward the levels they would have reached without the geoengineering. (Parker and Irvine 2018, 456)

Fourth, MCB lifecycle assessments reveal problems of carbon neutrality. If fossil fuels are employed for vessel propulsion and particle dispersion, the method will lead to additional GHG emissions and worsen climate change (NRC 2015b, 107). However, relying on existing commercial fleets and routes will limit excessive emissions while promising a net positive impact in terms of enhanced albedo in comparison to a no-intervention alternative. However, transitioning commercial fleets to fossil-free energies would guarantee carbon neutrality.

In conclusion, MCB feasibility and efficacy stay indeterminate (NRC 2015b, 121–122). Technological maturity is evaluated at TRL 4 (Low *et al.* 2022, 2). The insufficient knowledge of the interplay between the ocean and the atmosphere impedes modeling and future interventions. Feasibility assessments are further impaired by the limited understanding of the interactions between injected particles and cloud formation. The parameters for clouds in climate models remain too crude. Cloud formation appears difficult to monitor, especially following a field experiment, which diminishes the possibility of producing solid evidence.

Cirrus Cloud Thinning (CCT)

A less discussed proposal consists of "modifying the properties of high-altitude ice clouds, increasing the atmosphere's transparency to outgoing thermal radiation" (National Academies of Sciences, Engineering, and Medicine 2021, 2). The principle is to reduce cirrus clouds, which are cold and loaded with ice crystals residing in the upper troposphere and known for their forcing characteristics (Mitchell *et al.* 2011, 259). On average, cirrus trap more outgoing heat from the ground, that is longwave (infrared) radiation, than they reflect incoming shortwave solar radiation, which causes a net warming (Hamilton 2013a, 56–57; Storelvmo *et al.* 2013). Since the mechanism is to ease outgoing radiation, and not to block incoming radiation, some scientists distinguish CCT from other SRM methods such as MCB or SAI, rebranding it as "thermal radiation management" (Muri *et al.* 2014, 4,174) or "Earth radiation management" (Mitchell *et al.* 2011, 258; Winsberg 2021, 10112) instead of *SRM*. Despite the importance of the question, CCT is considered in this book as an SRM intervention.

CCT could take the form of seeding the upper troposphere, where the coldest cirrus reside, with ice nuclei exciting the formation of larger ice crystals and then hastening their fall. As a result, clouds will thin out, build at lower altitudes, and with diminished forcing characteristics. Furthermore, stimulated cirrus formation will diminish the atmospheric concentration of water vapor, which is the most common GHG (Lohmann and Gasparini 2017, 249). Different particles could be used for seeding. Bismuth triiodide (BiI_3) is favored due to its (apparent) safety, availability, lower costs than silver (the main alternative), and ice forming properties (Mitchell and Finnegan 2009, 3; Mitchell *et al.* 2011, 262; Muri *et al.* 2013, 4175; Storelvmo *et al.* 2013, 178; Storelvmo *et al.* 2014, 4). Additional proposals include seawater (Muri *et al.* 2013, 4,175) and dust particles (Lohmann and Gasparini 2017,

249; Storelvmo *et al.* 2013, 178). CCT could rely on commercial aircraft. The seeding material could then be mixed with the fuel or directly injected into the exhaust systems (reactors) (Mitchell and Finnegan 2009). Dispersion by drones constitutes another option (Mitchell *et al.* 2011, 262).

Cirrus' net global radiative impact, which depends on their altitude and optical thickness, is quantified to be approximately 5–6 W/m^2. CCT cooling potential might be able to offset the forcing induced by a doubling of atmospheric CO_2 (climate sensitivity) (Mitchell and Finnegan 2009). More sobering evaluations situate the effect around 2–3 W/m^2 (Lohmann and Gasparini 2017),[2] which could nonetheless compensate for a large part of the forcing imputable to industrialization estimated ca. 2.4 W/m^2 (Storelvmo *et al.* 2013). If undertaken broadly, and not limited to specific areas, CCT could decrease the average global temperature by 1°C (Muri *et al.* 2013, 4,189).

In the scientific and policy literature, CCT is far less discussed than MCB and SAI (NRC 2015b, 130–132). The technology could be judged as nascent, remaining at a stage dominated by model simulations and clouds chamber experiments, with a handful of small and short-lived field studies. With those caveats in mind, CCT offers benefits. First, the main seeding material (BiI_3) seems to be relatively inexpensive and nontoxic (Muri *et al.* 2013, 4,175). Second, the dispersion by commercial jetliners may constitute an advantage, especially when compared to SAI which will require the development of dedicated aircraft (Smith and Wagner 2018). Third, CCT demonstrates further merits over SAI (Mitchell and Finnegan 2009, 6), such as a limited reduction in global mean precipitations (Duan *et al.* 2018; Muri *et al.* 2013, 4,189). Lastly, CCT could be employed either broadly, for lowering the average temperature, or locally, as a targeted initiative. Seeding could be confined to areas particularly vulnerable to global warming, such as the polar regions (Mitchell and Finnegan 2009, 6; Mitchell *et al.* 2011). Such deployment could help slow ice loss, preserving the albedo, for instance in the Arctic Sea during summers (Lohmann and Gasparini 2017; Storelvmo *et al.* 2014).

One of the main obstacles to the technology maturation (TRL4 [Low *et al.* 2022, 2]) is the poor understanding of cirrus cloud dynamics (NRC 2015b, 131), which stems from scant *in situ* observations (Gasparini and Lohmann 2016, 4,888) and bounded modeling. Such lacunas worsen engineering insufficiencies (Gasparini *et al.* 2020, 10). This explains why the crux of the intervention, namely the optimal amount of seeding, is crippled with uncertainties. Insufficient ice nuclei concentrations could have negligible effects on the climate, while excess concentrations could aggravate global warming (Lohmann and Gasparini 2017; Storelvmo *et al.* 2013). Limited knowledge renders the potential for targeted interventions more difficult to assess. The nature, intensity, and probability of distant outcomes need to be clarified. The reaction of the Earth's system could substantially vary based on the location of intervention (Gasparini 2020 *et al.*). Cooling the Southern Hemisphere could help mitigate Sahelian droughts, while focusing on the Northern Hemisphere may produce significant positive impacts only when combined with seeding in the Southern Hemisphere (Muri *et al.* 2011, 4,189). However, other studies challenge those findings (Gasparini 2020 *et al.*, 5).

In conclusion, three main limitations to CCT deserve to be mentioned: First, the brief lifetime of the particles in the troposphere (a couple of weeks), while echoing the comparative advantage of MCB over SAI, suffers from the same drawbacks. A lasting impact will require continuous seeding without any interruption for more than a few days, while the potential damages of a termination shock will escalate in proportion to the duration and extent of the deployment (Parker and Irvine 2018, 459). Second, some seeding materials could be expensive, or represent hazards for human life and ecosystems (Mitchell and Finnegan 2009, 6). Third, specific regions may be directly or indirectly harmed by CCT. On the one hand, seeding in areas where cirrus cloud formations are rare could raise local radiative forcing and temperatures (Lohmann and Gasparini 2017, 249). On the other hand, models suggest possible changes in precipitation and convergence zones (Muri *et al.* 2011), which could accentuate rainfalls under tropical and high latitudes and droughts in the subtropical regions (Gasparini *et al.* 2020, 6), potentially impacting already vulnerable populations.

STRATOSPHERIC INTERVENTIONS

The most prominent and controversial (UNEP 2023) method in this category is SAI, "a strategy for increasing the number of small reflective particles (aerosols) in the stratosphere in order to increase the reflection of incoming sunlight" (National Academies of Sciences, Engineering, and Medicine 2021, 2). The stratosphere is the layer between eight and eighteen kilometers above the Earth's surface depending on the latitude and fifty kilometers (National Research Council 2015b, 66–67). Although diverse aerosols or their precursors could potentially be utilized (e.g., calcite [Keith *et al.* 2016]) using aircraft, balloons, naval canons, hoses, or drones, sulfates like sulfur dioxide (SO_2) or hydrogen sulfide (H_2S) (leading to sulfuric acid [H_2SO_4] formation) are the main contenders (NRC 2015b, 67–69, 77; Schäfer *et al.* 2015, 41; Shepherd *et al.* 2009, 29; Wagner 2021, 54–55). Sulfates may not be the most efficient aerosols, but they are widely available, and properly understood (Keith *et al.* 2016; Wigley 2014, 12). Thus, this section focuses on stratospheric sulfur injection (SSI), reflecting the current scientific and policy discussions (Tracy *et al.* 2022).

Branded the "Pinatubo option" by Ken Caldeira (Kintisch 2010, 57), SSI offers to mimic the dust release of a volcanic eruption. A key difference is that particles would be directly introduced in the stratosphere to extend their lifetime (Wigley 2006).[3] In the troposphere, aerosols have a shorter lifetime, from a few days to one month, while they can reside in the stratosphere longer, from twelve months (Visioni *et al.* 2018) to a couple of years (NRC 2015b, 66–67). The idea is hardly new (Kintisch 2010, 56). In his 1784 paper—*Meteorological Imaginations and Conjectures*—Benjamin Franklin conjectured about the cooling effect of volcanic eruptions. In 1974, Budyko evoked the possibility of spraying aerosols in the stratosphere for altering the climate. The proposition has been discussed in the following decades (e.g., Kellogg and Schneider 1974; Schneider 1996). More recently, researchers from the Lawrence Livermore National Laboratory (Teller *et al.* 1997, 2002) and others (e.g., Crutzen 2006; Wigley 2006) promoted the option.

Thus, SAI consists in spraying aerosols into the stratosphere at an altitude that varies according to the latitude of the intervention, since the troposphere is thinner at the poles (ca. 6 km) than at the equator (ca. 20 km). To maximize their longevity, particles need to be dispersed above the tropopause, the boundary between the troposphere and the stratosphere (Smith *et al.* 2022a). Insertions could take place in the lower stratosphere (ca. 20 km) or higher up (ca. 25 km). To optimize the lifetime of the aerosols, high altitude would be preferable, but it would require developing novel aircraft or alternative vectors,[4] more hazardous to operate (Smith *et al.* 2022a).

Low-altitude stratospheric injections appear to be feasible in a near future (ten to twenty years) (UNEP 2023) and relatively affordable, depending on the strategy. For instance, to halve the radiative forcing induced by the representation concentration pathway (RCP) 6.0 (emissions peaking in 2080, then declining), Smith and Wagner (2018) estimate the expenses to be US$2.5 billion/year for a fifteen-year implementation (2033–2047). The overall budget would be around US$35–38 billion, including the design, production, and operation of aircrafts capable of flying at an altitude of twenty kilometers while carrying a significant payload (25 t). This estimate does not, however, account for a monitoring system, a limitation shared by other cost appraisals (NRC 2015b, 97).

Smith (2020) expands the evaluation by comparing three strategies ("halving future warming, halting warming, and reversing temperatures to 2020 levels") under three emission pathways (RCPs 4.5, 6.0, and 8.5[5]). The annual budget fluctuates in relation to the RCP and the adopted strategy, from US$7 billion (halving global warming under RCP 4.5) to US$71.7 billion (reversing global warming under RCP 8.5). The middle-of-the-road proposal costs approximately US$30 billion (halting global warming under RCP 6.0). To give some perspective, according to the US Department of Defense, the planned US defense budget for 2023 was US$816.7 billion (Garamone 2022). In comparison, the SSI middle-of-the-road strategy would roughly be equivalent to the resources allocated to the US Navy for shipbuilding (US$32.6 billion). In addition to the wealthiest industrial states, the method would be available to major corporations such as Apple, which reported revenue of US$394.3 billion and a net income of US$99.8 billion in 2022 (Apple 2022).

In sum, SSI could be financially accessible to many public and private agents (Schäfer *et al.* 2015, 41; Smith 2020, 10), constituting one of the most cost-effective CE methods. Technological maturity also seems within reach, *only* necessitating adapting existing aircraft (mostly design and propulsion), although the technical feasibility of the adaptations is sometimes disputed (NRC 2015b, 95), which could explain why Low *et al.* (2022, 2) situate the TRL at 3. In any case, no crippling obstacle to rapid maturation emerges from the literature. Accessibility justifies worries about hypothetical unilateral or "minilateral" (Morrow and Svoboda 2016) deployments (de Coninck *et al.* 2018; Victor 2008), in which a single country or a coalition of a few countries decides with little or no international cooperation to initiate stratospheric injection. Such a prospect stems from a free driver effect, a concept coined in reference to free riding.[6] Free driving characterizes situations in which a course of action is so accessible that the problem is not to coordinate "making things happen," like in the case of carbon abatement, but to coordinate not "overdoing it," or not undertaking a potentially harmful path[7] (Wagner 2021, 16–17; Wagner and Weitzman 2012; Weitzman 2015, 1050). The possibility of unilateral/minilateral deployment might, however, be exaggerated since SSI may require stronger international cooperation and agreement than usually assumed (Halstead 2018, 70).

Computer modeling as well as previously observed volcanic eruptions suggest that SSI could rapidly cool down the globe (Keith *et al.* 2020; Soden *et al.* 2002). The impact will vary according to the specifics of the injection strategy such as the location and volume of aerosols (Irvine *et al.* 2010; Irvine *et al.* 2019; Irvine and Keith 2020; Keith 2013, 51–61), resulting in a maximal efficacy around 10–15 W/m² (Schäfer *et al.* 2015, 43). The intervention could be phased out immediately, with the effects disappearing within a couple of years. Affordability, maturity, rapidity, and efficacy explain why SAI is perceived as "the archetypal geoengineering technique" (Hamilton 2013a, 59). However, GHG accumulation may thin stratocumulus clouds, impairing SAI efficiency (Schneider *et al.* 2020).

Nonetheless, only a handful of research projects, spanning from modeling to modest experimentation, have been initiated (UNEP 2023). The most salient are the Geoengineering Model Intercomparison Project (GeoMIP),[8] the Stratospheric Aerosol Geoengineering Large Ensemble Project (GLENS),[9] the Stratospheric Controlled Perturbation Experiment

(SCoPEx),[10] and the Stratospheric Particle Injection for Climate Engineering (SPICE).[11] Testing remains restricted to computers and in the laboratory, without any scientific field experiments yet. Experiments and protocols have been cancelled, mostly because of public opposition, which illustrates SSI's low social acceptance, even in the context of a priori controversial climate alteration (Carlisle *et al.* 2022). In 2012, a SPICE field experiment in the UK devised for pumping water through a hosepipe maintained at an altitude of one kilometer by a balloon was called off due to public outcry and patent issues (Cressey 2012). In 2021, as part of the Harvard SCoPEx protocol, balloons were planned to be sent over the Swedish Arctic. Under public pressure, especially from the Saami Council, the Swedish Space Corporation put a halt to it (Fountain and Flavelle 2021; Oksanen 2023; Samiraddi 2021). In April 2022, a tiny rogue injection was undertaken by a US start-up company (*Make Sunsets*) without a solid scientific protocol (Osaka 2023; Temple 2022).

Injections could be continuous from several locations or be local and/or intermittent for optimizing benefits and minimizing harms (Irvine *et al.* 2017, 2019; Irvine and Keith 2020; Lee *et al.* 2021; MacMartin *et al.* 2012, 2018, 5; NRC 2015b, 83; Smith *et al.* 2022; Tilmes *et al.* 2017; Visioni *et al.* 2019).[12] For instance, sulfur injections focused on the 60°N (Arctic area) during springtime could maximize aerosol optical thickness in the summer, double arctic ice restoration in September, and limit negative side effects such as ozone loss[13] and sulfur pollution, restricting the latter to the dispersion zone (due to the influence of the Polar Vortex) (Walker *et al.* 2021). An additional advantage of high-latitude interventions could be to moderate the technical constraints on aircraft design since the tropopause is lower at the poles than in the tropical and temperate areas.

SAI offers the prospect of rapid net forcing reduction (Crutzen 2006). Based on data collected from volcanic eruptions, the temperatures could start declining within a few months (Keith *et al.* 2010; UNEP 2023, 11). As a result, SSI appears to be cheap, fast to implement, and fast to phase out since sulfate aerosols only remain for a couple of years in the stratosphere. Depending on the dispersion strategy, average temperatures could significantly drop, alleviating part of the burden imposed by global warming on human beings and ecosystems. In addition to retarding the Arctic Sea ice loss, other benefits include dampened exposure to heat-related health risks in cities, a less pressing need for crop adaptation, and protection of ecosystems against thermic increase. Moreover, land biosphere productivity might be enhanced in the tropical regions thanks to lower heat stress (NRC 2015b, 87–88). Sea-level rise could be fended off due to the slowed retreat of glaciers and limited oceanic thermal expansion (Wigley 2014, 13; Tracy *et al.* 2022, 7–8). More generally, the frequency of climate events like hurricanes, droughts, and floods may be attenuated in comparison to a pathway characterized by unabated carbon emissions, for example in the case of doubled CO_2 concentrations (Irvine *et al.* 2019; Robock 2020). So, properly designed, SSI could have far-reaching positive consequences.

However, SSI implementation presents significant challenges and risks, from the research and development (R&D) phase up to deployment. Some are specific to SAI while others are shared with SRM or CE methods. First, knowledge is lacking in many respects. Adequate monitoring capabilities are needed, especially for surveying volcanic eruptions, which are natural large-scale experiments (NRC 2015b, 92). Nonetheless, efficacy assessment would most likely require several decades of observation with constant injections for separating the signal from the noise, that is, the effects imputable to SSI from other variables (Hamilton 2013a, 67; Robock 2020, 63). In the case of limited deployment, any detection and attribution of the impact on regional temperatures and precipitations might remain elusive, even after 70 years of continuous observation (MacMartin *et al.* 2019).

The behavior of sulfate aerosols ought to be researched further in relation to their dispersion in the stratosphere and their optimal size (NRC 2015b, 69). Both variables are important because they influence particles' coalescence and hence their efficacy (Smith 2020, 3). Therefore, they should be properly modeled (NRC 2015b, 80). The optimal injection latitude and altitude for generating predefined local or global effects need to be clarified. Some impacts are not studied well enough, especially regarding ocean circulation and ecosystems (NRC 2015b, 98–99; Visioni *et al.* 2020; Zarnetske *et al.* 2021). Research is even more necessary because the regional climates will be altered, with severe consequences for populations (Keith 2013, 51–61; Robock 2016, 2020). While SSI could rapidly bring down global mean temperatures, possibly close to pre-industrial levels, it would not reduce GHG concentrations.[14] In the absence of ambitious mitigation and CDR, the intervention will generate new, potentially destabilizing, climates (Flegal and Gupta 2018, 49; Irvine *et al.* 2010; MacMartin *et al.* 2018; McLaren 2018; NRC 2015b, 84–85; Shepherd *et al.* 2009, 24).

The hydrologic cycle may be severely affected (Duan *et al.* 2018; Hamilton 2013a, 63; MacMartin *et al.* 2018). The observation of volcanic eruptions indicates that SSI will probably diminish evapotranspiration and evaporation, leading to local declines in precipitation and river runoff (Tilmes *et al.* 2013, 11,037). The hydrologic cycle could be weakened in comparison to a world without albedo enhancement. If it compensates for quadrupling CO_2 concentration, SSI will not re-establish the original hydrological conditions (Tilmes *et al.* 2013). Those projections may, however, be contested because they are often based on scenarios in which very ambitious interventions are undertaken with the goal of restoring preindustrial temperatures (Keith 2013, 53–54), while other strategies are available like halving the impact of anthropogenic emissions (Irvine and Keith 2020; Irvine *et al.* 2017, 2019; MacMartin *et al.* 2022).

SSI will affect soil moisture (NRC 2015b, 95) and cloud cover (NRC 2015b, 85), notably cirrus formations (NRC 2015b, 71), thinning them out and further cooling the planet (Kuebbeler *et al.* 2012). Global average precipitation may be weakened (NRC 2015b, 75), with altered rainfall patterns particularly impacting the Monsoon (Burns 2016, 11; Nalam *et al.* 2018; Robock *et al.* 2008). However, those projections are debated (Sun *et al.* 2020; Tilmes *et al.* 2013). Moreover, some of the negative effects may be shaved off by modulating injections based on the seasonality or latitude (McLaren 2018, 221). Beyond the Monsoon, the evidence remains elusive about how SSI could lessen the frequency and consequences of extreme events such as droughts, storms, and floods, by comparison with unabated climate change (Tracy *et al.* 2022, 6–7). Another area of uncertainty is the extent to which SSI could help preserve the sea ice (NRC 2015b, 87).

TEXTBOX 4.2 PROJECTIONS, UNCERTAINTY, AND TRADE-OFFS

The fact that the outcomes of SSI deployment cannot be perfectly, or even adequately, anticipated at a moment (t) is often interpreted as proof that the climate system is so complex that any ambitious intervention is bound to generate unpredictable harms. SSI would be "unreliable" (Hulme 2014), creating uncertainty gaps that science could not bridge. Contradictory projections about disruptions of the hydrologic cycle, human health, and ecosystems support the belief that science is myopic when confronted with the transformations carried by SSI. Therefore, models and injection scenarios would be of little value for policy.

Conflicting projections partly stem from the diverging injections' latitude, altitude, aerosol amount, seasonality, and goals (fully reverting, halving, and stopping global warming) across deployment scenarios. The resulting benefits, harms, and risks fluctuate accordingly, sometimes significantly. For instance, the use of extreme climate scenarios and interventions (e.g., canceling the warming resulting from doubled CO_2 concentrations), especially in early studies, has produced drastic regional disparities such as for the Monsoon (Flegal and Gupta 2018, 50). Modest strategies, however, generate less pessimistic projections (Keith and Irvine 2016). The variation of effects across models is not uniquely a consequence of uncertainty. It does not prove that climate models and science in general are of little use for public policy.

Nonetheless, it remains true that the impacts of even modest interventions cannot be fully predicted. CE outcomes depend on the complex interactions between climate forcing (GHG concentrations stemming from socioeconomic decisions and potential albedo enhancement) and responses. So future evolution is difficult to forecast, yet it does not necessarily invalidate science. It may advocate for a more focused, "mission-driven" research agenda, "which is defined by its explicit end goal of supporting informed future decisions regarding deployment" (MacMartin and Kravitz 2019, 1090), even if limited to laboratory and computer models (Lenferna et al. 2017). Although imperfect, projections are nonetheless crucial for identifying possible risks and trade-offs between different scenarios, for example, when choosing between unabated climate change, a world with a level (x) of mitigation but without SSI, and a world with a level (x) of mitigation and a level (y) of SSI (or mixed CE strategy). Impact assessments are needed, for instance, to correct the misconception that SSI can restore preindustrial temperatures and climate, as well as to demonstrate how it will create novel climates (Flegal and Gupta, 2018; Irvine et al. 2017; Keith 2013, 51; McLaren 2018).

At the end of the day, scientists and policymakers are responsible for protecting populations (and presumably ecosystems), especially vulnerable ones, against CE and climate change, which is embodied in the United Nations Framework Convention on Climate Change (UNFCCC)'s injunction to "prevent dangerous anthropogenic interference with the climate system." They must prepare for adaptation, redress, compensation, and so forth. If employed, SAI will redistribute climate burdens and benefits. The magnitude of those effects combined with the relative accessibility of the method could be interpreted as imposing a moral obligation of cautiously investigating and comparing deployment scenarios at different latitudes, altitudes, seasons, and so on, which supports increased research (Cicerone 2006; Long et al. 2015; MacMartin et al. 2016; Robock 2016).

SSI carries further risks and uncertainties for human beings and ecosystems (Robock 2008, 2016, 2020; Tracy *et al.* 2022; Visioni *et al.* 2020; Zarnetske *et al.* 2021). First, there might be health consequences related to the use of sulfur, such as a surge of respiratory complications (Tracy *et al.* 2022, 3). Sulfates may also weaken the recovery of the ozone layer (National Academies of Sciences, Engineering, and Medicine 2021, 42–44; NRC 2015b, 71.86), potentially increasing the UV-B radiation and the frequency of skin cancers (Eastham *et al.* 2018; Tracy *et al.* 2022, 4–5). Other aerosols might, however, minimize or suppress the damages to the ozone layer (Keith *et al.* 2016). Nevertheless, the fact that incoming sunlight will be more scattered could compensate for the effect of the ozone loss on UV-B and moderate the cancer threat (NRC 2015b, 70). On aesthetical grounds, SSI will whiten the sky, giving it a milky aspect (Kravitz *et al.* 2012; NRC 2015b, 97) and degrading astronomical observations (Robock 2020).

SSI presents risks to ecosystems and nonhuman living beings too (Trisos *et al.* 2018; Zarnetske *et al.* 2021). It may aggravate sulfate deposition and acid rain, and its distribution, affecting the soil pH and toxicity of previously pristine areas (Visioni *et al.* 2020). However, since the quantity of released particles could be lower than other anthropogenic sources, the effect might remain minimal (NRC 2015b, 90, 95), especially if ambitious mitigation is conjointly undertaken, cutting down the volume of sulfate aerosols imputable to fossil fuels. More refracted sunlight may also affect photosynthesis, with negative impacts on crops such as maize, soy, rice, and wheat (Proctor *et al.* 2018), although some species could thrive (NRC 2015b, 70). If bioenergy with carbon capture and sequestration (BECCS) has been deployed, the yield of energy crops might decline. This last aspect underpins the importance of understanding better the interactions between various CE interventions and other components of climate policies.

A crucial downside shared with other SRM methods is that the origin of climate change—carbon emissions and concentrations—is left untouched. Diminished radiative forcing may nonetheless have a favorable impact on carbon sources and sinks. For instance, permafrost thawing may be slowed down, helping to maintain vast amounts of methane stored underground, which will reduce the risk of positive feedback and runaway planetary warming (Wagner 2021, 52). In addition, decreased global average temperatures may bolster both ocean and land carbon intake (Tracy *et al.* 2022, 5). That being said, and keeping in mind the fact that those positive outcomes are debated, SSI does not directly lower sources of GHG or enhance sinks, contrary to CDR. It does not prevent ocean acidification, which threatens marine life (Tracy *et al.* 2022, 8).

The most serious worry resides not in the intervention, but in its disruption (NRC 2015b, 59–66). As mentioned for MCB and CCT, "termination shock" (Hamilton 2013a, 57), "termination effect" (Shepherd *et al.* 2009), or "SAI intermittency" (Baum *et al.* 2013) designate the adverse consequences that will follow an abrupt and uncontrolled interruption of SAI/SSI deployment. In the case of SSI, temperatures, precipitation, and ice loss will change more rapidly than under a business-as-usual scenario (Jones *et al.* 2013). The amplitude of the thermic rebound will likely be correlated to the intensity of the accumulated interventions (NRC 2015b, 43), with Earth warming potentially twenty times faster than the current pace of 0.2°C per decade (Burns 2016, 14; Matthew and Caldeira 2007). The harms to humans and ecosystems will be proportionate to the magnitude of the rebound, but not uniformly distributed. Many human beings, animals, and ecosystems will not have sufficient time for adapting or migrating. However, the seriousness of any termination shock will depend on the scale of the prior deployment. The adverse effects risk could be moderated by engaging in a modest injection strategy, phasing it out as soon as possible, and building redundant deployment infrastructure (Parker and Irvine 2018). In any case, the mere prospect of such catastrophic outcomes highlights the necessity of crafting and inserting, from the very start of any research project, "exit ramps" ("criteria and protocols for terminating research programs or areas" [National Academies of Science, Engineering, and Medicine 2021, 9]) and "exit strategies" (Schäfer *et al.* 2015, 100).

In addition to those risks and uncertainties, SAI and to a lesser extent other SRM methods like MCB and CCT raise significant issues of justice and governance that become especially acute when it comes to novel climates. Many of those issues are not created *ex nihilo* by SRM but aggravated in the form of compound vulnerabilities. CE harms may combine with damages from climate change and other disadvantages to generate "skewed vulnerabilities" (Carr and Preston 2017; Gardiner 2011; Preston 2012). Finally, challenges are distributive *and* procedural, that is related to the allocation of burdens and benefits, *and* to the way of conducting scientific enquiry and deciding SSI deployment (Svoboda 2017).

On distributive grounds, vulnerable populations, particularly those exposed to currently unfolding climate change, may be further harmed by CE interventions. Africa and Southeast Asia could be among the regions the most exposed to SRM owing to the alteration of the Monsoon regime. A recurrent objection is that risks associated with thermal changes and precipitation tend to concentrate in the least developed regions (Biermann and Möller 2019). Distributive consequences could be worsened by procedural injustices. Researchers and lay people from the Global South are excluded from scientific and policymaking discussions surrounding CE and SSI (Carr and Yung 2018; Hourdequin 2019; McLaren 2018; Winickoff *et al.* 2015). The research is dominated by scientists from the North, more specifically from the United States, the UK, Germany, Japan, and a few other countries, with only a handful of representatives from China and India (Cao *et al.* 2015). A further worry is that the pool of scientists directly involved and whose opinion matters is restricted to a narrow group, disparaged by critics as the "geoclique" (Kintisch 2010, 8; McKinnon 2019), which raises the probabilities of capture by private or technocratic interests (cf. chapter 5). The relative absence of perspectives from the Global South is reflected in the design of climate models used for evaluating the impact of SSI which focus on two variables: temperatures and precipitation, while additional parameters would be pertinent to include from the point of view of vulnerable populations (Hourdequin 2018; McLaren 2018). More generally, the interests and views of people from developing and emerging countries tend to be underrepresented, for example in public perceptions studies (Carr and Yung 2018; Suarez and van Aalst 2017; Sugiyama *et al.* 2020; Visschers *et al.* 2017; Whyte 2018; Winickoff *et al.* 2015).

Beyond the epistemological limitations of climate and CE models, the lack of visibility of Southern researchers creates procedural injustices in a context of "de facto governance" (Gupta and Möller 2019) of SRM research by bodies such as the IPCC, the Royal Society, the US NRC, and science academies from industrialized nations. Research is already controlled through bottom-up initiatives from individuals and institutions based in the North. In the broader scheme of things, such defective representativeness and responsiveness of scientific research and policymaking could undermine legitimacy, by undercutting the public acceptance of, otherwise necessary, climate interventions.

Although developing countries are on average left out of SAI research and governance, they will not forcibly oppose any R&D or deployment agenda.[15] They might support research. Everything considered some CE strategies might be advantageous to vulnerable populations in the Global South, especially when compared with unabated climate change (Horton and Keith 2016). The exclusion of the interests of affected people nonetheless generates a problem of legitimacy spanning from R&D to governance structures (McLaren 2018). At the end of the day, in addition to procedural injustices, such exclusion ultimately undermines the legitimacy, and arguably the acceptability, of ambitious, and potentially beneficial, SAI interventions. This constitutes a practical reason for including more researchers from the Global South or giving them the lead on SAI research (Rahman *et al.* 2018).

SPACE-BASED INTERVENTIONS[16]

The last category is the most hypothetical and neglected intervention in the literature. Despite conveying science-fiction undertones, it could nevertheless offer the most potent SRM method in the future (Sánchez and McInnes 2015). The idea of space intervention for SRM purpose has been floating around for approximately a century (Baum *et al.* 2022) and was recently revived (Early 1989). General assessment reports such as the Royal Society (Shepherd *et al.* 2009, 32–34) or the NRC (2015b, 127–128) cover the intervention, but not to the same extent as SRM alternatives like MCB or SAI.

Qualified as a "climate airbag" (Baum *et al.* 2022, 20), the method consists in placing in low Earth orbit (LEO) (altitude up to 2,000 km), geosynchronous orbit (GEO) (35,786 km), or at the Sun–Earth Lagrange point 1 (SEL1) (ca. 1.5 million km from Earth) a single massive object or a myriad of smaller objects that will absorb, reflect, or scatter incoming radiation. Those objects could be shields, shades, reflectors, mirrors, sails, prisms, Fresnel lenses, or dust, and other particles (Baum *et al.* 2022, 3; Fuglesang and Miciano 2021; NRC 2015b, 127–128; Roy 2022; Shepherd *et al.* 2009, 32). Dimming, scattering, or reflecting 1.8 percent of the solar radiation may be sufficient to fully cancel the warming induced by a doubling of atmospheric CO_2 concentrations (Angel 2006, 17,184).

A preliminary issue is to determine the appropriate deployment distance from Earth, knowing that not all orbital heights display equivalent stability (e.g., objects at SEL1 will drift due to radiative pressure [Early 1989]) (Baum *et al.* 2022; Sánchez and McInnes 2015). Orbits closer to Earth such as LEO or GEO are more accessible and cheaper (e.g., fuel use for launchers) than more remote ones such as SEL1. However, those areas are more crowded with various items such as satellites and debris, increasing the risk of collision. Moreover, massive objects orbiting at those heights will create shadows, while not maintaining their position in relation to the Sun, remaining over specific regions, which will cancel out their efficacy during the nighttime.

Another crucial issue regards the method of deployment. One of the most feasible and cost-effective options would be to launch a multitude of objects forming a cloud of, for example, one million tiny spacecrafts (Angel 2006), the placement of which will need to be adjusted and coordinated at the deployment point, for instance through self-propelled devices. The optimal site from which to launch the apparatus is still debated. If many proposals assume that it can and will be done from Earth, a deployment from space is theoretically conceivable. If undertaken from Earth, the deployment requirement for launchers and fuel will be massive. Therefore, presupposing that the Moon or other space objects like meteorites can be industrially exploited and mined, extraterrestrial sites could offer more efficient locations for launches (Ellery 2016).

A few potential advantages of space-based CE emerge from the scant literature. First, it could significantly reduce global average temperatures, possibly up to 2–3°C (Fuglesang and Miciano 2021, 270). Second, and unlike many CDR interventions, the land footprint will be quasi-null (Baum *et al.* 2022), leaving untouched natural resources such as water and land and threatening food security less. In sum, space-based CE offers to preserve Earth's scarce resources (Ellery 2016, 278). Third, contrary to other SRM methods such as SAI or CCT, the chemicals injected into the atmosphere will remain limited to rockets' exhaust gas. Those first two advantages underscore the lower risks of harmful interference in the biosphere in comparison to other CE interventions (Baum *et al.* 2022). Fourth, despite the necessity of constantly monitoring and maintaining the whole installation, with regular launches, the method could produce more enduring effects than SRM alternatives like MCB, CCT, or SAI. Fifth, the installation could also be scaled down or dismantled rather quickly, that is within a few weeks (Roy 2022, 371).

A particularity of space CE lies in its potential co-benefits (Baum *et al.* 2022). Investments in sun-reflecting technology, rockets, and infrastructure could hasten space exploration. In addition, the placement of orbiting photovoltaic mirrors could generate energy (Roy 2022, 372). The method could incentivize exploring, exploiting, and colonizing celestial objects such as the Moon and asteroids, for example for extracting resources. The National Aeronautics and Space Administration (NASA) and the European Space Agency are already engaged in the Artemis program[17] regulated by the Artemis Accords,[18] which is a cooperative venture aimed at establishing a Moon station in the 2030s for peaceful purposes. The China National Space Administration (CNSA) and Roscosmos—the Russian State Space Corporation—are working on a concurrent plan: the International Lunar Research Station.[19] The whole intervention could be part of a broader project of observing and monitoring Earth.

However, there are potential drawbacks and limitations. First, like other SRM methods, space-based CE does not address GHG emissions and concentrations, leaving the cause of dangerous climate change untouched and not tackling such harmful consequences as ocean acidification. It will not compensate for all anthropogenic changes. As for SAI, impacts may vary across the globe. Using a general circulation model, Lunt *et al.* (2008) observed, among other things, that space shades could cool the tropics, slightly warm higher latitudes, inducing a retreat of sea ice, and alter ocean circulation (thermohaline) as well as precipitation. Space interventions may turn back the clock to what the climate was during the Cambrian period, 500 million years ago. Such disparate impacts could be partly offset by modifying the orbits of the shades.

Second, space intervention is among the least mature CE methods, with a TRL evaluated at 2 (Low *et al.* 2022, 2). Optimistic forecasts situate full maturity, at the soonest, by mid-century. Experts widely agree that more research is required, especially on how various deployment strategies differently affect incoming radiation (Shepherd *et al.* 2009, 33) and climate responses (Sánchez and McInnes 2015, 22). Without breakthroughs on carbon-free propellant and considering the sheer number of rockets to be sent, space-based CE will emit vast amounts of GHGs. In addition, technical difficulties may cripple the intervention. The positioned shades, sails, or mirrors will be vulnerable to solar pressure, wind, and storms, as well as micrometeorites and other debris that could damage them or push them off orbit (Lior 2013, 407; Matloff *et al.* 2014, 177–178). Moreover, the placement of objects in the higher orbits will have to be constantly maintained, for preventing the sunshades from drifting away. Solutions could include self-propulsion, regular replacement by new objects, or setting up a particles belt such as for Saturn rings (Baum *et al.* 2022, 4). In sum, the method will most likely come too late as a bridging technology and will take too long to install for offering an emergency option.

Third, the sheer number of launches and rockets necessary to deploy space intervention may hinder economic viability (Fuglesang and Miciano 2021; Sánchez and McInnes 2015). Cost appraisals, from the R&D stage until hypothetical deployment, vary greatly, from US$1 trillion to US$20 trillion (Baum *et al.* 2022, 4). The lack of cost-effectiveness could be challenged (Roy 2022, 371–372). The costs per kilogram launched have steeply declined under the competitive pressures of companies such as SpaceX and Blue Origin. In any case, space-based CE might still offer an affordable strategy in case of need, since the overall costs represent a significant, but relatively low, portion of the annual world GDP (evaluated by the World Bank at US$96 trillion in 2021). But, again, those evaluations are crippled with uncertainties, such as the industrial capacity for producing sunshades' components in the extraterrestrial space (Fuglesang and Miciano 2021, 270).

Fourth, legal constraints might be insurmountable. Owing to possible weaponization, the 1977 Convention on the Prohibition of Military or Any Other Hostile Use of Environmental Modification Techniques (EMTs) (ENMOD) could be employed for blocking deployment, or even field experiments (Baum *et al.* 2022, 13). However, the Convention only prohibits the implementation of EMTs that aim at *deliberately* hurting a third party, tolerating space-based interventions that *might* result in unintentional harms. In addition, the colonization or exploitation of the Moon or any other objects could initiate international legal battles, impairing the feasibility of utilizing space-based installations for solar CE. For example, the article II of 1966 Outer Space Treaty (Treaty on Principles Governing the Activities of States in the Exploration and Use of Outer Space, including the Moon and Other Celestial Bodies) prohibits any national appropriation of the Moon and other celestial objects,[20] which complicates the claims for resources ownership that, on Earth, is territorially based. However, a few countries, like the United States with the *Commercial Space Launch Competitiveness Act* (2015),[21] already passed legislation allowing their citizens to acquire resources harvested on the Moon and objects. The international community is taking steps for ending the legal ambiguity. The United Nations Committee on the Peaceful Uses of Outer Space established in 2021 a Working Group on Legal Aspects of Space Resource Activities with the mandate of "[developing] potential rules and/or norms, for activities in the exploration, exploitation and utilization of space resources" (Hanlon 2022).

Finally, as with other SRM methods, there is a risk of a termination shock (Baum *et al.* 2022), that is an abrupt halt of the deployment causing a thermic surge, possibly exceeding the initial thermic compensation brought on by the intervention. The difficulty with preventing or responding to sudden interruption, especially in comparison to SAI, is that space is less accessible than, for instance, the stratosphere. Sending material and human resources in the required amount in the space is a more challenging endeavor than transporting them somewhere else on the planet.

SRM: A GENERAL ASSESSMENT

As for CDR, any evaluative overview is at risk of generalizing and over-simplifying, especially when discussions are so heavily skewed toward SAI. As for CDR, solar CE presents a broad spectrum of methods that often significantly diverge from each other. USB and crop albedo are local and rather mature solutions, whereas interventions like CCT and space reflectors are more ambitious in scope but with a more elusive feasibility. SAI attracts most of the moral criticisms, while USB appears to be uncontroversial. Nonetheless, despite such diversity, it is still possible to apply the evaluative categories presented in chapter 2 for extracting a few tentative conclusions. It should be noted that, as for CDR, deployment is already on the way for some methods. In addition, under the condition of being properly developed, other initiatives promise local or global efficacy in the short/mid-run. However, the more planetary the intervention, the higher the odds of disparate impacts and potential catastrophic outcomes. As for CDR, the goal of this section is not to present a definitive assessment on the diverse methods, but to raise some highlights regarding feasibility, permissibility, and preferability.

In a nutshell, a brief summary could be that the feasibility of many methods appears moderate to strong, while permissibility presents more of a conundrum that includes hard to resolve political and ethical issues such as disparate impacts or global catastrophic risks. SAI, and to a lesser extent space-based methods, is potentially very effective, but proportionally very controversial interventions. Due to its characteristics, SAI captures most of the reflections on feasibility and permissibility. If any ambitious SRM happens at a scope beyond specific urban or agricultural areas, it will most likely take the form of the injection of particles in the stratosphere. Therefore, the governance and regulatory vacuum are particularly concerning in regard to this precise method.

Feasibility

Current and prospective feasibility greatly varies within the SRM category. The more modest and targeted initiatives like surface solar CE are presently the most accessible. Cities like New York or Los Angeles already brighten pavements, roofs, and other urban elements (Lomborg 2020). Crop selection has been practiced for millennia and genetic modification for a few decades, so crop albedo does not offer any impassable barrier. Moreover, feasibility is not strictly correlated to how ambitious the method is. In under a couple of decades, one of the most accessible SRM interventions may be SAI, which represents the second most ambitious method after space-based CE.

As repeated, *maturity* significantly differs across the techniques, on average, the more local, the more mature. SRM methods range from nascent, but hypothetically very efficient, techniques like space reflectors (TRL 2) to fully demonstrated in real-life conditions, but more modest, as USB (TRL 9). The interventions that promise the deepest impact (MCB, CCT, SAI, and space-based CE) still necessitate more elaborated research spanning from computer modeling to field experiments (National Academies of Sciences, Engineering, and Medicine 2021). Nonetheless, the prospect of reaching maturity in the near future for SAI appears strong, which raises questions of democratic control, governance, and justice.

Again, a diversity of situations characterizes technological, physical, and economic *scalability*. Local initiatives like surface albedo do not need to be deployed broadly for producing benefits, namely, to tackle the UHI effect. Scalability is therefore less of an issue in those cases. However, if the goal is to modify the whole climate, USB and crop albedo would demand solid international coordination, which is an obstacle for building up planetary alteration. Tropospheric, stratospheric, and space-based techniques experience serious challenges to their scalability in terms of infrastructure. They necessitate more massive investments in R&D, monitoring devices and, potentially, for deployment (e.g., SAI, CCT, and MCB), sometimes technological breakthroughs are necessary for ramping up the intervention (e.g., rockets and extraterrestrial equipment for reflectors).

Compared to CDR techniques such as forestry and BECCS, the most promising SRM methods (USB, crop albedo, SAI, and MCB) do not seem to create extensive land-use conflicts. SAI, CCT, and MCB do not need vast amounts of land, water, or minerals apart from the aerosols employed for injections (SSI and CCT). Space-based CE stands apart in terms of the magnitude of investment. It would require the development of a dedicated industry, including extraterrestrial elements (e.g., rocket launchers, mining facilities, and repair stations), that would demand much more than, for example, retrofitting or designing new planes capable of flying in the lower stratosphere (SAI).

Assessing *efficiency* implies tackling a preliminary issue about the ends to pursue. If the CDR target is obvious—reducing carbon emissions and/ or concentration—the SRM goal is far less straightforward (Irvine *et al.* 2017, 100). It is, however, easier to delineate clearer goals for surface-based interventions. For instance, the point of USB is mostly to counter the UHI effect. So, its goal is to lower the heat stress of a given urban area. For the other large-scale methods, is the objective to cancel all global warming imputable to human beings, or to alleviate part of? If the latter, by how much? Or is it to cushion the harms carried by the Anthropocene, which includes variables other than temperatures?

As noted above, often the scenarios of deployment that are investigated through computer modeling are extreme. It is assumed that SRM, especially in tropospheric, stratospheric, or spatial forms, will aim at compensating for most or all accumulated GHG emissions during the Anthropocene (e.g., the doubling of CO_2 concentrations). Furthermore, global average temperatures are considered as *the* variable to be optimized, which is a reductive, or even counterproductive, approach to policy. First, global average temperatures represent an "empirical impossibility" (Hulme 2014, 39). As such they exist nowhere. Relying on them exclusively masks the extent of local fluctuations. Second, they offer unsatisfactory proxies for climate conditions. They fail to capture other variables as precipitation, ocean circulation, carbon sink uptake, and so on. Third, their use "obscures most of what matters, in terms of weather, to humans and the things that they are attached: rain to grow crops, wind to power turbines, cyclones from which to shelter, and the like" (Hulme 2014, 43). At the end of the day, they nurture the illusion that SRM would be about regulating the "global thermostat." Thus, public policy and research require adopting better suited, multifactorial, metrics, moving beyond thermal optimization. A preliminary step is to determine what state of the world solar CE should achieve, what risks should to be mitigated, by how much, and under which acceptable trade-offs. Efficiency assessments should embody reflections on permissible risks and uncertainty, as well as include procedural dimensions.

Therefore, SRM presents an "optimization problem" in which climate goals and "an 'objective function' that measures how closely the goals are attained" (Ban-Weiss and Caldeira 2010, 1) need to be clearly identified. In contrast to the "curiosity-driven research model" that has marked most of the SRM research to date, there have been calls to scientists for adopting a "mission-driven research program" (MacMartin and Kravitz 2019). This more focused strategy, characterized by "the existence of a mechanism of centralised coordination" for research (Morrow 2020a, 619–620), will translate into the adoption of a "design perspective" for CE research and policy (Irvine *et al.* 2017, 99). The argument is the nature and magnitude of SRM's impact depends on implementation decisions.

[M]any of the climate effects of geoengineering are design choices, presuming the ability to actually impose changes with specific characteristics. As such, statements about the climate effects of geoengineering in general are ill-posed; such statements require the context of specific climate objectives and an approach designed to meet them. (Kravitz *et al.* 2016, 470)

According to the argument, asking whether specific SRM techniques could be efficient or too risky is a sterile strategy. It makes more sense to assess for which purpose different SRM methods may be useful, which ventures into engineering and governance territory (MacMartin *et al.* 2016).

Permissibility

Like feasibility, permissibility varies within the SRM category. For instance, USB and crop albedo (without GMOs) appear more justifiable or acceptable than SAI. Again, it is important to refrain from laying a blanket judgment over SRM as a whole-based solely on stratospheric techniques. Some solar CE methods could be reasonably advocated for.

Different methods (should) prompt different assessments of their *intrinsic moral value*. Some do not seem to be particularly controversial. For example, it is difficult to see which USB traits would make it inherently problematic. That is true that the intervention is not natural in the sense of promoting existing natural processes such as rock weathering and OIF (although supporting such processes does not shield against the unnaturalness objection). At the same time, what could be intrinsically wrong about painting the pavement and roofs in bright colors? Crop albedo, if not engaging in genetic modification, appears uncontroversial too. What could be wrong about selecting higher reflectance cereals? It could be claimed that it depends on the impact on the surrounding ecosystems, which is a consequentialist argument that does not judge the intrinsic properties of an intervention, but its possible outcomes.

Other methods prompt negative assessments. Again, SAI epitomizes everything wrong with solar CE, and climate alteration in general. It would be "undesirable," "ungovernable," and "unreliable" (Hulme 2014, xii). Such rejection flows from potentially catastrophic outcomes (cf. below), but also inner features. As seen in chapter 2, CE methods are often judged as intrinsically immoral for three main reasons: unnaturalness, hubris, and playing God, all of which are advanced against SAI (Clingerman and O'Brien 2014; Donner 2007; Flegal *et al.* 2019; Hamilton 2013a; Hartman 2017; Katz 2015; Lenferna *et al.* 2017; Morrow 2014; Preston 2012; Thiele 2018). SAI would be unnatural because it disrupts the biochemical and atmospheric balances by injecting massive volumes of particles. Although SSI mimics volcanic eruptions, the scope and length of the intervention would turn it into something artificial. In addition, it would be motivated by hubris, that is, delusive beliefs about human capacities. Along with space-based CE, SAI would incarnate at best the intention of administering the planetary climate through a "global thermostat" (Hulme 2014). Finally, humans would pose as possessing divine attributes, such as controlling the natural elements or being able to forecast the full consequences of their actions, which would be specious and morally objectionable. MCB or CCT could be subject to similar criticisms. They pursue the objective of tampering with the climate, potentially at a large scale depending on the deployment strategy. A moderating factor for MCB is that it enhances natural processes (marine stratocumulus formation). Within SRM, space-based CE stands apart, representing an even more unnatural, hubristic, and God-playing project than SAI. The vectors (shades, sails, mirrors, and reflectors) do not mimic any "natural" phenomena, while the whole proposal is much more ambitious, and thus potentially more arrogant, than SAI.

In sum, the bad press of CE in general, and SRM in particular, partly stems from controversies surrounding SAI, but not only. Its perceived advantages—cheap, fast to deploy/interrupt, and efficient at reducing radiative forcing while remaining imperfect (Keith *et al.* 2010; Mahajan *et al.* 2018)—explain why it attracts most of the attention. In turn, through a hasty generalization, all SRM methods, and sometimes CDR interventions as well, are assumed to suffer from the same drawbacks. This results in a caricatural debate. Beyond those fallacies, it remains true that SAI poses relevant questions in terms of anthropogenic arrogance and the relation with the planet system.

As a category, SRM is widely identified as creating more *risks and uncertainty* than CDR (MacMartin and Kravitz 2019, 1089).[22] The potential impact on the carbon cycle, that is, land and sea carbon uptake through plant productivity and ocean carbon solubility, remains unclear (Cao 2018; MacMartin *et al.* 2016). Further, SAI, space CE, and to a lesser extent, MCB and CCT carry risks of global catastrophes (Tang and Kemp 2021), that is, "events that would significantly harm or even destroy humanity at the global scale" (Baum *et al.* 2013, 169). The menace differentiates those techniques from CDR, jeopardizing permissibility (Baum *et al.* 2013; Morrow 2014; Trisos *et al.* 2018). The colossal damage, especially to vulnerable populations, which might be triggered by altered climate regimes and weather conditions as well as by a termination shock constitute major concerns.[23]

A couple of remarks are in order. First, not all SRM methods are equally susceptible to producing global catastrophic outcomes. Surface interventions mostly remain exempt from this concern. Second, the harms stemming from termination shocks could be constrained by engaging in modest cooling (Parker and Irvine 2018; Tang and Kemp 2021, 10) and modulating the injections (seasonality, location, and amount of aerosols) (Irvine *et al.* 2019; Irvine and Keith 2020). Third, governance mechanisms implemented from the R&D stage could help prevent catastrophes, in particular the establishment of "exit strategies" (Schäfer *et al.* 2015, 100) and ramps[24] activated at predetermined "physical science check points"[25] (Cairns 2014; Callies 2019b; Diamond *et al.* 2022). This last safety valve underpins the importance of carrying on research, although limited to models and laboratory experiments (cf. textbox 4.2).[26] Finally, there could be precautions such as the redundancy of deployment means (aircraft fleet, boats, and space launchers) (Parker and Irvine 2018), even though the feasibility of such protective measures might traduce an overly optimistic, "Panglossian," view of international politics (McKinnon 2019).

At the end of the day, some methods might be deemed too dangerous, that is representing unacceptable gambles, for even being researched. However, save for some exceptions, any judgment of impermissibility based on risks and uncertainty ought to be comparative. If undertaken, SRM will not create harms out of the blue relative to a situation without climate change or to the current conditions. Imagine that in 2040 the world is facing a combination of large-scale permafrost thawing, followed by massive methane releases, and an ensuing collapse of the West Antarctica ice shelf and the disappearance of the Arctic Sea ice during the summer. Imagine further that the only intervention immediately accessible is SAI. SAI would present catastrophic risks, but so would runaway climate change. This hypothetical example does not vindicate researching or deploying SAI. It underlines that impermissibility judgments rooted on risks and uncertainty ought to be comparative. A CE method needs to be ruled out if, everything taken together, it carries more significant risks and uncertainty than a future baseline and feasible alternatives. Should such a baseline be unabated climate change? Such a scenario justifies researching and potentially employing even the most controversial SRM techniques like space reflectors. But what about if future climate change is somehow cushioned? How much mitigation is it realistic to assume for determining the threshold of risks and uncertainty beyond which SRM methods would be judged unacceptable?

SRM is more at risk of disparate regional impacts than CDR because solar irradiance is more unevenly distributed than carbon (Kravitz *et al.* 2016, 469). Such impacts raise severe matters of *distributive justice* and compensation. Again, techniques like USB and crop albedo, owing to their local scope and modest ambition seem less prone to trigger distributive issues. *A contrario*, SSI and space-based CE present higher probabilities of generating winners and losers, although some preventive measures might moderate such outcomes (cf. above). Models show a significant decrease in rainfall for SRM save CCT, compounded with heterogeneous thermal and precipitation impact (Duan *et al.* 2018), that could lead to an expansion of drylands (Park *et al.* 2019). Compensation is complexified by two factors. The first is the challenge of international cooperation on climate issues (cf. Introduction), and *a fortiori* on victims' compensation. Moreover, it will be difficult to attribute specific adverse events and damages with a sufficient level of confidence to particular interventions. Such attribution will require extensive monitoring systems (NRC 2015b, 136). Those hurdles impede the permissibility of engaging in the methods that produce the most disparate impacts since there is little to no guarantee that appropriate compensation schemes could be devised.

Distributive issues are compounded by *procedural concerns*. Populations might be exposed to SRM harms while not having significantly contributed to climate change at the first place, and without being properly consulted or having participated in the decisions that lead them to suffer disadvantages (Hourdequin 2018, 2019; Whyte 2012). In addition, insufficient mitigation may create situations in which SRM stands as one of the options of last resort or an emergency intervention for the most underprivileged populations (e.g., in low-lying Pacific Islands) who are among the least responsible for past emissions and the most at risk of SRM going awry. In other words, already disadvantaged groups may be placed in front of an ethical dilemma, a daunting gamble, due to (relatively) no fault of their own: to accept the deployment of very risky and uncertain techniques like SAI or space reflectors or to suffer the full blow of unabated climate change (Gardiner 2013a).

The fact that SRM methods that carry global or regional impacts like SAI, space reflectors, and possibly MCB and CCT may be researched or implemented without proper consultation and be literally forced on vulnerable communities highlight potential issues of domination. In the neo-republican sense, domination denominates the capacity of an agent to arbitrary interfere in the circumstances experienced by another agent (Pettit 1997). More advanced nations, in particular the United States and UK, would be able to impose their views on CE over the least developed countries and indigenous peoples, especially for SAI (Biermann and Möller 2019; Smith 2018), because they monopolize the financial and human resources for researching and deploying global solar CE. The moral issue is compounded by the historical role played by developed nations in initiating the Anthropocene, the negative impacts of which fall disproportionally on disadvantaged nations that are the least responsible for the ongoing change.

Therefore, the permissibility of given SRM projects is intimately tied to how well such projects will respect procedural constraints (consultation, participation, etc.) at the benefit of potentially affected parties notably vulnerable groups (National Academies of Sciences, Engineering, and Medicine 2021; Suarez and van Aalst 2017; Winickoff *et al.* 2015). It could be claimed that whether an intervention will lead to the actual domination of particular regions is contingent on how R&D and deployment will be carried out. In response, it could be argued that because of their much larger capacities for R&D and deployment, the most developed countries already dominate the least developed ones by simply having the power of imposing the terms of the debates and their choices. An additional assertion could be made that some SRM methods by their very nature embody instances of domination.

Except for USB and crop albedo without genetic modification, SRM *acceptability* appears to be lower than for CDR (Cummings *et al.* 2017; Delina 2021; Raimi 2021), especially in countries where SRM is viewed as tampering with nature (Visschers *et al.* 2017). However, the picture of the public perceptions of SRM is, again, overshadowed by the figure of SAI. Studies heavily focus on SAI (Burns *et al.* 2016; Carr *et al.* 2012; Corner and Pidgeon 2014; Cummings *et al.* 2017; Mahajan *et al.* 2018; Scheer and Renn 2014; Visschers *et al.* 2017), with only a handful covering MCB (e.g., Bellamy *et al.* 2017). Therefore, the overall picture of social acceptability is still fragmentary, and the acceptance of specific methods may be contaminated by what people think of SAI. Moreover, since familiarity with SRM and CE is particularly weak, framing effects heavily influence public perceptions.

About SAI, Burns *et al.* (2016) find four components that are widely present in public perceptions: unfamiliarity, artificiality, conditional support for research (on factors such as robustness, predictability, efficacy, and governance), as well as unclear evidence regarding the belief that engaging into SAI research would cause people to lower their ambition on mitigation. Generally, SAI is negatively viewed for various reasons discussed in this chapter (unnaturalness, playing God, risks, and uncertainty). However, those negative attitudes may not fully translate into principled opposition. Laypeople usually distinguish between deployment and research while not overwhelmingly rejecting the latter (Pidgeon *et al.* 2013). Respondents in the Global South appear more likely to accept SRM than in the North (Sugiyama *et al.* 2020; Visschers *et al.* 2017). Such acceptance of research and maybe deployment, is, however, marked by reluctance and resignation, especially among vulnerable populations (Carr and Yung 2018).

In sum, studies on public perceptions suffer from shortcomings and biases, in particular framing effects, that is the manner that the information is presented would tend to elicit specific answers. First, deep unfamiliarity with solar CE, introducing interventions as mimicking natural phenomena or using analogies with natural processes, may stir perceptions of naturalness (Burns 2016, 537). Second, answers would also display an acquiescence bias, that is respondents are influenced by how questions are devised (Mahajan *et al.* 2018). In other words, as long as the public does not become acquainted with SRM, acceptance will remain subject to spurious inferences and more or less overt manipulation, which may be exploited in one direction or another, that is in support or against specific methods.

The permissibility of individual SRM techniques will partially rest on *intergenerational justice,* that is how they distribute benefits and burdens across time, whether they place future generations in front of impermissible risks or in a worse situation than any feasible alternative that could have been initiated nowadays. For being acceptable, a technique needs to demonstrate that it does not unnecessarily cripple future generations who are already disadvantaged by not being part of the current decision-making process that will constrain the policy options open to them. Any decision to research-specific SRM interventions may create a path dependency and potentially lock-in in terms of the available technologies for tackling climate change, especially if they crowd out investments on mitigation, adaptation, and/or research programs on other CE interventions. More importantly, SRM deployment may coerce future generations into prolonging interventions that are harmful, but constitute overall a lesser evil than unaddressed runaway climate change or a termination shock.

SRM techniques which are applied nonstop and at a large scale convey a more serious risk of trapping future generations into continuous deployment. The longer the intervention, the greater the damages of an abrupt termination. Future generations may be forced to pursue implementation even if the adverse effects become significant. The higher efficiency of methods like SAI and space reflectors represent a sword of Damocles. The potential negative effects of a sudden interruption correlate to the extent of the positive impact (i.e., thermic compensation). To sum up, the more efficient a method is at masking part of global warming, the more harmful its interruption. Thus, the most efficient methods, which are stratospheric and space-based, will trap future generations into constant deployment.

Similar to CDR, *regulation and governance* are in their infancy.[27] There is no dedicated international or national scheme for regulating deployment or research for that matter. Therefore, the barriers to individual country initiatives are minimal, and the potential for "de facto governance" (cf. chapter 2) is high. A few existing conventions and treaties such as the Outer Space Treaty (1967) or ENMOD (1977) tangentially or indirectly relate to SRM. Additionally, a handful of United Nations resolutions and decisions directly tackle solar CE (UNEP 2023, 37–38). As seen in chapter 2, the Conference of the Parties under the Convention on Biological Diversity (CBD) issued a few decisions that, among other things, call the member states to limit CE research to "small scale scientific studies that would be conducted in a controlled setting" (decision X/33, [w]), avoid transboundary harms, and apply the precautionary principle, especially in regard to biodiversity.

The potential challenges presented by the different techniques are nonetheless dissimilar. Surface-based methods present fewer problems than space-based CE or SAI. This difference is acknowledged by many commentators who focus their criticisms on SAI, usually leaving the other interventions out of the picture. The common reproach is that SAI cannot be correctly governed "because there is no plausible and legitimate process for deciding who sets the world's temperature" (Hulme 2014, xiii). In general, SAI would be extremely complicated to govern fairly (Svoboda *et al.* 2011) and efficiently. The inclusion and participation of the affected parties as well as the compensation of the victims would stay elusive. To meet the basic requirements of justice, the argument is often made that SAI, from the research stage up to deployment, ought to be managed by the Global South with the support of the United States and other key climate engineers. A possible avenue could be to moderate domination through the transfer of CE knowledge and CE capacity building (Smith 2018). However, since such a perspective remains wishful thinking, SAI should then be subject to an international nonuse agreement (Biermann *et al.* 2022). At least, it is how the argument goes. However, doubts have been voiced about the enforceability of any moratorium (SRMGI 2013).

Preferability

As for CDR, a full-fledged comparison of the different SRM options exceeds the ambition of this introductory book. Moreover, due to the hybrid or portfolio nature of climate policies, a full-fledged preferability assessment would have to integrate other components, namely mitigation and adaptation. As put by Keith and Irvine (2016, 551):

> If the goal is to evaluate the physical consequences of SG [solar geoengineering] where other policy choices are held constant then the impacts of SG should be compared against the impacts of not deploying SG in a reference GHG scenario. If, on the other hand, the goal is a more integrated assessment then it would be necessary to consider the way other policies and behaviors may change if SG is implemented.

That being said, it is still possible to extract a few features from the direct comparisons among individual SRM techniques. If SRM techniques share the core principle of immediate albedo enhancement, they are not equivalent in terms of scope. Some are obviously local, such as USB and crop albedo, whereas other interventions are more global, or even planetary, such as stratospheric and space-based methods. The tropospheric interventions (MCB and CCT) fall somewhere in the middle: not really local, since their impacts may spill beyond the location of the implementation, but not global either, owing to the rapid dispersion of the effects on cloud cover. Thus, any climate policy which seriously considers the inclusion of SRM will have to clarify the ends it pursues, as per the design perspective highlighted above. Is the objective local or regional cooling? In that case, surface and perhaps tropospheric techniques seem to be superior. USB could constitute an example of a Pareto superior intervention, since it may not make anyone significantly worse off while benefiting many. Is the intervention part of a broader effort for curbing global warming or the more general effects of the Anthropocene? Then, stratospheric and/or space-based interventions would be preferable. Desired or preferable policy goals are not the only dimensions to consider when deciding on an intervention. There are also proper issues of governance and regulation that relate to how efficiently different methods or mixes of methods can be managed domestically and internationally. Defective international cooperation will definitely take a toll on the capacity to fairly and efficiently govern tropospheric, stratospheric, and space solar CE.

At the end of the day, the preferability of SRM methods cannot be evaluated in isolation. It should be replaced in the broader perspective of multi-layered climate policies with more finely tuned objectives than global average temperatures and precipitations. This imperative gives rise to all sorts of questions and issues that have remained dealt with only in a marginal manner in the literature. How will a given method interact with other policy components? Would it impede or reinforce some non-SRM initiatives, like mitigation or specific adaptive measures? Weighed against current emissions trajectories and other equally feasible alternatives, what are the benefits, risks, and uncertainties generated by the option under consideration, especially when included to a policy mix? How well are the rights of the affected parties respected in comparison to alternatives? Will there be any realistic possibilities of shielding the most vulnerable populations against harmful consequences or at least to offer some compensation/redress? Without running through all the evaluative dimensions that have been explored since chapter 2, it is important to maintain a continuity here by emphasizing that feasibility and permissibility criteria should constitute the basis of the preferability assessments. And, in so doing, there is the necessity of balancing different requirements, the details of which, again, exceed the scope of those pages.

NOTES

1.   The consequences of a sudden halt or failure of the geoengineering system. For SRM approaches, which aim to offset increases in greenhouse gases by reductions in absorbed solar radiation, failure could lead to a relatively rapid warming which would be more difficult to adapt to than the climate change that would have occurred in the absence of geoengineering. SRM methods that produce the largest negative forcings, and which rely on advanced technology, are considered higher risks in this respect. (Shepherd *et al.* 2009, 35)

2. Muri *et al.* (2014) provide a more modest evaluation of CCT's global effect, a reduction of global radiative forcing by 1.5 W/m² and average mean temperatures by 0.9°C.

3. While releasing 20 Mt SO_2 (Bluth *et al.* 1992), Pinatubo eruption's maximum forcing was evaluated ca. -2.97W/m² (Wigley 2006, 452) so enough for compensating for the net forcing during the Anthropocene.

4. Aircraft is the main injection vector discussed in these pages due to its projected superior cost-effectiveness (NRC 2015b, 82).

5. As seen above, RCP 6.0 represents a scenario in which carbon emissions peaked in 2080 and then start declining. In RCP 4.5, emissions reach a plateau in 2040, and then decrease, while in RCP 8.0, no plateau is reached during the twenty-first century. The latter is sometimes, misleadingly, branded as a worst-case scenario, or business-as-usual scenario (Hausfather and Peters 2020).

6. A free rider is "a person who receives the benefit of a good but avoids paying for it" (Mankiw 2021, 212).

7. The general setup underlying free-driving is that of a public good or bad (GoB). The GoB game differs from the canonical public good games—even those in which agents can both give or take contributions—in that 1) the provision cost is so low that agents can provide the good; and 2) a conflict over the total provision is present because agents are characterized by heterogeneous, single-peaked preferences. (Abatayo *et al.* 2020, 13,394)

8. http://climate.envsci.rutgers.edu/geomip/

9. https://www.cesm.ucar.edu/community-projects/glens

10. https://www.keutschgroup.com/scopex

11. http://www.spice.ac.uk/about-us/aims-and-background/

12. Aerosols do not stay at their injection location. They are carried by jet streams.

13. The use of other aerosol particles (e.g., calcite and $CaCO_3$) could prevent damages to the ozone layer.

14. However, the claim is sometimes advanced that SRM could marginally enhance land and ocean carbon uptake (Matthew and Caldeira 2007; UNEP 2023, 18).

15. Funding, usually nongovernmental, instruments for SAI research in developing countries have emerged, such as the *Degrees* (DEveloping country Governance REsearch and Evaluation) *Initiative* formerly known as the SRM Governance Initiative.

16. This section is indebted to Baum *et al.* (2022), who publish a reference study by collecting a significant number of experts' evaluations while identifying the key challenges for space-based CE.

17. https://www.nasa.gov/specials/artemis/

18. https://www.nasa.gov/specials/artemis-accords/img/Artemis-Accords-signed-13Oct2020.pdf

19. http://www.cnsa.gov.cn/english/n6465652/n6465653/c6812150/content.html

20. "Outer space, including the moon and other celestial bodies, is not subject to national appropriation by claim of sovereignty, by means of use or occupation, or by any other means."

21. https://www.congress.gov/114/plaws/publ90/PLAW-114publ90.pdf

22. Some indirect risks, like delayed mitigation, are shared by the two families of interventions.

23. Risks are not only physical, but also social (Baum *et al.* 2013; Morrow 2014; Tang and Kemp 2021). SAI deployment might not only cause damages through altered weather conditions (e.g., droughts, storms, and floods), but through societal collapse, privatization/militarization, international tensions, and so on.

24. "Exit ramps" designate "criteria and protocols for terminating a research program" (Diamond *et al.* 2022, 2).

25. "Physical science checkpoints" refer to "important, relatively self-contained topical area with open questions that must be resolved for a solar climate intervention method to be considered viable" (Diamond *et al.* 2022, 2).

26. For a brief overview of the different issues to be addressed, cf. MacMartin *et al.* 2016.

27. For a more detailed legal assessment, the reader can consult Gerrard and Hester (2018).

Chapter 5

Engineering the Climate

Arguments and Objections

In a context where mitigation is lagging behind the Paris Agreement targets, thereby ramping up the risks linked to unabated climate change, climate engineering (CE), understood as *anthropogenic interventions in the climate system that predominantly aim at altering radiative forcing locally or globally to slow down, stop, or reverse climate change and/or related adverse events,* may increasingly appear as necessary for averting dangerous climate change (United Nations Environment Programme [UNEP] 2023). However, carbon dioxide removal (CDR) and solar radiation management (SRM) do not open the same opportunities and do not raise identical issues from a policy perspective. Although the scenarios respecting the 1.5–2°C targets include CDR (IPCC 2015, 2018), those methods will be slow to reach required efficiency while introducing all sorts of risks (leakage and carbon lock-in). Ultimately, they may not be part of the solution, but "an unjust and high-stakes gamble" (Anderson and Peters 2016). SRM, especially under the form of stratospheric aerosol injection (SAI), offers a more immediate intervention that could be helpful for "gaining time" before properly mitigating and/or capturing carbon. However, it comes with severe risks, two major ones being the creation of novel climates and termination shock.

This chapter pursues four goals. First, it places discussions on CE in the broader setting of carbon-based economies that have not sufficiently abated their emissions, are having trouble doing so, and will probably face the same difficulties in the next decades. In short, humanity needs to bridge a mitigation gap. Thus, CE would represent a necessity. Despite its apparent triviality, such a reminder is important when considering that the harms, risks, and uncertainty carried by CE are often assessed independently of the Anthropocene's threats. This chapter begins with a recap of why the context matters when it comes to understanding why some CE methods are becoming attractive. Insufficient mitigation and adaptation constitute the primary reason in favor of CE research and potential deployment.

The second section examines the prominent claim that research and deployment are distinct, therefore justifying the former does not necessarily force to endorse the latter. Public perceptions reflect the empirical distinction (Pidgeon *et al.* 2013), with stronger support for research than for deployment, even in the case of SAI. The most common arguments in favor of the distinction and, in the end, of research and development (R&D) are given. Next, objections to the distinction as well as to research are evaluated. The section closes on governance and how it could address some concerns about CE research.

The third section elaborates on four core arguments about why and how CE could be integrated into climate policy. In line with the task pursued since the beginning of the book to disentangle the different methods, this section shows how those arguments disparately apply to CDR and SRM, as well as within those categories. It is demonstrated that the case for CE either as a first-best solution or as a second-best solution, if exclusive of conventional policy tools (mitigation and adaptation), is indefensible. In addition, the popular view that CE could provide essential emergency responses is challenged. At the end, the most solid justification appears to be the combination of CE with mitigation and adaptation in the framework of portfolio policies.

The fourth section, which offers a methodical review of the main objections against CE, makes the bulk of this chapter. CE interventions attract five categories of criticisms, related to efficiency, precaution, capture, politics, and morality. Each category hosts objections that are presented and evaluated in turn. It is shown how those objections may apply differently to the whole spectrum of climate alteration techniques.

MITIGATION GAP AND NECESSITY

Any discussion bearing on climate interventions needs to take into consideration the context. If CE increasingly appears to be legitimate, it is due to the nonideal circumstances characterized by the "mitigation gap," which designates the widening distance between the actual cuts in carbon (and other GHGs) emissions and the cuts that are required for respecting the Paris Agreement targets. Inept abatement was invoked by Paul Crutzen in his defense of stratospheric sulfur injection (SSI; "so far, attempts in that direction [Note: mitigation] have been grossly unsuccessful" [Crutzen 2006, 211–212]). The necessity argument posits that CE, or some methods, would have been rendered *necessary* by the incapacity of states to coordinate their actions to adequately tame emissions for "prevent[ing] dangerous anthropogenic interference with the climate system" (United Nations Framework Convention on Climate Change [UNFCCC]).

Perspectives are indeed sobering. Odds are piling up that abatement will not be sufficient for respecting the 1.5°C, and plausibly the 2°C, limit (Boyd *et al.* 2015; Rockström *et al.* 2017; Rogelj *et al.* 2023). "To meet this target, the world would have to curb its carbon emissions by at least 49% of 2017 levels by 2030 and then achieve carbon neutrality by 2050" (Tollefson 2018, 172). On the assumption that no aggressive mitigation will happen anytime soon, and even if all the national pledges (nationally determined contributions [NDCs]) promised under the Paris Agreement are fulfilled, the probability is high that the temperatures will soar over 1.5°C, possibly 2°C, in the next decades (Riahi *et al.* 2021, 1065; UNEP 2022).

Unless drastic cuts in greenhouse gas (GHG) emissions are implemented immediately, global mean warming is likely to exceed the Paris Agreement target of 1.5°C above the pre-industrial level within the next 10–15 years. Warming levels of 4–5°C could be reached by 2100. (UNEP 2023, 8)

However, swift aggressive mitigation might not be adequate for respecting the Paris targets. This uncertainty reinforces CE attractiveness, especially CDR (Burns 2016; Lin 2019; Minx *et al.* 2018). CDR/negative emissions technologies (NETs) are already included in most of the climate models and projections that offer fair chances of remaining within the Paris goals. They also begin to be (insufficiently though) discussed by policymakers (Buylova *et al.* 2021; Haikola *et al.* 2018; Himmelsbach 2018; Peters and Geden 2017). The latter recognize, at least implicitly, the necessity of deploying some CDR methods, mainly forestry and bioenergy with carbon capture and sequestration/storage (BECCS). As put by Peters and Geden (2017, 619), "the governments that signed and ratified the Paris Agreement accept the IPCC consensus that CDR cannot be avoided if ambitious climate targets like 1.5°C or 2°C are to be met." Furthermore, if CDR is interpreted as a mitigation measure, it could be claimed that the Paris Agreement imposes on the signatory states to research and develop CDR technologies (Mace *et al.* 2021). In particular, the industrialized countries, according to the principle of common but differentiated responsibilities for tackling climate change and its impact (UNFCCC, article 3 paragraph 1), might be considered as having an accrued obligation of leading CDR R&D and deployment (Honegger *et al.* 2021, 332).

In other words, there is the progressive realization that humanity has embarked upon an overshoot pathway of indeterminate proportions that may not be reversible in the coming decades. This realization motivates various recent international initiatives like the creation of the *Climate Overshoot Commission* formed in 2022 under the Paris Peace Forum.[1] Its explicit mandate is to assess, among other things, the possible impacts of and responses to an overshoot scenario. As mentioned in chapter 3, a key driver behind the overshoot is the energy infrastructure, existing and proposed, that may already commit humanity to exceed 1.5°C (Tong *et al.* 2019). Under those circumstances, CDR could compensate for the difficult-to-cut emissions, not only in power plants but also in air transportation and industry (e.g., cement factories) (IEA 2023; IPCC 2023, 60). Capturing carbon and other GHGs might become crucial, not only in the case of catastrophic scenarios.

> Even in a low overshoot scenario, carbon capture and storage and atmospheric carbon dioxide removal will be required to mitigate and compensate hard-to-abate residual emissions. Projects capturing around 1.2 $GtCO_2$ by 2030 need to be implemented, against the roughly 0.3 $GtCO_2$ currently planned for 2030. (IEA 2023, 3)

If the mitigation gap reinforces CDR legitimacy within integrated assessment model (IAM)'s and IPPC's scenarios, the picture is different for SRM. CDR appears more legitimate because it closely addresses insufficient carbon cuts, while SRM proposes to mask climate change by altering a single, dependent, variable (average global temperatures), with indirect effects on other variables (ocean circulation, precipitation, etc.). CDR could theoretically be deployed soon for avoiding an overshoot pathway or later and in a more ambitious manner for countering the harmful consequences of the overshoot. In contrast, the idea behind SRM is to temporarily shave off temperatures before ramping up mitigation and, potentially, CDR, or in emergency situations if temperatures get out-of-hand. The two kinds of CE methods do not offer identical strategies and so do not appear necessary for the same reasons.

In any case, the general sense of necessity surrounding CDR and SRM is reinforced by the fact that accumulated anthropogenic emissions commit humanity to climate changes over an extended period covering centuries if not millennia. Climate discussions take a short-sighted view by focusing on 2100. The challenge is not simply to stop releasing carbon sometime in a near future, but to interrupt emissions as soon as possible *and* to "clean up" part of what has already been emitted (CDR) as well as, perhaps, keep global average temperatures down for avoiding catastrophes (SRM) (MacMartin *et al.* 2018).

> [W]e note that 20–50% of the airborne fraction of anthropogenic CO_2 emissions released within the next 100 years remains in the atmosphere at the year 3000, that 60–70% of the maximum surface temperature anomaly and nearly 100% of the sea-level rise from any given emission scenario remains after 10,000 years, and that the ultimate return to pre-industrial CO_2 concentrations will not

occur for hundreds of thousands of years. If CO_2 emissions continue unchecked, the CO_2 released during this century will commit Earth and its residents to an entirely new climate regime. (Clark *et al.* 2016, 2)

This multi-millennia horizon is the backdrop of CE discussions. It underscores that GHGs *accumulation* is the essential problem to solve. In the absence of aggressive mitigation, CDR and/or SRM would have to be maintained over an extended stretch of time, as long as global temperatures will near the projected tipping points and until GHG concentrations are brought down to acceptable proportions, which assumes some agreement on what such levels could be. In sum, climate change and any hypothetical CE response commit humanity for a prolonged period of climate actions.

As seen in chapter 3, there are strong reservations regarding CDR scalability and efficiency at capturing enough carbon for tackling substantial overshoots (Anderson and Peters 2016; Bednar *et al.* 2020; EASAC 2018, 2022; Fuss *et al.* 2014, 2018; Mace *et al.* 2021). Even under moderate emission pathways, doubts exist about the possibility of remaining within the Paris targets with the help of CDR, without mentioning the prospect of an undesirable carbon lock-in (Anderson and Peters 2016; Asayama 2021; Dooley *et al.* 2018; Fajardy *et al.* 2019b; Otto *et al.* 2021). That explains the mounting pressure for including SRMs into the policy toolbox, even if it requires revising the Paris Agreement (Horton *et al.* 2016). Such a possibility stirs apprehension among researchers that, as for CDR, SRM could be embedded into climate scenarios without any substantial debate (Parker and Geden 2016).

If CE technologies are first and foremost considered because they promise to address the climate crisis, their legitimacy also results from decades of lobbying, starting with a 1992 report in the United States (Institute of Medicine, National Academy of Sciences, and National Academy of Engineering 1992), often at the expense of conventional mitigation. The current situation in which CE appears necessary could be interpreted as the achievement of an "invisible college" of CE promoters, namely "informal communication networks of approximately one hundred elite scholars from different affiliations who confer power and prestige to one another through citations, coauthorships, and common projects" (Möller 2021, 28). A consequence of the work of such invisible colleges may have been that the inclusion of CDR in scenarios, pathways, and models used in international climate politics occurred in a non-explicit manner, without prompting any public and contradictory debate. Moreover, responding to the political pressure of devising scenarios that respect the Paris Agreement, the IPCC introduced CDR into its assessment without an explicit mandate from policymakers (Parker and Geden 2016). CDR and BECCS, in particular, are criticized as parts of IAMs that produce increasingly unrealistic scenarios, that is divorced from actual and projected CDR implementation (Haikola *et al.* 2019).

In any case, a possible formulation of the necessity argument is as follows:

P_1: Through GHG emissions and accumulation humanity is interfering in a dangerous, potentially catastrophic, manner with the climate system.

P_2: So far, conventional mitigation has not been sufficient "to prevent dangerous anthropogenic interference with the climate system."

P_3: Without immediate aggressive mitigation, chances are slim "to prevent dangerous anthropogenic interference with the climate system."

P_4: Techniques of radiative forcing alteration (CDR and SRM) offer to either reduce or mask GHGs emissions and accumulation as well as their effects, that is to lower the chances of "dangerous anthropogenic interference with the climate system."

P_5: Humanity has the moral obligation "to prevent dangerous anthropogenic interference with the climate system."

C_1: Thus, Humanity has the moral obligation (i.e., it is necessary) to investigate techniques of reduction of radiative forcing for identifying those that are the most promising, in terms of efficiency, risk/uncertainty reduction, justice, and so forth.

C_2: As a consequence, the most promising techniques should be deployed.

As it stands, the argument may be too ambitious, allowing too much. It tends to demonstrate that an *obligation* applies to the research (C_1) and potentially deployment (C_2). It is possible to accept the sequence P_1–P_3 and challenge P_4, for instance by claiming that CE methods constitute distractions not tackling the origins of the problem, namely the use of fossil fuels. So, P_4 would be illusory. But rejecting P_4 is not necessary for invalidating the argument. It is also possible to agree to P_4 and P_5, but still refuse C_1. One could agree that CDR and SRM offer to lower or mask radiative forcing, but judge that overriding reasons exist for *not* investigating those techniques. Such reasons could be that the increased knowledge and the perspective of deployment will corrupt humanity by feeding delusional beliefs of grandeur (hubris), inhibit robust GHG cuts (mitigation deterrence and risk compensation), create impermissible gambles (too high risk/uncertainty), generate unjustifiable or non-compensable disadvantages (distributive concerns), and so forth.

A more modest manner to apprehend CE and the whole argument is to consider that researching CE might not be necessary, but to claim that an engineered climate could be better *in some morally relevant sense* than a world without CE (Morrow 2014, 124), which will constitute a *pro tanto* reason to engage in CE research and, possibly, deployment. However, it implies that other, *pro tanto*, reasons could justify to refrain from doing so. Settling the issue demands to undertake a comparative work based on the criteria seen in chapter 2 and shows, on a case-by-case basis, that a world in which a hybrid climate policy, containing a dose of CE, is preferable to a world without, for instance in terms of harm reduction. This amendment will not shield the argument against all the criticisms (e.g., hubris), but it is easier to defend.

THE DISTINCTION BETWEEN
RESEARCH AND DEPLOYMENT

In the ranks of the scientists supporting CE, a common argument is that research does not equate to implementation. Thus, they should be evaluated independently (Cicerone 2006; Frumhoff and Stephens 2018; Jamieson 1996; Keith 2017; Long *et al.* 2015; Morrow *et al.* 2009; Robock 2016). The usual corollary is that if deployment could be rightfully contested objections to research would be more difficult to sustain. The intent is to insulate research from criticisms typically aimed at deployment like CE being an impermissible gamble. Accepting research would not necessarily commit us to endorse implementation and, more importantly, it would constitute a far more benign activity.

At first sight, research and deployment seem indeed distinct. The former would be limited to formulating concepts and hypotheses, modeling, and experimenting, preferably in encapsulated environments like computers, labs, or small field tests. The latter would scale-up the technology for significantly and durably impacting variables like temperatures, extreme climate events, or carbon emissions/concentrations. Surveys of public perceptions of CE show that the distinction sounds convincing (Pidgeon and Spence 2017; Pidgeon *et al.* 2013). CE research is judged as more acceptable than deployment (Sugiyama *et al.* 2020), under the conditions that experimentation is guided by pure scientific motives, contained, with effects that are as less uncertain as possible while remaining reversible (Bellamy *et al.* 2017).

For some CE methods, the likelihood of neatly separating research from deployment might appears contentious. For instance, what is/are the difference(s) between SSI, MCB, or CCT field tests at a scale and over a period long enough to prompt detectable climate responses, and full deployment, especially if we consider that it could take several decades before isolating the impact of SRM interventions (Hamilton 2013a, 67; Hulme 2014, 62; MacMartin *et al.* 2019; Robock 2020, 63)? In addition, once non-spurious material criteria of distinction have been established, it should be demonstrated that such a distinction matters for permissibility (cf.

chapter 2), that is that it insulates research from objections usually addressed to deployment, for example, in relation to adverse effects on human health and ecosystems.

In that respect, two main criteria are predominantly used in the literature for differentiating between CE research and deployment: intentions and scale (Lenferna *et al.* 2017, 578). According to the first criterion, only the activities led by scientific inquiry would qualify as research. The sole purpose would be to investigate a given method, its feasibility, potential impacts, risks, and so on. It would be about testing specific engineered processes (sulfur injection and direct air capture), assessing natural mechanisms (marine cloud formation), and producing scientific knowledge on natural mechanisms (e.g., cloud/nuclei interaction). The goal would be to pass this knowledge on to decision-makers. In contrast, deployment would consist in reducing the radiative forcing, locally or globally, for altering climate and weather conditions. The motivation would be less to validate knowledge and hypotheses than to steer the climate system in a less harmful direction.

The second criterion is the scale of the activity and its consequences. Scientific experiments are assumed to have a much more restricted spatial footprint and produce more limited climate responses than deployments. For instance, the UNEP (2023, 24) separates three kinds of SRM activities according to their scale: "indoor SRM research," "small-scale outdoor SRM experiments," and "large-scale operational SRM deployment." In this taxonomy, large-scale field experiments are reduced to deployments. The scope is key for differentiating between research (indoor and small scale) and deployment (outdoor and large scale).[2] However, as seen for soft or targeted CE (chapter 1), deployment can also be local and targeted (e.g., surface brightening and the preservation of polar ice caps). At the end, some targeted deployments may be more modest in scale than large field experiments.

In any case, to buttress the distinction, it could be added that the two kinds of activities are institutionally separated. They do not mobilize resources from the same actors and sectors. Different organizations and individuals have responsibility for R&D on the one hand and deployment on the other hand. Scientists, universities, and public/private research institutions undertake studies and experimentations. The funding could be public, that is provided by states, but the ultimate responsibility, in countries that respect a modicum of academic freedom, rests on the institutions carrying research. However, even if researchers may and probably should be associated with CE implementation (e.g., by providing monitoring and impact assessment), they will not constitute the decision-makers. They will not write regulations. Governments and international organizations will endorse the duty to deploy and devise the supporting framework.

If we leave the institutional dimension aside since it is far less common among the formulations of the distinction, a simple formalization of the argument could be:

P_1: Research is different from deployment in terms of intentions and consequences.

P_2: Intentions and consequences are materially and morally significant.

P_3: Research carries more modest intentions and consequences.

C_1: Thus, research could be more easily justified than deployment.

C_2: Thus, research's worth and permissibility should be evaluated independently of deployment.

Four remarks are in order. First, as seen above, P_3 is not universally true. Some research projects might be as or more ambitious than some deployment initiatives. Second, the conclusion C_1 is not that research is *always* justified. Indeed, research could be more modest in terms of intentions and consequences than deployment while still carrying risks, uncertainty, harms, and rights violation significant enough to reject it, all things considered. As such, CE research is not, *prima facie*, vindicated, although it could be easier to find reasons to engage in it than in deployment. Research programs experience a lower burden of proof than deployment initiatives even though permissibility should be assessed on a case-by-case basis, grounded on the criteria that have been spelled out from chapter 2. Third, this argument is usually combined with the necessity argument to strengthen the call to distinguish the two stages and, ultimately, carry out some research projects. In other words, because of the mitigation gap, the necessity of undertaking research would be reinforced. Fourth, the conclusion C_2 stipulates that research should be evaluated independently of deployment. This claim introduces specific reasons in favor of research, kept isolated from reasons that are more tightly tied to deployment, which will be presented in the next section.

Reasons for the Distinction (and in Support of Research)

Many justifications have been advanced for initiating research without committing to deployment (Cicerone 2006; Flegal and Gupta 2018, 49; Frumhoff and Stephens 2018; Keith *et al.* 2010; Winsberg 2021). Those arguments usually keep research and deployment as strictly separated as possible. They point at the differences between the two activities, but also underline that engaging in research does not *necessarily* lead to, or even consider, deployment. This position could be interpreted as supporting more "curiosity-driven" (MacMartin and Kravitz 2019), or "investigator-driven"

(Morrow 2020a), programs, in which the scientific questions prompted by CE are more generally explored, with less emphasis on practical deployment issues than mission-driven approaches.[3] According to the distinction, research carried for its own sake, that is expanding scientific knowledge, brings benefits independently of any hypothetical implementation. In this section, three common reasons for research undertaken for its own sake are detailed. In the "Justifications for Climate Engineering" section, reasons that are specific to deployment are presented.

First, it would be essential to advance knowledge of CE (Cicerone 2006; Keith 2017; Long *et al.* 2015), particularly because it constitutes a new and controversial technology. Moreover, due to the mitigation gap, the necessity of building reliable knowledge would be even more pressing, especially for diminishing the harms of hypothetical implementation (Winsberg 2021). Such pressure is acknowledged by various scientific bodies, like the Committee on Geoengineering Climate, which was tasked by the US National Academies to produce several reports on CE (National Research Council [NRC] 2015a, 2015b): "as a society, we have reached a point where the severity of the potential risks from climate change appears to outweigh the potential risks from the moral hazard associated with a suitably designed and governed research program" (NRC 2015b, 184). Faced with the statistically significant prospect of catastrophes, reluctance would get more difficult to justify, raising the odds of unilateral, or minilateral, deployment. The perceived expected utility of CE research compared to the perceived expected utility of climate change would become more favorable as the mitigation widens (Winsberg 2021). Thus, this growing temptation would impose a duty of investigating, CE interventions, especially because governance and regulatory safeguards are not in place yet, opening the perspective of premature deployment.

The imperative would be urgent in the case of SAI (Keith *et al.* 2010; NRC 2015b, 9–11) since it constitutes both the most "tempting" and "alarming" intervention (Winsberg 2021, 1113). As seen in chapter 4, SAI indeed promises to be fast to deploy/interrupt and cheap (Smith 2020; Smith and Wagner 2018), with immediate impacts on temperatures, but high chances to create novel climates with significant disparate outcomes (Flegal and Gupta 2018; Irvine *et al.* 2017; Keith 2013, 51; McLaren 2018). Moreover, the method may soon become affordable to private and public entities including rogue agents. One can imagine an extremely undesirable situation in which SAI would appear as the last-resort response to a crisis whereas decision-makers would only have access to fragmentary evidence as well as untested models. If any implementation should prove too risky or unpredictable, it is better to know as soon as possible, especially for governance purposes (Halstead 2018; Keith *et al.* 2010; Nature 2021). Finally, allowing some controlled public research, that is financed by state agencies, could help the creation of a scientific community with areas of shared understanding and norms (Winsberg 2021).

In addition to SAI, research will likely be beneficial for other SRM interventions, for example enhanced cloud formation for marine cloud brightening (MCB) and cirrus cloud thinning (CCT). If conducted properly, modeling possibly coupled with limited outdoor testing may provide the opportunity to reduce uncertainty (i.e., unknown or unpredictable adverse events). The claim could be made that research should not be restricted a priori to computers and laboratories. A reason could be that, in the absence of field experiments, CE research solely relies on "heuristic tools" like scenarios and climate models, the assumptions of which may remain unquestioned (Talberg *et al.* 2018b). Research is also crucial from an engineering point of view. For instance, it can help in designing albedo alteration techniques that maximize the gains and lower the risks. However, this first argument loses its force if it appears that research is incapable of tackling major uncertainties, most notably the disparate impacts (Flegal and Gupta 2018, 52).

In the case of CDR, research may produce knowledge that could serve to establish more reliable climate models and pathways than the ones currently used, for example in the Intergovernmental Panel on Climate Change (IPCC) reports. Because of the rising importance of methods such as BECCS and forestry in IAMs, it would be vital to figure out the feasibility and potential of those capture and sequestration processes as well as others like direct air capture with sequestration/storage (DACS). Some techniques such as ocean iron fertilization (OIF) or enhanced weathering may involve introducing pollutants in the environment, so the impacts on local ecosystems should be studied. But due to the prominent role played by BECCS and forestry in models, the most pressing task is one of optimization, that is investigating the conditions under which interventions could reach higher levels of capture and storage efficiency as well as devising robust instruments for measurement, reporting, and verification (MRV) (cf. chapter 3).

Second, CE research could also generate valuable scientific and engineering co-benefits. A solid program may further climate and environmental knowledge, as pointed out in chapters 3 and 4. For instance, research on OIF could contribute to a richer understanding of the life cycle of phytoplankton. Chances are nontrivial that studies on MCB will produce a finer-grained comprehension of cloud–aerosol interactions and cloud formation. Programs could also lead to a better grasp of the impact of the fluctuations of surface water temperatures on ocean circulation, or the thermic exchanges between ocean and land. On the engineering side, an MCB project necessitates precise monitoring tools (e.g., satellite imagery and survey aeroplanes), which could prove useful in additional domains. At the end of the day, CE research may lead to more accurate climate modeling. That progress could serve other purposes than the strict enhancement of marine cloud albedo. Instruments for surveying SAI field experiments could be utilized for volcanic eruptions and to detect any rogue injection (NRC 2015b, 189). As noted in chapter 4, research devoted to space CE could further space exploration and exploitation. Additional co-benefits could be generated by programs into other CE methods.

A last argument consists in advancing the intrinsic value of the freedom of (scientific) enquiry (Cicerone 2006, 224; UNEP 2023, 5).[4] A possible formulation is that the activities involved in CE research, namely observing climate processes (e.g., volcanic eruptions, phytoplankton, and geological carbon sequestration); elaborating hypotheses, scenarios, and pathways; running those hypotheses through computer models; or testing them indoors and outdoors; and so forth are worthwhile in themselves. They would constitute the expressions of a core value that is important in democratic societies based on technological progress: the liberty for scientists to not be restrained or censored.

However, if such values are undeniable when generally stated, the actual worth of specific projects depends to a large extent on what is investigated, for which purpose, and through which means (Gardiner 2010, 288–290). First, it could be claimed that, considering the emergency, mission-driven research is preferable to projects led by pure scientific curiosity (MacMartin and Kravitz 2019, 1090), which underlines the significance of the "design perspective" (cf. previous chapter). It is partly a question of efficient allocation of resources under conditions of urgency. Second, scientific tools and protocols could be used for the best or the worst as illustrated by eugenics or Nazism. Strict constraints on the "freedom of enquiry" may be judged desirable when it comes to investigating biological or nuclear weapons. Science could serve human progress as well as more sinister purposes. It could be replied that genuine science, by definition, excludes malevolent intents. However, the necessity for such qualification highlights that the freedom of enquiry cannot claim an absolute value independently of its intentions and means. Such constraints clearly manifest in the fact that any engineering or scientific activity is subject to deontological rules. For instance, even if medicine strives to preserve human health, which is a priori laudable, it is nonetheless subject to medical ethics, which severely restricts human experimentation for instance. The ideal of scientific freedom is thus conditional on fundamental principles.

As an illustration, Long *et al.* (2015, 31) propose a "checklist for funding research" on climate alteration made of five items. First, research should reduce the uncertainty and risks or improve the efficiency attached to a given method. Second, the dangers carried by a specific program should be minimized. Third, research should be led in a transparent manner, open to public scrutiny. Fourth, to limit the influence of vested interests, projects ought to be independently reviewed. Finally, regulatory assessments need to check whether the project respects domestic and international legal standards. As discussed below, in the section dedicated to research governance, other proposals have been advanced, like the Oxford or the Tollgate Principles, confirming that scientific freedom cannot, and must not, be absolute and should be subject to normative, that is moral, social, and legal, constraints.

Reasons against the Distinction (and Research)

A possible objection is to deny or downplay the distinction between research and deployment for better opposing the former (Biermann 2021; Biermann *et al.* 2022; Hamilton 2013b, 2014; Hulme 2014, 60–68). The divide could be viewed as less airtight as postulated (McLaren and Corry 2021), or as being less significant, for example morally, than assumed. A variant is to argue that research and deployment carry risks and uncertainty, perhaps not identical, but similar enough for refusing to relax standards when judging the former. SRM, especially SAI, is the focus of those objections.[5]

A crude objection, omnipresent in non-expert discourses, is to dispute the possibility or pertinence of discriminating between research and deployment. However, the more modest the research program, the less convincing the objection. Further, it is incontestable that computer modeling of enhanced albedo belongs to research, which significantly differs from spraying aerosols in the stratosphere.[6] A more robust objection is to pinpoint that *some* research activities may not be substantially distinct from deployment in terms of scope and impacts. For example, what would distinguish large-scale field experiments involving ample sulfur injections during a period prolonged enough for prompting clear signals, that is climate responses, from full deployment? A similar objection could be brought forward against MCB, CCT, and space-based CE. In short, SAI, among other methods, "cannot be tested without full-scale implementation" (Robock *et al.* 2010, 530). Field testing will require an extended time, normally several decades (some evaluations mention seventy years [MacMartin *et al.* 2019]), for gathering data that substantiate SAI efficacy or risks. Under those conditions, any outdoor research activity intended to prompt climate responses display attributes analogous to those of deployment. It may be replied that other dimensions than climate response can be tested, like aerosol behavior at different altitudes. This would suggest that the actual controversy does not bear on the possibility of separating between research and deployment per se but on where to draw the line (McLaren and Corry 2021, 13).

The third objection is perhaps the most serious. While the distinction between research and deployment may be accepted on conceptual grounds, its material effectivity and normative significance could be disputed. The claim is that chances are nontrivial that CE research will drift toward deployment, which will diminish the practical pertinence of the divide. R&D would spur a (morally reprehensible) slippery slope, at least for some methods (Andow 2023; Gardiner 2010; Frumhoff and Stephens 2018, 4; Hamilton 2013b; Hulme 2014, 68; Jamieson 1996; Ott 2012). Agreeing to research would lead to accepting deployment. Sliding could happen at any point in the sequence running from modeling to large outdoor experiments. It could be launched by factors like the "cultural imperative" to undertake any endeavor simply because it is possible, the vested interests and lobbying of a geoclique, sunk costs, or institutions dedicated to the regulation and promotion of CE activities (Andow 2023, 3; Callies 2019b, 675).

The objection combines two claims: empirical and moral. The former stipulates that initiating a research program creates incentives for pursuing the program further into implementation (Jamieson 2013, 535). The claim could be understood as any research *inexorably* heads to deployment, which is too strong. Research can stay strictly confined to computers and laboratories. A more modest formulation is to consider that *some* research projects, because of inherent features, may lead to actual deployment, for instance because they require massive and enduring field testing (e.g., with SSI). The concern is reinforced by the ambiguous stance adopted by some proponents of CE research who would defend deployment (Hamilton 2014, 18). However, the equivocal attitude of some scientists does not preclude that other scientists could abide by a stricter compartmentation (e.g., Smith *et al.* 2023, 41). The moral claim affirms that sliding from research to deployment would be reprehensible due to negative consequences like harms to humans, animals, or ecosystems, the violation of moral imperatives (e.g., rights).

Thus, the objection rests on the presumption that for a series of occurrences $[O_1 . . . O_i . . . O_n]$, agreeing to O_i strongly predisposes to accept O_{i+1}, and so on, until the moment when the frontier of impermissibility is crossed. At the core, the argument articulates two anxieties: whether initiating R&D on a given method will lead to disastrous deployment and whether moral concerns not to deploy will prevail in case of foreseeable catastrophes or deep uncertainty (Andow 2023, 15). The empirical claim leans on the presumption that such concerns will not be convincing enough for preventing from sliding down the slope. A possible response, however, is to insist that investigating various CE methods does not *necessarily* commit to implementing them. Accepting O_i does not equate or compel to accept O_{i+1}. The lack of governance structure may also impede rather than stimulate CE research and deployment, generating more of an "uphill struggle" than a slippery slope (Bellamy and Healey 2018).

Even if we admit the materiality of the slippery slope, protocols can always be halted, and projects canceled. To that effect, safeguards need to be devised for stopping research in an orderly and timely manner, that is, before irreversible harms unfold. This underscores the importance of "exit ramps," that is, "criteria and protocols for terminating research programs or areas" (National Academies of Sciences, Engineering, and Medicine 2021, 9).[7] "Check points" (Diamond *et al.* 2022), "way stations" (Bunzl 2009), or "stage gates" (Callies 2019b) may be established for pausing research at predetermined intervals and forcing to gauge feasibility and permissibility. At those points, efficacy, potential scalability, and monitoring prospects of the CE method could be reviewed for deciding whether to pursue the program. Those points could be integrated in the protocol of evaluating the technological readiness levels (TRLs) (cf. chapter 2). The protocol could be strengthened by involving the public, and not only experts, through consultations, focus groups, or even substantial participation to the stage assessments (Callies 2019b).[8] Assuming that public engagement reduces the odds of proceeding with too risky technologies, procedural measures may help to reinforce research legitimacy by rendering the path from R&D to deployment less slippery. In addition, as noted in chapter 2, those procedural devices and CE research in general could gain in legitimacy by including vulnerable populations. Some projects such as Stratospheric Particle Injection for Climate Engineering (SPICE) offer some early examples of citizens' consultation and participation (Pidgeon *et al.* 2013).

Governing Research

The potential use of exit ramps for preventing slippery slopes underscores the necessity of governance, defined as "the resources, information, expertise, and methods needed for the control of an activity in order to advance the potential societal benefits provided by SRM [Author's note: or CDR], while managing associated risks" (SRMGI 2013, 29). As mentioned in chapter 2, CE suffers from a lack of governance and regulation structures in relation to deployment, but also in relation to research (Burger and Gundlach 2018). The consequence is that scientists and engineers operate in a relative vacuum or, more precisely, under "de facto governance" (Gupta and Möller 2018) or "governance-by-default" (Talberg *et al.* 2018a). CE research is subject to fragmentary domestic and international legal instruments, treaties, and institutions, such as the decisions of the Convention on Biological Diversity (CBD), the London Convention and Protocol, or the Convention on the Prohibition of Military or Any Other Hostile Use of Environmental Modification Techniques (EMTs) (ENMOD). Moreover, institutions and instruments devoted to CE are underdeveloped, with no dedicated regulatory and governance framework in any major country or internationally.

In response to this lacuna, different official bodies have been advocating for more substantial research governance (National Academies of Sciences, Engineering, and Medicine 2021; NRC 2015a, 2015b). Practical initiatives have flown from the academic community under the form of principles. Those could be qualified as "*self*-regulation" or "*self*-governance" (in opposition to *external*) since they originate from the very community that needs to be regulated. Moreover, except if those principles are endorsed by state authorities, which could enforce them, their efficacy entirely depends on the good will of researchers. The Oxford Principles represent one of the earliest and most discussed initiatives. Its five principles are intended to cover both research and deployment.

Principle 1: Geoengineering to be regulated as a public good.

While the involvement of the private sector in the delivery of a geoengineering technique should not be prohibited, and may indeed be encouraged to ensure that deployment of a suitable technique can be effected in a timely and efficient manner, regulation of such techniques should be undertaken in the public interest by the appropriate bodies at the state and/or international levels. (Rayner *et al.* 2013, 502)

This directly tackles the end goal of CE research (the "public interest") and the possibility of capture (cf. below). It attempts at minimizing the odds of research being carried out for the benefits of private/corporate interests, that is, of a minority. Such odds are nontrivial due to the prospect of global disparate impacts and local harms generated by methods like SAI, CCT, MCB, OIF, or ocean-enhanced weathering. This first principle subjects any project to the obligation of being guided by the "public interest," and therefore to be regulated as a "public good," terminology that is widespread among CE researchers (Long *et al.* 2015, 30; National Academies of Sciences, Engineering, and Medicine 2021, 6). However, there is the need for precisely defining the "public interest." Is it formed of the interests of all-affected parties? If so, the public interest may be prone to internal tensions, containing incompatible expectations. Does it aggregate the interests of the *majority* of the affected parties? In that case, the vulnerable minorities are at risk of not having their interests properly taken into account, for example, in destitute areas and in the Global South, whose voices are usually inaudible. To circumvent such risk, principle 2 imposes conditions of public engagement. At a more fundamental level, the assimilation of CE to a public good (to which the Oxford Principles refrain) is controversial (cf. textbox 5.1).

TEXTBOX 5.1 IS CLIMATE ENGINEERING
A GLOBAL PUBLIC GOOD?

Among the supporters of CE research, the claim is common that CE is, or should be considered as, a global "public good" (Barrett 2007, 20, 38-41; Fajardy *et al.* 2019; Flegal *et al.* 2019; Heyen *et al.* 2019; Keith 2017; Morrow 2014b; Rayner *et al.* 2013; Weitzman 2015), which is contested by some (Gardiner 2013b; Gardiner and Fragnière 2018, 145–152). The debate calls for conceptual clarification about what constitutes a public good. Although several answers are possible, the social sciences typically follow the definition crafted by economics as "goods that are neither excludable nor rival in consumption" (Mankiw 2021, 795). Non-excludability means it is impossible, or extremely difficult, to bar agents from consuming the good. Non-rivalry means that the amount of public good consumed by one agent does not reduce the amount available to other agents. Archetypical examples present in the literature include national defense, lighthouses, public education and so on. Despite its limitations, the public good rationale is useful for argumentative purpose as it frames any issue as one of underprovision while advancing the necessity of state or international governance (Bodansky 2012, 655; Gardiner 2013b, 515). Whereas a public good generates spillover (or externalities) effects (Dworkin 1985, 221–233), potential beneficiaries can eschew contributing to its provision, that is, free ride. To be optimized, the provision of public goods would then need to be regulated somehow by some apt authority through forced contribution.

Are CE techniques non-rival and non-excludable? It depends on the technique. CDR seems to fit the description (Fajardy *et al.* 2019, 5). In principle, carbon extraction and sequestration uniformly benefits everyone since atmospheric carbon concentrations are brought down globally. The benefits (lowered CO_2) enjoyed by some does not deduct any amount available to others. However, some initiatives might generate local harms, for example, pollution with ocean-enhanced weathering, land-use conflicts with forestry, and BECCS, so it means that overall net benefits could be reduced for some groups. On the other hand, the picture is more equivocal for SRM. Most of the methods do not match the definition (Gardiner 2013b, 516). USB and crop albedo will have small-scale effects, which could make them *local* public goods, but interventions such as SAI, MCB, CCT, and space CE will produce large-scale disparate impacts and new climates, with (significant) net winners and net losers. Moreover, the intervention by a country may prevent other countries from deployment in a more advantageous manner. Finally, for SAI central issue is not free riding, but free driving, which may lead not to under-provision, but to over-provision (Bodansky 2012, 665; Wagner 2021, 16–17; Wagner and Weitzman 2012; Weitzman 2015). Nevertheless, it could be claimed that the technical definition of a global public good, in terms of non-excludability and non-rivalry, does not require universal benefits (Morrow 2014, 96). The general impact could be mixed, or even negative (Flegal *et al.* 2019, 414). In a laxer fashion, a global public good could be understood as an activity or an institution that could significantly affect the collective well-being of a "public." In sum, CE could match this loose description, despite not being potentially equally beneficial, or even not beneficial at all.

Branding CE as a global public good has the effect of posing the question of regulation and imposing normative constraints. It requires regulation to be undertaken for the common interest of humanity, present and future. At the extreme, it might justify excluding private initiatives and establishing state control over all the activities, from R&D to deployment. However, making CE state-run may be undesirable and, for the most part, unfeasible. Without reaching such extremity, since (potentially) all humans could be affected (positively or negatively), all humans have (potentially) a stake, and thus should have a voice, in how those methods could be researched, developed, and deployed. If one remains skeptical toward qualifying CE deployment as a public good, it could still be claimed that CE *research* nonetheless qualifies as such. The increased knowledge, higher ability to monitor (rogue) deployment, and so on would positively contribute to humanity. Again, this could justify public control, or at least substantial oversight from state institutions. At the end of the day, less than CE, it seems that the genuine public good is an atmosphere with low GHG concentrations characterized by a stable climate, which has not crossed tipping points, and in which the frequency of catastrophes is limited.

Principle 2: Public participation in geoengineering decision-making.

Wherever possible, those conducting geoengineering research should be required to notify, consult, and ideally obtain the prior informed consent of, those affected by the research activities. The identity of affected parties will be dependent on the specific technique which is being researched—for example, a technique which captures carbon dioxide from the air and geologically sequesters it within the territory of a single state will likely require consultation and agreement only at the national or local level, while a technique which involves changing the albedo of the planet by injecting aerosols into the stratosphere will likely require global agreement. (Rayner *et al.* 2013, 502–503)

This principle acknowledges the procedural constraints bearing on CE research and deployment. As seen before, such constraints are particularly imperative regarding vulnerable groups. Public engagement can come in various shades: from consultation to co-decision. Furthermore, participation requirements apply to the general public, but also to scientists working in institutions usually not implicated in the main CE R&D projects, for example, in the Global South (Winickoff *et al.* 2015). Such requirements may not be all or even most of the time implementable in satisfying ways, which could incline one to judge norms-building attempts as being delusional. It could be responded that ironing out shared principles is not a futile enterprise. It is vital for setting standards of what morality or justice requires. The constraints imposed by principles might lead to effective regulation in a not-so-distant future. More importantly, regulation needs to start somewhere, as with the Asilomar principles (Svoboda 2017, 119). As noted in chapter 2, participation is essential for several reasons, among which, as a mark of respect toward affected parties, as a detection tool for hidden risks and vulnerabilities, and for securing acceptance. When implementing this second principle, the challenge is to go beyond paying lip service to public engagement and introduce it in a meaningful manner within research programs (Frumhoff and Stephens 2018; Pidgeon *et al.* 2013).

Principle 3: Disclosure of geoengineering research and open publication of results.

There should be complete disclosure of research plans and open publication of results in order to facilitate better understanding of the risks and to reassure the public as to the integrity of the process. It is essential that the results of all research, including negative results, be made publicly available. (Rayner *et al.* 2013, 503)

This principle addresses the difficulty of evaluating CE due to the complexity of the climate processes involved, but also the fact that most of the knowledge is produced within small circles. Thus, research outcomes need to be more widely accessible, especially to disadvantaged populations. This could help to rebalance asymmetric relations of domination through "power equalization," which consists in this case to give access to the results of research to groups and nations so they can elaborate their own programs (Smith 2018, 352–358). Though, for disclosure to play its role, the beneficiaries should have the capacity and resources to use this knowledge, for example, for developing monitoring tools to detect rogue deployment or their own research strategy. A few initiatives aim at supporting research in the Global South such as the Developing Countries Impacts Modeling Analysis for SRM (DECI-MALS) (Wagner 2021, 88). However, a possible criticism is that knowledge dispersion increases the threat of rogue or uncoordinated implementation since more entities will have access to up-to-date scientific and engineering knowledge. In short, it would represent an instance of CE proliferation.

Principle 4: Independent assessment of impacts.

> An assessment of the impacts of geoengineering research should be conducted by a body independent of those undertaking the research; where techniques are likely to have transboundary impact, such assessment should be carried out through the appropriate regional and/or international bodies. Assessments should address both the environmental and socio-economic impacts of research, including mitigating the risks of lock-in to particular technologies or vested interests. (Rayner *et al.* 2013, 503)

This principle tackles the issue of the concentration of capabilities in a few institutions mostly located in the Global North. The main concern addressed is the risk of research teams being stuck with developing inefficient, harmful or unfair CE methods because of vested interests, lobbying, sunk costs, corruption, and so on. If seriously undertaken, independent assessments could drastically lower not only the odds of such inefficient, harmful or unfair allocations of resources, but also of slippery slopes. For being apt, principle 4 requires protocols to be crafted in a manner that render them responsive to external reviews through, as mentioned above, "check points," "way stations," "stage gates," and so on. In the same vein, David Keith proposes an adversarial design, in which a blue team of scientists investigate the benefits of a method and a red team focuses on the drawbacks (Keith *et al.* 2010). Independent assessments, as public engagement, ought to have a substantial impact on how research is undertaken.

Principle 5: Governance before deployment.

> Any decisions with respect to deployment should only be taken with robust governance structures already in place, using existing rules and institutions wherever possible. (Rayner et al. 2013, 503)

The last principle goes beyond research by highlighting the necessity of not implementing CE before the regulatory infrastructure is built up. Again, it may not be feasible, especially considering that some techniques like SAI, due to their relative accessibility and their global impacts, would require a modicum of international cooperation for being adequately regulated. Another obstacle to effective structure of governance lies in the calls for bans or moratoriums. Such pushes could deter initiatives that aim at establishing accountable research infrastructure. It is nonetheless true that prohibition is also a means of regulation and that a moratorium does not prevent building overseeing institutions. Moreover, banning some methods could be fully justified. In short, there are various modes of governance (Gupta *et al.* 2020). That being said, the overshadowing threat of a blanket ban on all, or most, CE methods might impede the development of consistent governing institutions.

Although the Oxford Principles constitute the most discussed initiative, they are not the only or most full-fledged approach (e.g., Asilomar Scientific Organizing Committee 2010; Hubert 2021; Morrow *et al.* 2009). Despite their merits, they can be challenged with solid reasons. According to Gardiner and Fragnière (2018), they would be too vague and procedural, for example, when it comes to the provisions for public consultation and participation. As a result, Gardiner and Fragnière propose a substantially revised version (the Tollgate Principles), which, injects more demanding ethical requirements (Textbox 5.2).

TEXTBOX 5.2 THE TOLLGATE PRINCIPLES

1st Tollgate Principle (Framing): Geoengineering should be administered by or on behalf of the global, intergenerational and ecological public, in light of their interests and other ethically relevant norms. (. . .)

2nd Tollgate Principle (Authorization): Geoengineering decision-making (e.g. authorizing research programs, large-scale field trials, deployment) should be done by bodies acting on behalf of (e.g. representing) the global, intergenerational and ecological public, with appropriate authority and in accordance with suitably strong ethical norms (e.g. justice, political legitimacy).

3rd Tollgate Principle (Consultation): Decisions about geoengineering research activities should be made only after proper notification and consultation of those materially affected and their appropriate representatives, and after due consideration of their self-declared interests and values. (. . .)

4th Tollgate Principle (Trust): Geoengineering policy should be organized so as to facilitate reliability, trust and accountability across nations, generations and species. (. . .)

5th Tollgate Principle (Ethical Accountability): Robust governance systems (including of authority, legitimacy, justification and management) are increasingly needed and ethically necessary at each stage from advanced research to deployment. (. . .)

6th Tollgate Principle (Technical Availability): For a geoengineering technique to be policy relevant, ethically defensible forms of it must be technically feasible on the relevant timeframe.

7th Tollgate Principle (Predictability): For a geoengineering technique to be policy relevant, ethically defensible forms of it must be reasonably predictable on the relevant timeframe and in relation to the threat being addressed.

8th Tollgate Principle (Protection): Climate policies that include geoengineering schemes should be socially and ecologically preferable to other available climate policies and focus on protecting basic ethical interests and concerns (e.g. human rights, capabilities, fundamental ecological values).

9th Tollgate Principle (Respecting General Ethical Norms): Geoengineering policy should respect general ethical norms that are well-founded and salient to global environmental policy (e.g. autonomy, justice).

10th Tollgate Principle (Respecting Ecological Norms): Geoengineering policy should respect well-founded ecological norms, including norms of environmental ethics and governance (e.g. sustainability, precaution, respect for nature, ecological accommodation).

(Gardiner and Fragnière 2018, 166–168, emphasis added).

In sum, independently of any hypothetical deployment, research will need to be regulated from the onset and governed one way or another, at least for enforcing bans or moratoriums (Burger and Gundlach 2018). A glance at domestic and international regulatory apparatus feeds the concern that the legitimate institutions that could perform such a duty are not in place yet. Besides, it could still be claimed that a modicum of public research is necessary, even on the most controversial interventions, to identify the threats, but also devise appropriate monitoring tools for controlling possible deployments by public and private actors.

At the end of the day, policymakers need reliable data spanning the various CE techniques for making informed decisions on whether to allow or support further research and, ultimately, permit or proscribe deployment. Moreover, hybrid or portfolio policies will presuppose optimizing a combination of mitigation, adaptation, and at least a grain of CDR (deemed to bridge the mitigation gap). Acknowledging the nonequivalence of the methods, each of them ought to be independently evaluated without sweeping judgments. Regulation from the R&D stage ought to be taken seriously. The crux of the issue is not whether to launch and regulate research, but which project could be launched, by whom, and under which conditions. Clearly, that section has left many central questions untouched, for example, who, or which institutions, should govern research? Should they be domestic and internationally coordinated?

And more fundamentally, how to build and guarantee the legitimacy of such institutions?

JUSTIFICATIONS FOR CLIMATE ENGINEERING

This part presents four arguments for integrating CE into climate policy. It complements previous attempts at mapping out arguments in favor of CDR and/or SRM (e.g., Baskin 2019, 96–100; Buylova *et al.* 2021). As already noted, two flaws often muddy the discussions. The first is to reduce the analysis to a criticism of SAI. To eschew this pitfall, it is essential to insist on the variety of methods when assessing the justifications for (this section) and objections against (the next section) CE. The second flaw is to evaluate interventions as stand-alone options that could be implemented independently or in substitution of hybrid policies. The detachment of CE from conventional policy (mitigation and adaptation) expresses either an error of omission (not remembering that CE cannot suffice for preventing climate catastrophes) or an error of judgment (not understanding that CE cannot suffice for preventing climate catastrophes). The first two arguments below (A_1 and A_2) suffer from that flaw. Finally, these arguments constitute ideal types, that is, stylized and simplified reconstructions of more complex and diverse lines of thought.

Argument 1 (A_1) stipulates that CE, or some CE methods, is *the first best*, that is, the most preferable option. A basic formulation could be as follows:

P: All things considered, CE, or some methods, would be superior to conventional mitigation (and adaptation) when it comes to "prevent dangerous anthropogenic interference with the climate system" (UNFCCC)

C: Thus, CE, or some methods, should be pursued at the exclusion of mitigation (and adaptation).

The argument derives a normative conclusion (C) from an empirical premise (P). CE would represent a better, that is, cheaper, more cost-effective,[9] or more socially acceptable policy option (Baskin 2019, 97) (Often, the techno-fix objection [cf. next section] interprets the support for CE research and/or deployment as a first-best advocacy). According to the objection, instead of tackling the source of the problem, through aggressive mitigation, decision-makers would get distracted by the belief that CE offers the most efficient, acceptable, response to dangerous climate change (Preston 2012, 26–27).[10] Entities connected to the fossil-fuel industry and free-market think-tanks (Baskin 2019, 98–100; Hamilton 2013a, 72–82) push forward some version of A_1, namely not changing the energy infrastructure while developing carbon capture and sequestration/storage (CCS) for instance, at the risk of initiating a carbon lock-in (cf. chapter 3).

To our knowledge, no expert supports A_1, most likely because the argument is unsound, P being untrue. Chapters 3 and 4 show that no CE method is ready or could be readied fast enough, alone or in combination with other methods, to avoid Anthropocene-related catastrophes. On the one hand, no CDR intervention alone or combined with other interventions, in the absence of drastic and immediate mitigation, can guarantee to remove sufficient volumes of GHGs, and swiftly enough for preventing climate disasters. Without even considering the necessity to lower GHG concentrations, CDR simply cannot catch up with current emissions. Moreover, if tipping points are crossed, it will be extremely difficult, most likely impossible, for CDR to compensate for positive feedback loops (e.g., methane emissions, collapse of the West Antarctica ice sheet, and Amazonia dieback) (Shue 2017). At best, such methods as BECCS and DACS need more time for reaching maturity and scalability. More feasible alternatives like forestry cannot secure sufficient and durable CCS.

On the other hand, it is hard to see how initiating SRM *without* mitigation could be superior to any policy mix containing substantial conventional mitigation. Even under its most mature form (SSI), SRM can only mask part of anthropogenic induced warming while leaving other variables untouched (e.g., increased ocean acidification). With SAI (and potentially MCB and CCT), new harmful regional climate regimes will be imposed on vulnerable populations. Moreover, there is a dangerous gap between the lifetime of atmospheric carbon (from a few centuries to several millennia) and the lifetime of stratospheric aerosols (a couple of years) (Pierrehumbert 2019). In the absence of drastic mitigation, Albedo enhancement is not a sustainable "solution."

Argument 2 (A_2) is a variant of A_1. Sometimes interpreted as defending CE as a "plan B," A_2 posits that CE as a whole, or some set of CE methods, constitutes a second best after mitigation. A possible formulation could be the following:

P_1: In the absolute, conventional mitigation is the first-best option
P_2: However, conventional mitigation is unfeasible and/or undesirable on practical grounds.
P_3: Under those circumstances, conventional mitigation should be abandoned.
C: Then CE, or some methods, could offer a substitute "to prevent dangerous anthropogenic interference with the climate system" (UNFCCC).

P_2 supports the weight of the whole argument. Conventional mitigation would be unfeasible due to contextual factors such as the difficulty of international cooperation for cutting GHGs, vested interests in fossil energies, lack of time for properly decarbonizing economies, and so forth. This pessimistic view is sometimes advanced by researchers, for example, Lowell Wood: "Mitigation is not happening and is not going to happen" (quoted by James Fleming [2007, 48]). In addition, initiating mitigation to the extent required for averting catastrophic climate change could be deemed undesirable since it would demand tremendous financial resources, causing economic stagnation and condemning many to poverty, thus undermining the goals of sustainable development and poverty eradication, as stipulated by the UNFCCC and Paris Agreement.

The difference with A_1 is slight, but still significant. A_2 supports engaging in a suboptimal solution (carbon capture or albedo modification) not by choice, but due to necessity, as *per* some "contingency plan" (Pamplany 2019, 3080). This explains why CE is often introduced, discussed, and criticized as a "plan B" (Corry 2017; Fragnière Gardiner 2016; Goodell 2010; Gordon 2010; Hulme 2014, 99–105; Michaelson 2013; Olson 2012, 29; Pierrehumbert 2019; Shue 2017, 110).

> We are talking about Plan B, because Plan A seems so expensive that a few key players remain intent on blocking it. Plan A is best; Plan B may be the best we can do. And of course, as we have said many times, the two are not mutually exclusive. A shift of nomenclature from the technological mechanisms of a climate change strategy to the policy nature of that strategy helps clarify the issue in place, and properly shifts attention from means to ends—or at least from proximate means (technological intervention) to meaningful means (regulation, management, or mitigation). (Michaelson 2013, 87)

Although Michaelson makes it clear that plans A and B are not mutually exclusive, plan B is often used in a more ambiguous manner, leaving open the issue whether CE precludes or supplements plan A (Stilgoe 2015, 122). To that respect, A_2 provides a formulation of the former through P_3, whereas the latter, as in Michaelson's quote, constitutes the basis of a distinct argument, which defends combining CE, or some methods, with mitigation and adaptation. This last formulation is the core of A_3, detailed below. Finally, plan B could be read as an emergency or contingency response in case plan A, mitigation, fails (Baskin 2017, 97). However, the fact that (some instances of) CE could offer a second-best option is distinct from emergency considerations. For example, the deployment of BECCS and DACS at the expense of mitigation could be chosen, not as a first best, and not as a decision taken in urgency, but as the execution of a less satisfying, regrettable, nonetheless more realistic option. So, the emergency frame does not perfectly reflect A_2.

To summarize, under the premise P_3 of excluding mitigation, A_2 suffers from the same shortcoming as A_1. On the one hand, A_2 overlooks that CE cannot form a potent substitute to aggressive mitigation if the goal is to avert catastrophic climate change.[11] On the other hand, A_2 apprehends CE in isolation, separated from other policy instruments, and not as a potential part of a hybrid strategy. As A_1, A_2 presents a false dilemma[12] between conventional climate policy and CE, while other possibilities are available, such as combining CE with mitigation and adaptation.

Argument 3 (A_3) stipulates that CE, or some set of CE methods, is a necessary part of a *portfolio, hybrid, or symphony approach along mitigation and adaptation*. A simplified formulation could be that:

P_1: Conventional mitigation, along with some dose of adaptation, is the first-best option.

P_2: However, neither conventional mitigation nor CE, on their own, could guarantee with enough certainty and under acceptable conditions "to prevent dangerous anthropogenic interference with the climate system" (UNFCCC) and/or to respect additional moral constraints such as sustainable development.

P_3: Each of those tools carries particular benefits when it comes to tackling "dangerous anthropogenic interference with the climate system" (UNFCCC) and/or to respect other moral commitments.

C: Thus, those instruments, or some of them, should be used in conjunction, within a hybrid framework.

As per A_3, mitigation, possibly along with adaptation, retains priority. However, CE would offer a fallback or failsafe option, not across the board (basis of A_2), but for some specific parts of the policy (Buylova 2021; Keith 2000, 274). The expression "portfolio" (NRC 2015a, 2015b) hints at the idea, borrowed from risk management, that assets' diversification attenuates risk exposure. As a diversified portfolio of stocks is less prone to totally lose its value, a diversified portfolio of policy tools is less likely to fail. A_3 relies on the possibility of balancing the different components in a beneficial fashion, namely, to lower the prospect of catastrophes. A more ambitious interpretation of A_3 is to consider that adding (some) CE methods to mitigation and adaptation will improve the chances of optimizing policies (Schäfer *et al.* 2015, 40), like in a "symphony" (Long 2017).[13] However, multiplying tools does not guarantee optimal outcomes. Mutually impairing effects could emerge across the various instruments, such as SAI impeding forestry interventions (cf. chapter 4). More fundamentally, the notion of optimum needs to be defined. Does it refer only to climate variables (temperatures, precipitations, etc.) or does it include impacts on human communities and ecosystems? Is it a Paretian notion,[14] according to which no one should be made worse off as the result of the hybrid policy while at least one agent should become better off, which is demanding? Or does it follow laxer conditions, such as Kaldor–Hicks criterion, which allows for situations in which some are made worse off if and only if they can be compensated for their losses?

A variant of the portfolio frame is mobilized for justifying a controlled overshoot (cf. chapter 3). It begins by positing that mitigation, alone or coupled with adaptation, carries the nontrivial possibility of not lowering climate risks significantly and swiftly enough for eschewing "dangerous anthropogenic interference." Another claim is then added stating that if mitigation is drastically ramped up for maximizing the chances of preventing catastrophes, it will most likely be at the expense of core moral commitments such as poverty alleviation and sustainable development (Callies and Moellendorf 2021). Therefore, all things considered, it would be better to rely on fossil energies for a period while fostering carbon-free technologies (Nordhaus T. 2018). In that scenario, although deploying CE alone is not desirable per se, in conjunction with mitigation and adaptation, some CE methods could still offer a shred of efficiency that will both allow delaying full decarbonization and minimize harms on the most vulnerable who rely on cheap, fossil, energies. This is the locus of the buying time argument. Time could be gained by either sucking up carbon out of the atmosphere (Shackley and Thompson 2012; SRMGI 2013, 19; Wigley 2006), which could be helpful for compensating for difficult-to-decarbonize sectors like energy generation, air transportation, or agriculture (Buylova 2021, 12; Tamme and Beck 2021), or by shaving temperature peaks through albedo enhancement (MacMartin *et al.* 2018; Tilmes *et al.* 2016). *In fine*, CE methods could act as bridging technologies in the long road toward safer, carbon-free, societies, concretely allowing for an overshoot.

Argument 4 (A_4) is one of the most discussed and controversial arguments (Flegal and Gupta 2018, 52). It formulates the potential necessity of deploying CE, or some methods, for averting an imminent catastrophe. A possible formulation could be the following:

P_1: Conventional mitigation is the first-best option, along with some dose of adaptation.
P_2: Thus, conventional mitigation and adaptation should be pursued.
P_3: However, nontrivial chances exist that the climate spirals out of control due to positive feedback loops and tipping points, presenting an emergency.

P_4: CE, or some methods, could efficiently deal with the emergency and prevent "dangerous anthropogenic interference with the climate system" (UNFCCC)

C: In the prospect of a climate emergency, it could be justified to engineer the climate.

Contrary to A_3, A_4 does not plan to implement CE as part of the default option. Deployment would stay conditional on the occurrence of an emergency, loosely defined as "an imminent 'catastrophe' or 'crisis' brought about by runaway climate change" (Bellamy and Lezaun 2017, 406). Moreover, the argument assumes that CE, or some methods, represents the lesser of two evils, that is, prospective CE deployment carries lower expected losses than climate catastrophe (which reasoning follows a maximin logic). Under those circumstances, investing into R&D would contribute to "arm the future" (generations) in front of potential disasters (for a critique, cf. Gardiner 2010).

Framing CE as an emergency response is controversial. First, the concept of what could constitute an emergency is subject to definitional and practical disagreements, especially if venturing into what could constitute an emergency compelling enough to justify CE deployment (Gardiner 2010, 291–295; Hulme 2014, 21–26). The concept of a (compelling) emergency is open to interpretations. Beyond the conceptual hurdles, proclaiming a global state of emergency might be complicated by contextual considerations. Some regions might face looming catastrophes, while others might not experience unusual threats or even benefit from the changing climate. States, particularly those capable of CE implementation, will not have identical preferences for optimal temperatures, which will generate collective action problems (Emmerling and Tavoni 2018; Heyen *et al.* 2019). Moreover, for being efficient, that is, globally accepted and not spiraling out into counter-CE or sabotage attempts, the characteristics of the emergencies that could justify deployment, as well as the conditions for such deployment, need to garner international agreement. In addition to those crucial questions, it should be noted that framing a situation as an emergency constitutes a powerful, yet dangerous, rhetoric. It could be invoked for suspending part of the regular norms (Gardiner 2013, 29; Horton 2015), potentially leading to a "tyranny of urgency" (Stilgoe 2015, 199). More prosaically, it can be used for forcing public acceptance (Bellamy and Lezaun 2017, 406), as for other frames (cf. chapter 2) (textbox 5.3).

TEXTBOX 5.3 FRAMING THE CLIMATE ENGINEERING DISCUSSION

In the literature, the reasons for and against CE, or some methods, often rely on analogies or convoke images and concepts that act as frames, nudging perceptions and judgments in one direction or another. Positive instances present CE as a plan B, a contingency plan, a buy time/stopgap device, an emergency/last resort/last-chance response, an insurance policy, a fallback option, the lesser of two evils, and a desperation solution, among others. CE could also be framed negatively as playing God, a technofix, a Promethean project, a slippery slope, a moral hazard, so on. Often used in a lax manner, those frames are notorious for their ambiguity, being prone to contradictory interpretations. They are mobilized for rhetorical purposes, for example, as positive analogies (e.g., insurance and lesser evil) for convincing the necessity of researching and developing CE, or some methods, or negatively for persuading that CE, or some methods, would represent impermissible threats (e.g., moral hazard and playing God). Such strategic uses have not gone undetected (Bellamy and Lezaun 2017; Buylova *et al.* 2021; Corner and Pidgeon 2014; Gardiner 2010, 2013b; Hale 2012; Hamilton 2013a; Pamplany *et al.* 2020; Pierrehumbert 2019; Preston 2012, 30; Scott 2012; Shue 2017; Talberg *et al.* 2018b).

Without analyzing all of those frames, the insurance analogy is noticeable due to its prominence among arguments that advance CE R&D (Preston 2013, 30; Scott 2012; Shepherd *et al.* 2009). Loosely formulated, it posits that investing into CDR and SRM would guarantee that humanity would benefit from a fallback option or plan B (other analogies) in case of an emergency (another one) or in case of more general need. CE, however, does not neatly fit the description or the intent of an insurance policy. At the core, insurance is a term used for designating risk-pooling mechanisms (Bernstein 1998; Landes 2013; Moss 2002). Members of the pool, as policyholders, will insure themselves against adverse events by paying premiums, that is, sums that entitle them to (partial) compensation in case of losses. CDR boasts of removing and storing carbon away from the atmosphere in order to decrease the risk of catastrophes and also to keep climate variables like temperatures and ocean acidification in check. SRM offers to enhance albedo for reducing global warming. Based on this account, it is unclear in which sense CDR and SRM could constitute an insurance policy. Neither promises to compensate those who will suffer from the Anthropocene. Moreover, the odds are nontrivial that CE deployment may harm populations, many of which are already suffering from the Anthropocene (skewed vulnerabilities). After the damage has been inflicted, compensation may come, or not. It is a separate issue, independent of CE techniques per se. If compensation takes place, that will stem from a distinct mechanism from CE. Further, an insurance policy does not offer to cancel out or reduce risk exposure. It cushions the consequences of such exposure when concrete losses materialize. CE promises something different. CDR offers to lower the odds of adverse events (i.e., climate harms and catastrophes) by altering the original source of those events, GHG emissions and accumulations. SRM proposes to lessen a secondary driver, which is a variable dependent on GHG concentration, namely temperatures.

At the end of the day, the use of frames raises two questions: how descriptively accurate are they? And how do they enlighten the discussion? First, they could be evaluated *qua* concepts, by determining their internal consistency, and so on. On those grounds, it is likely that most of those frames will prove weak, arguably misleading, like with insurance (for other examples, cf. Gardiner [2010] and Hale [2012]). Regarding the second question, it is possible that those analogies bring something of value to the analyses on CE feasibility and permissibility. That should be decided on a case-by-case basis. Do specific frames "elicit the relevant discussions" (Scott 2012, 158) or do they contribute to obscuring what is at stake? To return to insurance, it could be claimed that, although the frame is not descriptively accurate, nevertheless apprehending CE from the angle of risk pooling highlights crucial components of climate politics, for example about international solidarity and its boundaries. In front of the Anthropocene, should human beings pool global catastrophic risks? If so, how to devise instruments for protecting populations, not only against climate risks, but also the ensuing losses? In situations where the identification of CE impacts may remain ambiguous (e.g., SAI), thus rendering difficult or impossible to clearly separate what is due to natural variability from what is imputable to the Anthropocene or to imputable to the Anthropocene or to CE, insurance mechanisms could nonetheless offer compensatory options.

The allusion to tipping points (Lenton 2021; Lenton *et al.* 2019) performs a critical function in the argument (Hulme 2014, 14). Accordingly, emergencies could be located at those junctions beyond which "a small amount of extra climate forcing, usually linked to global warming—for example, greenhouse gas forcing—triggers a qualitative change in part of the climate system" (Lenton 2021, 325). Despite their intuitive appeal, tipping points might be difficult to employ (Lenton 2019). Uncertainty about the distance before reaching such points may persist until the last moment. They could be crossed without noticing. Once detected, the disastrous changes (e.g., permafrost thawing, forests dieback, and loss of ice caps) may be well underway and irreversible. Finally, to avoid tipping points, humanity may be required to initiate "pre-emptive" CE, that is, deploying it well before an emergency emerges, which will face the same obstacles (inertia and defective cooperation) than conventional climate policy.

The emergency rationale offers only limited support to CDR since any substantial attempt at capturing and durably sequestrating carbon will take several decades or centuries for noticeably impacting GHG concentrations. If emergency is understood as a situation in which nearing catastrophes mandate immediate action, CDR cannot represent a solution. Of course, developing and deploying CDR will, in the long run, lower the odds of disasters, but it cannot, as such, provide emergency responses. On the other hand, the argument could partly justify engaging in SRM R&D. However, if SAI could potentially constitute an emergency initiative (Lenton 2019), it remains unclear whether other SRM techniques could. Surface albedo interventions (urban surface brightening [USB] and crop albedo) may dispense some local relief in case of runaway temperatures, for example, heatwaves, but with little to no global impact. Tropospheric methods (MCB and CCT) could be theoretically deployed as a matter of urgency, assuming they will be mature at the time of need. However, they will demand constant dispersion without any interruption lasting more than a few days, under the threat of termination shock. They do not offer perennial responses, especially when considering that carbon stays much longer in the atmosphere than the injected particles (Pierrehumbert 2019). Finally, to be implemented, space CE will require a fully dedicated industry, possibly extraterrestrial infrastructure, as well as a staggering number of rocket launches, among other things. Of course, such infrastructure could be prepared beforehand, in prevision of a potential emergency, but under the current conditions and until the mid-century, this method cannot respond to an urgent situation. In sum, for the time being, only SAI could offer a viable approach, under the assumption that the technology is rapidly developed over the coming years.

In conclusion, A_1 and A_2 share the crucial flaw of assuming too much from CE and from its capacity to counter the Anthropocene in the absence of robust mitigation (and adaptation). Both support strategies that severely exacerbate the prospect of disasters. On its side, A_4 justifies CE methods only and only if they have the potential of aptly countering emergencies. For the time being, the argument can only be used to vindicate stratospheric injection as a global response, although more modest techniques such as USB and crop albedo might also provide some local relief in case of temperatures spiraling out of control. On the surface, A_4 may appear as the safest option, that is, the one minimizing harm, since it confines CE to the margins of climate policy, as a last-resort initiative. However, there are substantial shortcomings. Any emergency deployment will require a broad agreement among CE-capable agents on what constitutes an emergency, under the sanction of initiating a cycle of engineering, counter-climate engineering (counter-CE thereafter) as presented in the next section. An even broader agreement with other potentially affected nations would be preferable for reducing the odds of conflict.

In addition, it will impose the necessity to detect urgencies that are underway in due time. *In fine,* A_3, CE as part of a portfolio policy, looks the strongest because of the diversification strategy, although this does not imply that A_3 is free of ambiguities and concerns. Among other things, it may justify embarking upon an overshoot, at the costs of disastrous consequences if afterward mitigation and CDR cannot compensate for the emissions over the budget (cf. chapter 3).

OBJECTIONS TO CLIMATE ENGINEERING

This last section offers an overview of the major objections brought up against CE, regrouped into the following categories: efficiency, precautionary, capture, political objections, and moral corruption (table 5.1). In addition, the robustness of each objection is briefly assessed. The goal is to acquaint the reader with the concerns climate alteration is raising. The taxonomy faces limitations, though. The compartmentation between the different objections is not airtight. They may overlap or support each other. For example, slippery slopes could be interpreted as instances of carbon lock-ins. Moral corruption is often a central component of the technofix criticism. Despite those limitations, it is hoped that this section offers an as complete as possible overview of the reasons to oppose CE.

Efficiency Objections

The first category challenges CE efficiency. Those objections could be described as asserting that CE, or specific methods, will most likely fail to adequately lower radiative forcing for "prevent[ing] dangerous anthropogenic interference with the climate system" either because they will not remove carbon in sufficient quantities and rapidly enough, or because they

Table 5.1   Objections to Climate Engineering

Efficiency Objections	Precautionary Objections	Capture Objections	Political Objections	Moral Corruption
• Weak Cost-Effectiveness	• Accrued Risks/Uncertainty	• Private Interests	• Inequalities/Vulnerabilities	
• Crowding Out	• Compound Pollution	• Technocratic	• Domination	
• Defective Cooperation	• Lock-In	• Militarization	• Democracy	
• Technofix	• Slippery Slope	• Rogue Deployment		
	• Risk Compensation			

will not enhance the albedo in a timely, significant, enduring, and non-harmful manner.

The Weak Cost-Effectiveness Objection. Since CE requires massive investments, it will be more cost-effective to *simply* cut emissions to avoid disasters (e.g., Mann 2021, 158). Everything else being equal, it is cheaper to refrain from emitting carbon than to release it and then extract it or mask its forcing. The view is unchallenged within the scientific community that mitigation is the first best, under the conditions that agents sufficiently reduce their emissions. Still, the crux of the issue is not whether conventional mitigation is the first best, but whether, on its own, it offers realistic perspectives to "prevent dangerous anthropogenic interference with the climate system" under the current circumstances. Again, the widening mitigation gap fuels pessimism. The more GHGs amass in the atmosphere, the less defensible the claim of the superior efficiency of the mitigation-only strategy over hybrid ones (e.g., containing CDR) for avoiding catastrophes like heatwaves, droughts, floods, or sea-level rise. As time passes and due to the extended atmospheric lifetime of carbon, the center of gravity of efficient policy moves away from emission cuts toward concentration reduction, which is precisely what some CDR methods promise. As a result, CDR methods are less vulnerable to the inferior cost-effectiveness objection than SRM interventions, which support mitigation only tangentially. The possibility of consolidating carbon abatement with CE highlights the false dilemma at work when presenting the issue as a stark choice between mitigation and CE. To reiterate the point made in chapter 2, it is preferable, *all things considered,* to assess efficiency across the board, that is, by examining also hybrid policies that embody mitigation, adaptation, and some modicum of CE (e.g., forestry, DACS, and USB). Furthermore, any evaluation needs to include a comparison of different policy mixes based on their cost-effectiveness for reducing specific threats, for example, heatwaves or floods, which is the approach explicitly endorsed by the design perspective (cf. chapter 4).

The Crowding Out Objection. Another objection stipulates that CE represents a dangerous distraction from carbon abatement. By investing into CE, either vital resources are diverted from mitigation, or the sole prospect of climate manipulation undermines mitigation efforts and efficiency[15] (Biermann 2021; Biermann *et al.* 2021; Burns *et al.* 2016; Carvalho and Riquito 2022; Corner and Pidgeon 2014; Fleming 2010; Fuss *et al.* 2014; Hamilton 2013a; Otto *et al.* 2021; Robock 2008b). Resources are indeed scarce. Any investments in SAI, DACS, BECCS, reforesting, or OIF could have *hypothetically* served to bolster energy efficiency and fossil-free energies. However, nothing guarantees that, if not spent on CE, (most of) those resources would have been used for carbon cuts. Furthermore, investments in mitigation and CE are not mutually exclusive, even if those decisions are made under strict budget constraints. It is not a simple zero-sum game. There might be co-benefits. At the end of the day, the objection rests on an empirical claim that may be difficult to evaluate: CE will significantly drive out financial support for abatement. Also, the objection is more about mitigation efficiency than CE efficiency. Speculatively, CE capacity of robbing resources from mitigation might boost its comparative cost-effectiveness.

The Defective Cooperation Objection. The claim is that CE inefficiency stems, not from intrinsic features, but from deficient cooperation. In that respect, CE faces two distinct issues. The first is a classic public good problem in which collective benefits can only be generated if a coalition large enough contributes to their provision, but in which participants have strong incentives to free ride, that is, let other participants shoulder the provision. The shortcoming cripples many CDR methods. Everything else being equal, large-scale CDR would be globally beneficial through climate risk minimization. However, those benefits would require that forestry, land sequestration, enhanced weathering, CCS, BECCS, and so on be pursued simultaneously by a multitude of agents. Under the current circumstances, it is improbable that a single country, or a small coalition, could extract enough carbon for "prevent[ing] dangerous anthropogenic interference with the climate system." Thus, deficient cooperation hinders the provision of CDR, as it happens for mitigation. Moreover, some methods may generate negative externalities, like BECCS with food prices, or impair other CE initiatives, like SAI with forestry or energy crop cultivation. Tackling those effects will necessitate a minimum of coordination among states and private entities, even prior to deployment while incentives for free riding will be high.

The second cooperative defect pulls in the opposite direction, namely, over-provision, through *free driving*. This is particularly salient for SAI and possibly tropospheric and space methods. As shown in the previous chapter, SAI may become soon accessible to various agents, who will be able to alter the climate without international cooperation (Wagner 2021, 16–17; Wagner and Weitzman 2012; Weitzman 2015, 1050).[16] However, it is likely that those agents will hold diverging climate preferences, which will lower the probabilities of a Paretian optimal deployment, which benefits at least one agent without making anyone else worse off. Some countries will prefer to keep temperatures where they are, while others would rather warmer or cooler conditions. The issue will escalate as more actors build SAI capacity.

In addition, global SRM initiatives may spark "counter-climate engineering" (Abatayo *et al.* 2020; Bas and Mahajan 2020; Biermann *et al.* 2021, 4; Emmerling and Tavoni 2018; Heyen *et al.* 2019; Horton 2013, 174), which is "the use of technical means to negate the change in radiative forcing caused by SRM deployment" (Parker *et al.* 2018, 1058). Parker *et al.* (2018) identify two forms of counter-CE: "countervailing" attempts, that is, injecting a forcing component in the stratosphere to compensate for the cooling particles, and "neutralizing" attempts, that is, injecting aerosols that destroy the cooling particles. Potential consequences include retaliation, escalades, and conflicts (Abatayo *et al.* 2020; Bellamy and Healey 2018, 7). Besides SAI, other CE methods are exposed to counter-CE, for example, MCB, CCT, OIF, and enhanced weathering. However, the prospect of CE/counter-CE spiraling out of control may incentivize countries to cooperate (Heyen *et al.* 2019), for example, through an international nonuse agreement. The looming threat of CE/counter-CE draws attention to governance, lending support to calls for arrangements ranging from early regulation and structures of governance (Gupta *et al.* 2020; Jinnah *et al.* 2019) to moratoriums and bans on deployment/research (Biermann *et al.* 2021).

The Technofix Objection. The term "technofix" (Weinberg 1967) designates "the solution to an intractable social problem that results from reframing it as a social problem to framing it as an engineering puzzle" (Scott 2012, 159). While originally positive, the denomination has become mostly derisive. *Qua* technofix, CE would serve as an excuse for disregarding indispensable reforms and indulging into extra emissions (Carr *et al.* 2012, 177–179; Hamilton 2013a; Hulme 2014; Preston 2012, 26–27; Scott 2012, 159; Wibeck *et al.* 2015). CE would also trivialize the Anthropocene by turning it into a simple technical issue, without questioning the consumerist ideology at the heart of the climate challenge (Hamilton 2011, 17). More radical formulations go one step further by lambasting the central role played by technology in our societies (Scott 2021, 158).

It is still possible to try to contrast "good" from "bad" technofixes, keeping in mind that one thing is to determine whether CE_1, in itself, is a good technofix, while it is quite a different thing to decide whether CE_1 is a good technofix in conjunction with other initiatives within a broader, hybrid, policy. The latter is a more delicate task. It requires integrating elements such as the magnitude of mitigation, adaptation, and the interactions between the various policy instruments. With this caveat in mind, Daniel Sarewitz and Richard Nelson (2008) offer three criteria for distinguishing legitimate from illegitimate technological fixes.

> I. The technology must largely embody the cause—effect relationship connecting problem to solution. (Sarewitz and Nelson 2008, 871)

A good technofix provides a "successful" solution in a complex context. This criterion raises a problem definition: is the issue global climate change or warming? If the former, SRM per se do not provide good technofixes, although the evaluation could be sharpened on a case-by-case basis, for example, USB could be a relevant adaptation measure for addressing the urban heat island (UHI) effect. CDR appears more robust since it targets the root cause of the Anthropocene. However, not all methods yield equally realistic capacities for durably capturing carbon. CCS is probably the most promising, followed by DACS and BECCS. There is room for discriminating further among the proposals. More importantly, the problem could be refined as one of minimizing climate harms while blocking or reversing the Anthropocene. In that case, more SRM and CDR methods may offer valuable solutions as a complement to mitigation and adaptation.

> II. The effects of the technological fix must be assessable using relatively unambiguous or uncontroversial criteria. (Sarewitz and Nelson 2008, 871)

The second criterion undercuts the legitimacy of SRM as a relevant technofix. Although local interventions could be straightforward to gauge (e.g., lowered urban average temperatures and heatwaves frequency), the influence of global albedo modification (e.g., SAI) will remain elusive for decades (MacMartin *et al.* 2019). In addition, SRM impact measurement is controversial, for example, the priority given to temperatures and precipitation over other variables (McLaren 2018) (cf. chapter 4). For space CE, solar dimming might be easier to quantify, although the ensuing effects might stay indistinguishable from natural variability, as for SAI. Yet, it could be claimed that, at the exclusion of surface-based techniques, SRM remains difficult to review, for the time being. On the other hand, the quality of the assessment of CDR fluctuates according to the method. If evaluating the performance of fully engineered devices (CCS, BECCS, and DACS) appears robust, the case is less clear for methods like enhanced weathering, OIF, or forestry. This underscores the importance of MRV (cf. chapter 3). More generally, the capacity of quantifying forcing performance does not change if the technology is considered in isolation or within a portfolio.

> III. Research and development is most likely to contribute decisively to solving a social problem when it focuses on improving a standardized technical core that already exists. (Sarewitz and Nelson 2008, 871–872)

A good technofix is characterized by R&D not devoted to pursue pure knowledge production or radical innovation, but to the amelioration of existing processes or instruments. In other words, the route from knowledge creation, diffusion to implementation, should be shortened. The criterion seems to favor technologies with strong maturity and scalability prospects. Also, mission-driven projects ought to be prioritized over curiosity-based ones (cf. above). However, it is unclear whether it rules out fundamental research that could possibly lead to breakthroughs in a distant future. In any case, the investigation of fully engineered CDR techniques (CCS, BECCS, and DACS) appears vindicated since the objective is to improve existing devices and technologies (CCS), similarly for forestry, enhanced weathering, biochar, USB, SAI, and, partially, MCB. Studies into space CE could qualify as an acceptable technofix if focusing on technologies already in use for launchers, reflectors, or spatial exploration. The status of methods like CCT or OIF is more ambiguous because the core technology or processes could be judged as more speculative. Finally, the evaluative framework (isolated CE methods versus hybrid policies) does not fundamentally alter the assessment.

In sum, it is difficult to disparage CE as offering "just" technofixes, whether the problem is identified is either climate change or global warming. Clearly, potential efficiency varies across techniques. Only a few, most notably fully engineered CDR, seem to live up to all of the criteria. A limitation to such an assessment is that CE ought to be examined within a portfolio framework. Regarding the most radical part of the objection, namely, the rejection of CE based on the contestation of the prominent role played by technology in modern societies, nothing much can be added since such a line of argumentation implies dismissing most of the CE interventions *qua implementations of technological solutions*. It could still be claimed that some methods eschew the criticism due to their low-tech features, for example, forestry. Moreover, the boundaries of the appropriate use of technology could be questioned. For example, does USB constitute an unacceptable call to technology? Does the critique justify refusing any form of technology, which could be considered as absurd and dangerous?

Precautionary Objections

This category regroups all the objections to CE based on the climatic, social, or political risks and uncertainty it would generate. To some extent, all objections rest on *some* loose version of the precautionary principle, which is crudely defined by the UNFCCC in its article 3.3.

> The Parties should take precautionary measures to anticipate, prevent or minimize the causes of climate change and mitigate its adverse effects. *Where there are threats of serious or irreversible damage, lack of full scientific certainty should not be used as a reason for postponing such measures*, taking into account that policies and measures to deal with climate change should be cost-effective so as to ensure global benefits at the lowest possible cost. (emphasis added)

The precautionary principle (or measures) constitutes the cornerstone of international climate politics, along with the prevention of "dangerous anthropogenic interference with the climate system" (UNFCCC, article 2) and the right/obligation to sustainable development (UNFCCC, article 3.4). However, the principle represents less an internally consistent concept than a broad category accommodating various, sometimes contradictory, formulations (Steel 2015; Sunstein 2005). Beyond physical threats, CE is often perceived as being ungovernable (Hulme 2014), thus likely to create harms and damages that could not be curbed or compensated for.

The Accrued Risks/Uncertainty Objection. A key concern is that CE, or some methods, would necessarily, or most likely, produce new risks, uncertainties, or augment those existing in an unacceptable manner (Biermann *et al.* 2021, 2; Hamilton 2013a; Hulme 2014). As formulated by Keith (2000, 77), the objection claims that:

> Geoengineering entails "messing with" a complex, poorly understood system; because we cannot reliably predict results it is unethical to geoengineer. Because we are already perturbing the climate system willy-nilly with consequences that are unpredictable, this argument depends on the notion that intentional manipulation is inherently worse than manipulation that occurs as a side effect.

Indeed, the previous chapters showed that CE methods generate risks and uncertainty of their own. But the probability of adverse events as well as uncertainty, in themselves, do not form a compelling objection. Any climate action or combination of mitigation and adaptation carries its lot of risks and uncertainty. Not doing anything, aggressively mitigating, not mitigating enough, heavily or not adapting, and so on is harmful, often in unpredictable ways. As evoked by Keith, the accrued uncertainty/risks objection uses the deliberate nature of CE for grounding the moral unacceptability of CE. However, as seen in chapter 1, the divide between "accidental Anthropocene" and "intentional climate change" may be less solid, with less far-reaching consequences than usually assumed.

The question of intentions aside, a more robust version of the objection is that CE, or some methods, inflates risks and uncertainty to an impermissible degree in relation to a given benchmark. The benchmark could be constituted by feasible, CE-free, alternatives. The precautionary principle would prohibit from engaging in R&D and/or deployment for those methods. Due to the prospect of a termination shock, SAI, and to a lesser extent MCB, CCT, and space CE seem to fall under the critique (Baum *et al.* 2013; Hamilton 2013a; NRC 2015b, 59–66; Trisos *et al.* 2018), despite some disputed measures proposed for minimizing risks and uncertainties, for example redundant infrastructure (McKinnon 2019; Parker and Irvine 2018). However, if methods like SAI appear to present odds of global catastrophes, it cannot be generalized to CE as a whole. Indeed, in which sense would the climate system become more hazardous or unpredictable in an unacceptable manner due to forestry, biochar, land enhanced weathering, CCS, DACS, or USB, especially compared to a world with no CE, but insufficient abatement?

More importantly, the UNFCCC conception of the precautionary principle does not rule out CE. Literally, it imposes to "anticipate, prevent or minimize the causes of climate change and mitigate its adverse effects," a formulation flexible enough to allow *some* CE. Moreover, it stipulates that "[w]here there are threats of serious or irreversible damage, lack of full scientific certainty should not be used as a reason for postponing such measures," which could be interpreted as an injunction to initiate research on the methods that present fair chances to prevent serious or irreversible damages. Unfortunately, a full-blown analysis would require more elaboration. For the time being, it is enough to note that nothing in what precedes implies that no convincing version of the objection could be ironed out. The conclusion of this brief review is that the objection is more demanding than usually assumed. It is not sufficient to claim that CE will increase risks and uncertainty for rendering it irrelevant as a component of climate policy. It should be shown that the CE methods under consideration come with unacceptable accrued risks and uncertainty, especially when gauged against feasible alternatives.

The Compound Pollution Objection. A variant consists in pointing that CE will pile up another layer of pollution on the top of the GHGs (among other forms of pollution), which would lead to increased risks and uncertainty and/ or be intrinsically bad (which last part falls outside the compound pollution objection). The following quote captures those two reasons.

> You are adding pollution to the existing pollution, creating the same instability. Pollution for a pollution problem; deliberate climate change as a solution to climate change is not just insane, but it is, as Einstein said, repeating the mindset that got you into the crisis in the first place. And what is that mindset? A mechanistic worldview. The idea of mastery and control, of engineering a solution for everything living. (Vandana Shiva quoted by Oksanen [2023, 6])

The criticism relies on the identification of actual pollutants, which is not possible for all CE methods. Many CDR interventions seem immune to the objection, like forestry, CCS, and DACS. BECCS could end up contaminating the soil and groundwater, not because they represent pollutants in themselves, but due to chemical fertilizers used for energy crops. So, compound pollution is contingent on the deployment conditions. As mentioned, SAI is vulnerable to the critique, presupposing that sulfur particles are demonstrated to be harmful and that no safe substitute exists. Space CE could also contribute to pollution through the exhaust gases of the rockets and orbiting debris. Depending on the seeding material, MCB and CCT might fall under the criticism. But SRM surface-based interventions, assuming that crop albedo does not use genetically modified organisms, do not threaten of augmented pollution.

At the end, except as a general analogy, the scope of the objection seems limited. The contaminating character of many CE methods might be difficult, or impossible, to demonstrate. Even the main target of the critics, namely SAI, might not require hazardous particles since substitutes to sulfates may be developed. The pollution over pollution could be read as claiming that "CE is bad because, *qua* pollution, it represents a harmful, hubristic, anthropogenic interference into the climate." Then, the core of the objection lies less in the pollution than in the very fact of meddling with "nature" with an arrogant attitude.

The Lock-In Objection. CE would initiate irresistible forces that would trap humanity into continuous, possibly perpetual, deployment (*CE lock-in*) and/or fossil fuels use (*carbon lock-in*), the latter echoing the risk compensation objection below. Those forces could be fed by sunk costs and escalating commitments (cf. chapter 3), vested interests, institutional and regulatory structures, and so on (Cairns 2014). For the objection to convince, the mechanism(s) causing the lock-in, that is, path dependence, needs to be demonstrated. It should be shown how and why a specific CE method would necessarily or most likely trap humanity. It is not enough to simply assume that CE per se, no matter the method, threatens a lock-in.

The objection is often mobilized against both CDR and SRM, but not in the exact same terms. SRM would be mainly prone to CE lock-in, and a bit less to carbon lock-in. Due to the menace of termination shock, once initiated it would be perilous to halt albedo modification. As seen in chapter 4, not all methods equally carry such a risk. If SAI is particularly exposed as well as, theoretically, MCB, CCT, and space CE are, surface-based techniques that do not display such a risk, or if they do, chances of catastrophic outcomes are low. For instance, the pursuit of USB does not present any serious danger to human health, ecosystems, or the climate per se. On the other hand, because they mask part of the accrued radiative forcing, SRM interventions are at risk of carbon lock-in, that is, continued GHG emissions. However, except if some stable SRM apparatus could be devised, for example, in space, constant deployment will remind humanity of the transient nature of the albedo enhancement, which may constitute a powerful motive for decarbonizing.

CDR is equally exposed to both lock-ins, with variations depending on the method. Yet, carbon lock-in is the main concern (Anderson and Peters 2016). CCS and BECCS fuel strong worries in that respect because they owe their very existence to fossil fuels (cf. chapter 3) (Asayama 2021; Mann 2021, 158; Vergragt *et al.* 2011). To make things worse, the suspicion is widespread that these methods are promoted by corporate actors because they could extend the life of the fossil infrastructure and industry while serving greenwashing purposes (Mann 2021, 153; Schenuit *et al.* 2021). In general, the exposure of a method to the objection depends on its potential for capturing carbon, the more efficient a method is perceived, the higher the lock-in risk. This explains why the CDR methods that are viewed as offering less enduring solutions are less subject to the critique. Forestry represents a specific case because of its apparent naturalness and co-benefits, which lessens lock-in concerns.

The force of the objection rests on factors that could be summed up as follows: First, does the deployment of a technique present a non-negligible risk of lock-in? An affirmative answer is not, however, sufficient for asserting with absolute certainty that there is an issue. Second, how deep is the alleged lock-in? As discussed in chapter 3, the problem lies less in the existence than in the intensity of a given lock-in (Shackley and Thompson 2012). Everything else being equal, easily escapable lock-ins are not contentious. That is the reason why mechanisms could be designed for preserving some flexibility during R&D and deployment, for example, by keeping open alternative technologies and policy options (Cairns 2014, 651). Moreover, regular evaluations of technological choices could be devised. Third, how objectionable is the alleged lock-in? Presumably, there are more and less perilous lock-ins. Maintaining forestry interventions is not morally equivalent than keeping injecting sulfur in the stratosphere.

The Slippery Slope Objection. This objection was already introduced in the section on CE research above. It can be assimilated to or interpreted as a type of lock-in (Burger and Gundlach 2018, 279; Cairns 2014, 652), namely debuting CE R&D or deployment would impose pursuing it despite evidence of inefficiency and/or dangerousness. Four variants can be distinguished:

V_1: Accepting CE research will force accepting deployment (The formulation seen above).

V_2: Accepting research on (or deployment of) harmless methods (e.g., forestry or USB) will force accepting research on (or deployment of) more harmful forms (e.g., SAI).

V_3: Embarking upon *temporary* deployment will force accepting *indefinite* deployment, like with SRM methods subject to termination shock or with CCS (Hulme 2014, 69).

V_4: Accepting CE will force accepting any instance of climate manipulation. As formulated by Keith (2000, 277), "[i]f we choose geoengineering solutions to counter anthropogenic climate change, we open the door to future efforts to systematically alter the global environment to suit humans."

The soundness of all variants rests on the same core mechanism. For a slope to be slippery, compelling reasons or factors must push from the initial, acceptable, stage to the final, unacceptable, one. The existence and compelling nature of the causation should be demonstrated. V_2 is therefore weak. It is unclear how consenting to forestry or DACS commits anyone to agree to SAI or space CE. V_3 utters the CE lock-in concerns as seen above, so there is no need to elaborate further. V_4 paints a dreadful picture: humans triggering all sorts of weather and climate modification simply because they can. To constitute a serious objection, it should be shown that such drifting is the direct consequence of CE R&D and deployment. More generally, as indicated during the discussion on research and deployment, checkpoints or safeguards could be set up for not sliding down the slope. Ultimately, the salience of the objection depends on how CE is regulated. This represents an institutional issue that is partly distinct from the methods themselves.

The Risk Compensation Objection. Also branded as "moral hazard" (Shepherd *et al.* 2009, 37), "mitigation obstruction" (Morrow 2014c), or "mitigation deterrence" (McLaren *et al.* 2019),[17] the objection can be interpreted in various, sometimes weak, manners (Hale 2012). Leaving aside those conceptual debates, risk compensation describes situations where "people respond to risk-reducing innovations by behaving less cautiously" (Morrow 2014c, 2). Applied to CE, the criticism is that by initiating R&D or deployment, climate change-related threats will be regarded as less serious, thus eroding commitments to abatement (Raimi *et al.* 2019). In the end, the "dangerous anthropogenic interference with the climate system" will be aggravated. The European Science Academies Advisory Board expresses that worry in relation to CDR methods and its impact on risk perception among political decision-makers (EASAC 2018, iv).

The objection rests on two claims. The first, empirical, argues that the knowledge of the possibilities offered by CE would cause a drastic reduction in the efforts to cut down GHGs. A dilemma then emerges for the policymakers: properly informing the public about CE at the risk of mitigation deterrence or not fully informing citizens about CE perspectives under the threat of violating some procedural requirements (e.g., transparency). The solidity of the empirical claim is, however, challenged (Merk *et al.* 2016; Reynolds 2015).[18] The ultimate effect could be the opposite: the rising prospect of CE could incentivize individuals and institutions to intensify their mitigation efforts (Fabre and Wagner 2020). The second, normative, claim posits that a situation with lower, i.e. less than optimal, mitigation but with CE will be *necessarily* inferior in some morally meaningful way to a situation with higher mitigation, independently of the other policy components. To be convincing, this last claim ought to be supported by strong reasons (Halstead 2018; Morrow 2014c), which is not the case yet.

Capture Objections

Those objections share the concern that CE, or some methods, could be seized by unlawful or illegitimate interests, possibly right from the research stage. Such illegitimate interests could be private (firms, individuals), military, technocratic, or stemming from rogue actors.[19] Captures are problematic for two kinds of reasons. The first, consequentialist, stipulates that any capture carries non-negligible chances of negative outcomes. It significantly increases the odds of CE R&D or deployment being inefficient, supporting particular interests at the expense of the common good (e.g., to "prevent dangerous anthropogenic interference with the climate system"), or intentionally harmful. A second kind of reasons is procedural and constitutes the core of the critique. Chances are nontrivial that CE would be seized by illegitimate interests, rendering it mostly unresponsive to legitimate ones, for example, of the affected parties, and democratic control. Captures would thus be wrong in themselves, independently of the consequences, beneficial or not, for the affected parties. Hence, even in the case of a benevolent capturer, who would steer CE for the good of humankind, CE would remain objectionable due to the vulnerability to capture.

The Private Interests Objection. It posits that probabilities are high that CE would be at least lobbied by powerful private actors for serving their own agenda (Blackstock 2012; Frumhoff and Stephens 2018, 4; Hamilton 2013a, 72–106; Szerszynski *et al.* 2013, 2813–2814). Lobbying may lead to the privatization of CE, for example, through instrumentalizing property rights (Burger and Gundlach 2018, 289) or a "patent land-grab" (Chavez 2015), or to plain corruption (Transparency International 2022). Such a prospect is a major driver behind the public opposition to the corporate involvement into SRM research (Pidgeon and Spence 2017, 5; Sugiyama *et al.* 2020, 7–8). But what are the odds of CE methods being seized by private interests? As a matter of fact, those interests already play a key role in forestry, CCS, BECCS, and DACS. The development of carbon (credits) markets and taxes strengthens the perspective of CDR profitability, which will incentivize private actors to ramp up their investments. Would it be forcibly bad? The objection cannot be that *any* private involvement represents an unacceptable capture. That would be an abusive generalization. Sectors like health care, infrastructure, and transportation are filled with examples of involvement of private investors that does not crowd out public institutions or run against the common good. The objection requires drawing the frontier between desirable and undesirable private involvements, which raises the broader issue of how to articulate private and public initiatives. Nonetheless, the claim could be advanced that any increased private involvement comes with such detrimental consequences as fake reporting or carbon credits trafficking. However, in principle, any illegal conduct could be tackled by regulation and control, especially through independent MRV. More fundamentally, objecting to an activity on the grounds that it could be used to commit illegal actions is convincing only insofar that the odds of such unlawful prospects as a direct reason for conducting the incriminated activity are shown to be significant.

In the matter of SRM, the probabilities of capture are limited by the absence of a market for albedo enhancement, that is, the lack of financial incentives. To some extent, it represents an instance of public goods (cf. textbox 5.1) in which inhabitants of a city or region will benefit from the surface or tropospheric-based interventions under the form of reduced thermic pressure, with the possibility of not contributing to SRM provision, which deters private investments. Thus, it constitutes a textbook case for public provision, municipal in the case of USB. The situation is similar for space CE, at the notable difference that it is heavily constrained by international law. Space is a global common, activities up there can only be carried for the interest of humankind and no country can own celestial objects. However, equipment shipped to space belongs to the sender, so private actors might place in orbit reflectors, sails, or scatterers in a configuration that favors their interests. The main worry is about SAI. Despite the absence of a market for aerosol injection, the demand already exists and will likely grow. Private or public entities, for example, governments of Pacific low-lying islands, may become tempted to finance SAI. In that case, and assuming that no part of the world collapses into anarchy, states will be able to regulate or prohibit CE interventions. Nonetheless, the danger lingers of a state which, by imprudence or complicity, allows private actors to deploy without proper oversight. Then, the capture will result, not from private initiative per se, but from defective state supervision.

The objection could be understood as prompted by mistrust, or outright hostility, toward the private sector. Leaving aside arguments that oppose free enterprise as a matter of principle, it is still possible to agree that corporations are mostly driven by profits, and not the public good (concept that should be precisely defined). However, any private involvement into CE will not necessarily lead to a capture or undermine climate policy. Furthermore, companies have contributed to solving global environmental issues in the past, as with *The Montreal Protocol on Substances That Deplete the Ozone Layer* (1987) (Glaberson 1988). While the unfolding crisis raises legitimate suspicion about the genuine motivation of the fossil-fuel industry, it cannot be generalized to all sectors. Thus, the objection should be interpreted less as a rebuttal of CE per se than an important reminder of the necessity of regulation. Instruments for that purpose already exist, like laws, open patents, regulatory bodies, independent control, licencing, and so on (Grasso 2022).

The Technocratic Objection. Technocrats represent another potential source of capture (Hamilton 2014, 20; Szerszynski *et al.* 2013, 2812; Stephens and Surprise 2020). Academic experts or public officials could rob citizens of their roles in the debate about CE permissibility, escape democratic oversight, and bend CE in favor of some particular interests. The mistrust faced by public institutions and experts fuels the critique: the less trustworthy the institutions/experts, the more they are suspect of manipulating CE. However, the legitimate, democratic, regulation of climate alteration by officials and experts should be separated from instances of illegitimate capture. The frontier should be based on some conception of the "common good," which would require to define how it is understood. Nonetheless, as mentioned above, even CE programs led with a solid and widely agreed conception of the common good (e.g., reducing risks uniformly and offering acceptable compensation to the losers) remain opposable if they do not respect procedural constraints of consultation and participation. Methods that mobilize complex techniques or necessitate heavy-handed regulation may be more prone to technocratic capture (e.g., space CE or DACS). As for the private capture objection, the concern is then less about CE per se than how it is managed, how strictly experts and public officials are monitored, and so on. Ultimately, the probability of capture is negatively correlated to the robustness of the system of checks and balances in force. Well-functioning democratic societies rooted on solid public accountability are probably less vulnerable than authoritarian regimes in which officials are shielded from public scrutiny and in which experts are subject to political leaders.

The Militarization Objection. CE would carry notable risks of military usage (Frumhoff and Stephens 2018, 4; Hulme 2014, 25; Kellogg and Schneider 1974; Schneider 1996; Stephens and Surprise 2020, 4). As seen above, CE interventions may cause counter-CE because of defective cooperation, possibly escalating to retaliation and geostrategic instability. The militarization objection advances one step further by positing the nontrivial possibility of CE being employed as a weapon. Previous attempts at altering weather conditions for military purposes seem to vindicate the concern (e.g., the Popeye Operation in Vietnam). However, the number of techniques vulnerable to the objection is limited. It is difficult to see how most of CDR methods could be weaponized. How could forestry, biochar, algaculture, CCS, BECCS, or DACS inflict damages on an adversary? Maybe OIF or enhanced weathering may serve to pollute an enemy's territorial waters, but that will not result from using CE as a *carbon capture and sequestration* device. On the other hand, SRM interventions appear, at first sight, more susceptible to weaponization. The vision of malevolent governments spraying sulfur, seeding stratocumulus, or thinning out cirrus for causing harmful temperatures or precipitation fluctuations strikes the mind. However, as chapter 4 shows, tropospheric and stratospheric methods cannot guarantee that the outcomes will be circumscribed enough for transforming SRM into efficient weapons (Halstead 2018, 69). The impacts will be difficult or impossible to contain within a region. Chances will be high of inflicting damages to allies, neutral states, or even the deployer itself. It does not imply that SRM interventions could not become precise enough in the future for entering into the military arsenal, but the prospect remains hypothetical. The same goes for space-based CE.

The Rogue Deployment Objection. As per the free-driver effect, it is usually assumed that CE, or some methods, could be deployed by an agent or a small coalition without the assent of other actors or without prior consultation (Fleming 2010, 185–186; Keith *et al.* 2010; NRC 2015b, 152; Smith 2020; UNEP 2023, 5, 19; Victor *et al.* 2009, 71–72), which is a public concern (Bellamy *et al.* 2017, 198; Pidgeon *et al.* 2013, 454). To some extent, unilateral implementation could count as a positive feature since it gives the opportunity for bypassing defective international cooperation in the provision of global public goods such as reduced GHGs' concentrations. Such a positive feature is reinforced in the case of CDR methods that do not carry such transboundary negative externalities as pollution (which excludes OIF and enhanced ocean weathering). However, whether a unilateral deployment could be beneficial, everything considered, does not address the core of the rogue deployment objection, which is procedural. What makes the free-driver effect so alarming is the threat of unilateral, or minilateral, intervention carried by public or private actors without the consent of legitimate, domestic, and international institutions, and more generally of the potentially affected parties. The perspective is especially feared in the case of solar CE, even more so for SAI.

Rogue deployments have already happened, for example, the fertilization of small ocean patches like in British Columbia (cf. textbox 3.2) (Tollefson 2012) or the release of sulfur particles in Mexico (Osaka 2023). The public opposition to unilateral actions, however, lessens the odds of rogue initiatives (Michaelson 2013, 105–106), although the prospect remains credible. For instance, stratospheric or space-based CE would only require a single motivated agent with adequate capabilities and little concern for domestic, international law, or procedural rules for engaging in rogue SRM. But the overall menace appears to be overstated (Baum *et al.* 2022, 13; Halstead 2018). Even for SAI, as seen for counter-CE, the success of any unilateral deployment will depend on cooperation, or at least tacit acceptance, of various countries (Horton 2013). In other words, the incentive structure on the international scene promotes the logic of multilateralism that, in turn, impedes rogue deployment.

Political Objections

This category gathers objections that point at the ill-fated political implications of CE. Those objections relate to how societies are organized, how the interactions between their members are regulated, as well as how goods are distributed. Concretely, the claim is that CE R&D and/or deployment will create/accentuate inequalities/vulnerabilities, create/accentuate relations of domination, or will undermine democracy.

The Inequalities/Vulnerabilities Objection. The disparate impacts produced by CE, or some methods, would worsen existing inequalities and/or vulnerabilities or generate new ones to some impermissible extent (Biermann *et al.* 2021, 3; Carr and Preston 2017; McLaren 2018, 2021; Preston 2012; Stephens and Surprise 2020; Suarez and van Aalst 2017). Inequalities refer to the unbalanced distribution of some goods at the domestic or global level. In that respect, CE, or some methods, will disproportionally affect some groups in an objectionable manner, especially those who are already vulnerable. Vulnerabilities describe the lack of capabilities to face risks. Without a shadow of a doubt, some CE methods will create or accentuate inequalities and/or vulnerabilities. At first sight, CDR's core impact, the reduction of GHG concentrations, will be evenly spread throughout the planet. Moreover, biochar, CCS, or DACS do not seem to directly cause or aggravate unbalances, or at least not in an unacceptable manner. But it cannot be concluded that CDR as a category does not present challenges. The odds are significant that large-scale forestry and BECCS will put a tremendous pressure on resources like land and water, potentially impairing food security. On the other hand, SRM interventions like SAI, and most likely MCB, CCT, and space CE, will generate disparate impacts. Some may particularly burden specific regions, for example, by drying out lands or increasing the frequency of floods or storms. However, USB seems shielded against the criticism.

The objection cannot be that CE is unacceptable due to the inequalities and vulnerabilities it creates or intensifies per se. Any climate policy, whether based on aggressive, moderate, or no mitigation, creates or intensifies inequalities and vulnerabilities. Instead, the argument is that the accrued inequalities and vulnerabilities are morally objectionable, perhaps because, for instance, they overburden already disadvantaged groups (Gardiner 2011, 119). Yet, it is possible to disagree about the overall projected impact of specific methods on populations. For example, Horton and Keith (2016) claim that the "global poor" as a category will benefit from solar CE. However, the objection is not addressed by waving overall projected benefits. Overall positive outcomes do not preclude extreme variations in the distribution of benefits and harms in which some groups end up in precarious situations, worse off than under any feasible alternatives. Deflecting the objection may require demonstrating that the distribution of disproportionate impacts remain within acceptable limits, but, more importantly, it necessitates showing that inequalities/vulnerabilities could be adequately "taken care of" by appropriate institutions through redress, compensation, or redistribution.

The magnitude of the disparate impacts and the possibility to moderate them are essential. Often CE methods are rejected precisely because of the non-negligible odds of excessive unbalanced impacts that appear difficult, or impossible, to be tackled afterward. Under the current international circumstances characterized by defective cooperation, the transboundary disparate effects of methods like enhanced ocean weathering, OIF, tropospheric, stratospheric, and space-based SRM render more complex *ex post* (after deployment) compensation. A deeper trouble is that, depending on the method, it may be arduous, or even impossible, to impute climate harms and catastrophes to CE (Hulme 2014, 101–104; MacMartin *et al.* 2019; Robock 2020), undermining liability attribution (Burger and Gundlach 2018, 290). Nonetheless, global insurance mechanisms could allow setting aside issues of attribution and focus on compensation, although they raise informational (e.g., moral hazard) and cooperation issues (e.g., establishment of an international insurance scheme). In sum, CE methods with significant transboundary effects might be impossible to govern in an acceptable manner, that is, while cushioning enough the inequalities and vulnerabilities it creates.

The Domination Objection. This objection stipulates that CE should be opposed because it is a technology that is dominating in itself or could serve dominating purposes (e.g., through the capture by specific interests, as exposed above). According to a commonly accepted definition within political theory, domination characterizes relations in which an agent has the capacity to arbitrarily interfere with another agent's choices (Pettit 1997). In the case of CE, the argument is often advanced that CE, or some methods, from the research stage, reinforces the subjection of the Global South to the Global North (Bierman and Möller 2019; Frumhoff and Stephens 2018; Preston and Carr 2018, 314–318; Smith 2018; Winickoff *et al.* 2015). More generally, through CE R&D, industrialized nations are building capacity to arbitrarily interfere within the life of vulnerable groups, without their consent, for example, through unilateral or minilateral deployments. This could then degenerate into "predatory" or "parochial" CE (Gardiner 2013a).

If it may be difficult to determine how CDR could be intrinsically dominating, possibilities exist that CDR implementation ends up in arbitrary interference, for example, by expropriating farmers of their lands for reforestation or energy crop cultivation, or forcing communities to accept massive, potentially dangerous, carbon sequestration sites nearby. SRM is perhaps more at risk of being intrinsically dominating or becoming a vehicle for arbitrary interference, as expressed by the capture objections. Nonetheless, it could be underlined that CE is not inherently dominating. It depends on the material and institutional conditions under which CE is conducted, which justifies designing CE, right from the research stage, in such a way that reduces domination (Morrow 2020). In particular, "power accountability" or "power equalization" tools could be devised (Smith 2018). The former consists of creating or reforming oversight institutions (e.g., the UNFCCC) for giving a voice to vulnerable countries and groups. The latter is to support vulnerable countries or groups in strengthening their CE capacities, so they can initiate their own projects, or threaten industrialized countries with counter-CE if their interests are neglected. The former is preferable to the latter for two reasons. On the one hand, it does not rely on CE capabilities proliferation. On the other hand, it is focused on building centralized decision-making structures and international coordination that could prove useful for tackling the Anthropocene. In sum, less dominating manners may exist for implementing CE R&D as well as deployment, but they require a modicum of international cooperation.

The Democracy Objection. The objection goes beyond claiming that CE, or some methods, represents a conundrum for regulation and governance, to assert that climate alteration threatens democracy. The concept of democracy itself is open to variations. It can be understood as participative, representative, liberal, direct, and so on. To some extent, all conceptions share the fundamental trait of the political power being theoretically exercised by the people, most often through elected representatives. Democratic regimes come in diverse flavors. The dominant conception is the *liberal* one. The democratic, popular, rule is completed by liberal features that protect individuals from abuses of power from the majority and the state. In the liberal democratic context, the objection could be understood as claiming that CE, or some methods, would threaten either popular sovereignty or individual rights and freedoms. Some of the objections already evoked above represent potential infringements to the liberal democratic order. For instance, the capture of CE by private, technocratic, or military interests means that the people have lost control over the R&D or/and implementation of climate alteration. Another example is constituted by the inequalities and vulnerabilities objection. CE, or some methods, may create imbalances that could be difficult or impossible to redress or compensate for, undermining the liberal component.

Yet, the democratic objection could be understood as claiming that the consequences of (some) CE methods would strain liberal democratic institutions so much that they may be in danger of collapsing. A substantial review of the objection would require longer developments than the passing mention offered here. For the time being, it is enough to point that the objection could be answered by demonstrating that institutions possess the resources for responding to those challenges in a rather efficient and morally acceptable manner. Another interpretation is to consider that CE, or some methods, are by their very nature incompatible with democracy. For example, the claim could be that CE, or some methods, embody authoritarian tendencies that will translate into undermining liberal democracies. If the critique is difficult to sustain against (most of) CDR methods, it may be advanced against SRM techniques like SAI for various reasons (Szerszynski *et al.* 2013). For example, SAI's global scope may call for autocratic rule or exacerbate anti-democratic tendencies. The uncertain and disparate effects coupled with attribution limits could strain institutions based on fair redistribution of resources, opportunities, and welfare. SAI might be prone to capture. Those reasons for considering CE as carrying a mortal risk for democracies have been challenged (Callies 2019b; Horton *et al.* 2018). Without elaborating further, the core of the counter-objection is to underscore that technologies do not impose a particular political regime (e.g., authoritarianism) while insisting that existing domestic and international institutions may possess the resources and capacities for tackling the challenges posed by SAI/SRM.

Moral Corruption

At the end of the day, giving in to CE, independently of the benefits and risks, would represent a deep moral transgression. The objection can take several forms. As seen in chapter 2, CE could be judged intrinsically bad. By initiating CE, humans would display excessive arrogance (Hamilton 2013a; Thiele 2019), fall prey of the hubristic belief that they can control the climate, or "the global thermostat" (Hulme 2014), akin to some divine entity (Hartmann 2017). It is not necessary to develop that objection further. However, CE may not only affect one's moral character. The implications could be more far-reaching. Moral corruption may also result from burdening other people with the consequences of one's choices. Through climate change and manipulation, the current generation is in the position, maybe unique in History, of significantly deteriorating the living conditions and drastically corseting the choices of the future generations (Gardiner 2006, 2011, 45–48). The corruption would then spread at two levels. First, current generations, especially affluent ones, by postponing necessary mitigation, would transfer to future generations an issue that has worsened. Second, they would engage in moral rationalization, that is, reframing one's unethical actions into ethical ones. They would pretend that delaying abatement is the right thing to do, or that it is morally laudable/acceptable. Rationalization takes the form of either assuming that the future generations will be better equipped than the present one to tackle climate change (e.g., by attributing high values to the social time discount rate [STDR], cf. chapter

2) or suggesting that the Anthropocene is less severe than projected (e.g., by the IPCC). Those assumptions act as self-serving excuses for defending that current emissions are, at the end of the day, "okay." CE would support rationalization by providing additional justifications for continuing to emit GHGs, and presumably in much larger quantities than deemed safe ("CDR will permit reaching net-zero emissions" and "SRM will offer a bridging technology for ramping up carbon abatement"). As a result, future generations will most likely face soul-crushing dilemmas, for example, enduring horrendous climate conditions or taking the chance of a termination shock by conducting SAI or space CE. Under such circumstances, the existence of the future generations could be irrevocably "marred" by the pressure to deploy, that is, definitely tainted by the lingering feeling to have committed maybe a lesser, but an evil gesture anyway (Gardiner 2010, 300–302). To repeat, not all techniques present a Faustian bargain, but all emerge as solutions in a context of past and present generations self-indulging in perpetuating actions whose costs will be passed onto the next generations.

NOTES

1. https://www.overshootcommission.org

2. Indoor SRM research investigations have involved theoretical analysis, social science research, computer simulations using climate and Earth System models and laboratory experimentation. Small-scale outdoor experimentation might emit limited quantities of material over a limited time to examine critical and poorly understood SRM-related processes in the real atmosphere with negligible climatic impact. Operational SRM deployments would likely be of planetary scale and need to last for decades or more to be effective. It should, therefore, be possible to define a level beyond which an SRM experiment would no longer be small-scale. (United Nations Environment Programme 2023, 5)

3. Mission-driven projects do not necessarily, or even likely, commit to deployment. For instance, Morrow (2020a, 629, 632) makes a further distinction inside the category between information and deployment-oriented programs. It could, however, be objected that any research project centered on tackling practical scientific and engineering issues related to a given method, even if aimed only at creating knowledge, will likely create incentives and vested interests to embark upon deployment. The objection, which has the form of a slippery slope (cf. below), is an empirical claim that needs to be demonstrated for the method under discussion.

4. The invocation of scientific freedom can have tangible consequences regarding which activities are deemed to be governed/regulated and how they should be. For instance, the argument is used by the panel of researchers put together by the UNEP (2023, 25) for opposing any stringent regulation for indoor experimentation/research.

5. The initiative *Solar Geoengineering Non-Use Agreement* is representative of the rejection of the method that extends to research dimensions. https://www.solar-geoeng.org

6. Several classifications of CE research stages are possible (e.g., Burger and Gundlach 2018, 291). This book employs the following sequence: computer modeling, laboratory/indoor experiments, small-scale outdoor experiments, large-scale outdoor experiments, and then deployment, which departs from the research category.

7. 'Exit ramps' (. . .) should be an explicit part of the program, with mechanisms to terminate a research activity, for example, if it is deemed to pose unacceptable physical, social, geopolitical, or environmental risks or if research indicates clearly that a particular SG technique is not likely to work. (National Academies of Sciences, Engineering, and Medicine 2021, 9)

8. For exit ramps to play their role, there is a need to clarify the conditions under which the slippery slope could be activated, and to devise mechanisms that protect the decision-makers involved in the ramps. Agents involved in the research process need to be shielded against cognitive biases, heuristics, or fallacies such as sunk costs or escalating commitments, vested interests, lobbying, as well as the various forms of pressure coming from private companies, research institutions, or the general public (e.g., in case of perceived emergency).

9. This claim is contestable (Hulme 2014, 19; Schneider 1996, 295).

10. Adhering to the view that some social problems could be fixed by technology does not commit to rule out mitigation and adaptation. The point made in the paragraph above is simply that A_1 could be interpreted as positing that CE, on its own and without the support of mitigation/adaptation, constitutes a viable fix to the Anthropocene.

11. All the arguments discussed in this section assume that:

1. Anthropogenic climate change is real.
2. If unabated, it will cause global catastrophes.
3. Thus, to avoid all or some of those catastrophes, GHG emissions and concentrations need to be cut.

Rejecting one of the premises or the whole argument equates adhering to some version of climate skepticism, that is, the view that climate change is either a hoax, or not as dangerous as claimed by scientific studies. Incidentally, adhering to climate skepticism invalidates the main reason to mobilize CE at the first place: countering the impact of climate change or some of its effects. Someone could endorse climate skepticism and still be willing to engage into CE. But then it will be for other purposes than tackling the Anthropocene and its consequences.

12. The false dilemma fallacy "reflects incorrect thinking because it presents a problem or issue as having only two possible solutions when in fact there are more" (Culver 2019, 346).

13. While the first interpretation embodies a maximin rule (favoring the option that maximizes the worst outcome), the second is compatible with the maximization of net benefits (cf. chapter 2).

14. Pareto efficiency stipulates that "a situation is efficient if no change is possible that will help some people without harming others" (Frank *et al.* 2022, 132).

15. The latter formulation echoes the risk compensation objection (cf. below).

16. The menace of unilateral deployment is contested (Horton 2013, Michaelson 2013).

17. Those denominations are often viewed as not being equivalent (Lin 2013).

18. Mitigation deterrence has also been raised against adaptation measures. And, as for CE, the empirical evidence remains ambiguous, at best (Carrico *et al.* 2015).

19. Capture objections could be conceived as one of the three categories under "regulatory drift" (Wolff 2019, 12). The two other categories are "regulatory complacency" when regulators become less vigilant and more lenient because regulation did not show initial problem and "regulatory gifting" when regulators transfer regulation to the CE industry.

Conclusion

A central ambition of this book has been to contribute to a more elaborated perception of the *anthropogenic interventions in the climate system that predominantly aim at altering radiative forcing locally or globally to slow down, stop, or reverse climate change and/or related adverse events*. For doing so, the circumstances formed by the persisting human influence—the Anthropocene—needed to be presented (introduction) before examining what is meant by climate engineering (CE; chapter 1). Once clarified the context and nature of CE, the next crucial step for supporting informed public debates was to identify evaluative (feasibility, permissibility, and preferability) criteria for assessing CE (chapter 2). Those criteria should, however, not be applied to climate alteration in general, but to particular techniques. Thus, chapters 3 and 4 delved into the details of carbon dioxide removal (CDR; chapter 3) and solar radiation management (SRM; chapter 4). Finally, the book closes on a critical review of the arguments that could be used for supporting or opposing CE, from research to deployment.

Independently of their academic value, those chapters highlight the requirements when it comes to form a judgment about whether to embark upon a specific climate alteration path. Because, no matter one's opinion on the feasibility and permissibility of the various CE methods, the debate about whether to research, develop, and/or deploy them is going to happen. The statement is both empirical and normative. Empirical since discussions on most of the methods *will likely* take place, and they already do. For example, BECCS and forestry are embodied into climate models, to the point that Paris targets seem unreachable without them, and voices demand transparency on their role. Normative because discussions *should* happen as a matter of democratic governance.

The claim that the debate is going to take place says nothing about who should be involved and how. Exchanges are already occurring within academic circles through articles, reports, books, and conferences, but they stay too confidential. On the contrary, they need to open up even more to researchers from the Global South as well as scholars from social sciences and humanities: lawyers, economists, political scientists, sociologists, and philosophers. Furthermore, discussions cannot remain confined to academic circles. They need to embrace a broad audience: policymakers, advocacy groups, and citizens. For sure, the technicity of the topic may deter participation, but this is the reason for multiplying works that simplify the material for a lay audience. The stakes are too high, and it could be extremely damaging for science, the climate and democracy that a minority monopolizes verdicts about what/how to research, develop, and/or deploy, without public participation and contributions from a large spectrum of experts in various disciplines.

Calls for participation should, however, not stay incantatory. An institutional infrastructure ought to support and circumscribe public engagement, which taps into the fields of governance and public policy. Furthermore, not all kinds of participation are desirable. Besides capture by specific interests, another pitfall for debates is to be swamped with caricatural depictions of CE methods as well as of the arguments mobilized for and against CE. Exchanges have to be informed, which is not trivial in an age of populist contestation of science and expertise. For instance, conspiracy views about chemtrails may derail assessments of SAI or CCT. More generally, the *equivalence claim*, which bundles all CE methods together for dismissing or endorsing them, needs to be dropped. If this book is premised on the pertinence of assembling SRM and CDR under a common label (CE), it also treads carefully among the differences between USB, SAI, MCB, CCT, BECCS, forestry, DACS, or CCS. As mentioned in the previous pages, CE, CDR, or SRM is often praised or rejected *en bloc* without nuance or attention to the details. The case would not be that preoccupying if humanity was not facing a formidable threat of catastrophic global risks that will endure for millennia. The magnitude of the task should be reckoned with, which requires reviewing the full range of policy tools, including carbon capture and albedo modification techniques. Such a reckoning is even more imperative because not only the existing generations, but the future ones, will shoulder the costs of decisions made nowadays.

The Anthropocene may be seen as commanding the adoption of a pluralistic approach. Pluralism of the policy instruments first. Without a priori committing to deploying any CE method, a broad array of tools should nonetheless be appraised. Then pluralism of the objectives. Climate policy goals ought to be diverse. Although the ultimate objective remains to cut GHG emissions as fast and as much as possible, abatement does not exhaust the content of feasible and desirable strategies. There is the need to adapt as well as to tackle the accumulation of GHGs in the atmosphere. Then, beyond temperatures and precipitation, other dimensions should be considered like ocean circulation, cryosphere, disasters, poverty alleviation, rights, inequalities/vulnerabilities, and so on. The true engineering or management aspect of climate policy may reside not so much in the hubristic attempts at tampering with the climate, but in tailoring hybrid interventions across various dimensions.

In that sense, the exceptionalism claim—CE represents a rupture within the history of humanity and, therefore, should be opposed for this precise reason—is damaging in several ways. First, it maintains a general state of denial vis-à-vis the role that human beings have already been playing as a climate force for millennia. Second, as a result, it trivializes the Anthropocene and exaggerates the contrast with CE. Third, such a contrast nurtures the misleading view that humanity would face a dilemma: either an arrogant full-blown experimentation with the Earth or a modest mitigation-only policy. Such a representation is damaging in the sense that it justifies disregarding the intermediary steps that could be undertaken in terms of hybrid policies, targeted interventions, and so forth. The point is not to deny that solid reasons may exist for building a morally relevant distinction between the Anthropocene and CE (e.g., based on intentions, although the case is less straightforward than it seems), but to warn against exaggerating the differences between the Anthropocene and CE. To some extent, CE results from the Anthropocene. Whether one endorses or opposes climate alteration as a general principle, it is difficult not to see it as the crowning moment of the anthropogenic interference within the Earth's system.

In fine, as Emile de Girardin, a nineteenth-century French journalist and politician, wrote: "to govern is to plan ahead; and not doing so, it is to run to ruin." Climate policy is notorious for its insufficient, somewhat defective, governance, a key reason being the lack of foresight from governments. This insufficient anticipation, coupled with culpable inertia, has been observed at work in mitigation since, at least, 1992 Rio Conference. Adaptation is also suffering from inadequate preparation worldwide. It would be a tragedy if, because of such a lack of foresight, humanity is confronted with circumstances in which climate alteration would be necessary for "prevent[ing] dangerous anthropogenic interference with the climate system" (United Nations Framework Convention on Climate Change [UNFCCC]), but in which the required methods would not be mature enough. A perhaps even worse situation could be that those methods are ready, but without any substantial governance and regulatory mechanisms apparatus that could help to moderate the issues of justice, risks/uncertainty, and many more that will emerge. Lessening such a risk among others, through knowledge diffusion, has been one of the tasks this book has been pursuing.

Bibliography

Abatayoa, Anna Lou, Valentina Bosetti, Marco Casarid, Riccardo Ghidonif, and Massimo Tavonic. 2020. "Solar Geoengineering May Lead to Excessive Cooling and High Strategic Uncertainty." *Proceedings of the National Academy of Sciences of the United States of America* 117, no. 24: 13393–13398. https://doi.org/10.1073/pnas.1916637117.

Adam, David. 2021. "How Far Will Global Population Rise? Researchers Can't Agree." *Nature* 597: 462–465. https://doi.org/10.1038/d41586-021-02522-6.

Akbari, Hashem, Surabi Menon, and Arthur Rosenfeld. 2009. "Global Cooling: Increasing World-Wide Urban Albedos to Offset CO2." *Climatic Change* 94, no. 3–4: 275–286. https://doi.org/10.1007/s10584-008-9515-9.

Alcalde, Juan, Stephanie Flude, Mark Wilkinson, Gareth Johnson, Katriona Edlmann, Clare E. Bond, Vivian Scott, Stuart M. V. Gilfillan, Xènia Ogaya, and R. Stuart Haszeldine. 2018. "Estimating Geological CO2 Storage Security to Deliver on Climate Mitigation." *Nature Communications* 9: 2201. https://doi.org/10.1038/s41467-018-04423-1.

Alexander, Larry, and Michael Moore. 2021. "Deontological Ethics." In *The Stanford Encyclopedia of Philosophy*, edited by Edward N. Zalta, Winter. https://plato.stanford.edu/archives/win2021/entries/ethics-deontological/.

Allen, Myles. 2019. "Why Protesters Should Be Wary of '12 Years to Climate Breakdown' Rhetoric." *The Conversation*, April 18, 2019. https://theconversation.com/why-protesters-should-be-wary-of-12-years-to-climate-breakdown-rhetoric-115489.

Anderson, Kevin, and Glen Peters. 2016. "The Trouble with Negative Emissions." *Science* 354, no. 6309: 182–183. https://doi.org/10.1126/science.aah4567.

Anderson, Ray G., Josep G. Canadell, James T. Randerson, Robert B. Jackson, Bruce A. Hungate, Dennis D. Baldocchi, George A. Ban-Weiss, Gordon B. Bonan, Ken Caldeira, Long Cao, Noah S. Diffenbaugh, Kevin R. Gurney, Lara M. Kueppers, Beverly E. Law, Sebastiaan Luyssaert, and Thomas L. O'Halloran. 2011. "Biophysical Considerations in Forestry for Climate Protection." *Frontiers in Ecology and the Environment* 9, no. 3: 174–182. https://doi.org/10.1890/090179.

Andow, James. 2023. "Slippery Slope Arguments as Precautionary Arguments: A New Way of Understanding the Concern about Geoengineering Research." *Environmental Values.* https://doi.org/10.3197/096327123X16702350862737.

Angel, Roger. 2006. "Feasibility of Cooling the Earth with a Cloud of Small Spacecraft Near the Inner Lagrange Point (L1)." *Proceedings of the National Academy of Sciences of the United States of America* 103, no. 46: 17184–17189. https://doi.org/10.1073/pnas.0608163103.

Ansell, Christopher, and Jacob Torfing. 2022. "Introduction to the Handbook on Theories of Governance." In *Handbook on Theories of Governance*, edited by Christopher Ansell and Jacob Torfing, 3–16. Cheltenham and Northampton, MA: Edward Elgar Publishing.

Apple. 2022. "Apple Reports Fourth Quarter Results." October 27, 2022. Last accessed, January 19. https://www.apple.com/lv/newsroom/2022/10/apple-reports -fourth-quarter-results/.

Arrhenius, Svante. 1896. "On the Influence of Carbonic Acid in the Air Upon the Temperature of the Ground." *Philosophical Magazine and Journal of Science* 5, no. 41: 237–276. https://doi.org/10.1080/14786449608620846.

Arrow, Kenneth. 2012. *Social Choices and Individual Values.* 3rd ed. New Haven: Yale University Press.

Arrow, Kenneth, Maureen Cropper, Christian Gollier, Ben Groom, Geoffrey Heal, Richard Newell, William Nordhaus, Robert Pindyck, William Pizer, Paul Portney, Thomas Sterner, Richard S. J. Tol, and Martin Weitzman. 2013. "Determining Benefits and Costs for Future Generations." *Science* 34, no. 6144: 349–350. https://doi.org/10.1126/science.1235665.

Asayama, Shinichiro. 2021. "The Oxymoron of Carbon Dioxide Removal: Escaping Carbon Lock-In and yet Perpetuating the Fossil Status Quo." *Frontiers in Climate* 3: 673515. https://doi.org/10.3389/fclim.2021.673515.

Asilomar Scientific Organizing Committee. 2010. *The Asilomar Conference Recommendations on Principles for Research into Climate Engineering Techniques.* Washington, DC: Climate Institute. http://www.climateresponsefund.org/images/Conference/finalfinalreport.pdf.

Autin, Whitney J., and John M. Holbrook. 2012. "Is the Anthropocene An Issue of Stratigraphy or Pop Culture?" *GSA Today* 22, no. 7: 60–61. https://doi.org/10.1130/G153GW.1.

Baldwin, Robert, Martin Cave, and Martin Lodge. 2010. "Introduction: Regulation – the Field and the Developing Agenda." In *The Oxford Handbook of Regulation*, edited by Robert Baldwin, Martin Cave, and Martin Lodge, 3–16. Oxford: Oxford University Press.

Ban-Weiss, George A., and Ken Caldeira. 2010. "Geoengineering as an Optimization Problem." *Environmental Research Letters* 5, no. 3: 034009. https://doi.org/10.1088/1748-9326/5/3/034009.

Barrett, Scott. 2007. *Why Cooperate? The Incentive to Supply Global Public Goods.* New York: Oxford University Press. https://doi.org/10.1093/acprof:oso/9780199211890.001.0001.

Barry, Roger G., and Richard J. Chorley. 2010. *Atmosphere, Weather and Climate.* 9th ed. London and New York: Routledge.

Bas, Muhammet, and Aseem Mahajan. 2020. "Contesting the Climate: Security Implications of Geoengineering." *Climatic Change* 162, no. 4: 1985–2002. https://doi.org/10.1007/s10584-020-02758-7.

Baskin, Jeremy. 2019. *Geoengineering, the Anthropocene, and the End of Nature.* Cham: Palgrave Macmillan. https://doi.org/10.1007/978-3-030-17359-3.

Bastin, Jean-François, Yelena Finegold, Danilo Mollicone, Marcelo Rezende, Devin Routh, Constantin M. Zohner, and Thomas W. Crowther. 2019. "The Global Tree Restoration Potential." *Science* 365, no. 6448: 76–79. https://doi.org./10.1126/science.aax0848.

Bauer, C., K. Treyer, T. Heck, and S. Hirschberg. 2018. "Greenhouse Gas Emissions from Energy Systems, Comparison, and Overview." In *Encyclopedia of the Anthropocene*, edited by Dominick A. Dellasala, Michael I. Goldstein (editors in chief), and Scott Elis (volume editor), 473–484. Oxford and Waltham, MA: Elsevier.

Baum, Chad M., Sean Low, and Benjamin K. Sovacool. 2022. "Between the Sun and Us: Expert Perceptions on the Innovation, Policy, and Deep Uncertainties of Space-Based Solar Geoengineering." *Renewable and Sustainable Energy Reviews* 158: 112179. https://doi.org/10.1016/j.rser.2022.112179.

Baum, Seth D., Timothy M. Maher Jr., and Jacob Haqq-Misra. 2013. "Double Catastrophe: Intermittent Stratospheric Geoengineering Induced by Societal Collapse." *Environment Systems & Decisions* 33, no. 1: 168–180. https://doi.org/10.1007/s10669-012-9429-y.

Beauchamp, Tom L., and James F. Childress. 2009. *Principles of Biomedical Ethics.* 7th ed. New York: Oxford University Press.

Bednar, Johannes, Michael Obersteiner, and Fabian Wagner. 2020. "On the Financial Viability of Negative Emissions." *Nature Communications* 10: 1783. https://doi.org/10.1038/s41467-019-09782-x.

Beerling, David J., Euripides P. Kantzas, Mark R. Lomas, Peter Wade, Rafael M. Eufrasio, Phil Renforth, Binoy Sarkar, M. Grace Andrews, Rachael H. James, Christopher R. Pearce, Jean-François Mercure, Hector Pollitt, Philip B. Holden, Neil R. Edwards, Madhu Khanna, Lenny Koh, Shaun Quegan, Nick F. Pidgeon, Ivan A. Janssens, James Hansen, and Steven A. Banwart. 2020. "Potential for Large-Scale CO2 Removal Via Enhanced Rock Weathering with Croplands." *Nature* 583: 242–248. https://doi.org/10.1038/s41586-020-2448-9.

Bernstein, Peter L. 1998. *Against the Gods: The Remarkable Story of Risk.* New York: Jon Wiley & Sons.

Bellamy, Rob, and Peter Healey. 2018. "'Slippery Slope' or 'Uphill Struggle'? Broadening out Expert Scenarios of Climate Engineering Research and Development." *Environmental Science and Policy* 83: 1–10. https://doi.org/10.1016/j.envsci.2018.01.021.

Bellamy, Rob, and Javier Lezaun. 2017. "Crafting a Public for Geoengineering." *Public Understanding of Science* 26, no. 4: 402–417. https://doi.org/10.1177/0963662515600965.

Bellamy, Rob, Javier Lezaun, and James Palmer. 2017. "Public Perceptions of Geoengineering Research Governance: An Experimental Deliberative Approach." *Global Environmental Change* 45: 194–202. https://doi.org/10.1016/j.gloenvcha .2017.06.004.

Bhat, Prerana. 2021. "Carbon needs to cost at least $100/tonne now to reach net zero by 2050: Reuters poll." *Reuters,* October 25, 2021. https://www.reuters.com/ business/cop/carbon-needs-cost-least-100tonne-now-reach-net-zero-by-2050-2021 -10-25/

Biermann, Frank. 2021. "Dangerous to Normalize Solar Geoengineering Research." *Nature* 595: 30. https://doi.org/10.1038/d41586-021-01724-2.

Biermann, Frank, and Ina Möller. 2019. "Rich Man's Solution? Climate Engineering Discourses and the Marginalization of the Global South." *International Environmental Agreements: Politics, Law and Economics* 19, no. 2: 151–167. https://doi.org/10.1007/s10784-019-09431-0.

Biermann, Frank, Jeroen Oomen, Aarti Gupta, Saleem H. Ali, Ken Conca, Maarten A. Hajer, Prakash Kashwan, Louis J. Kotzé, Melissa Leach, Dirk Messner, Chukwumerije Okereke, Åsa Persson, Janez Potočnik, David Schlosberg, Michelle Scobie, and Stacy D. VanDeveer. 2022. "Solar Geoengineering: The Case for an International Non-Use Agreement." *WIREs Climate Change* 13, no. 3: e754. https://doi.org/10.1002/wcc.754.

Blackstock, Jason. 2012. "Researchers Can't Regulate Climate Engineering Alone." *Nature* 486: 159. https://doi.org/10.1038/486159a.

Bluth, Gregg J. S., Scott D. Doiron, Charles C. Schnetzler, Arlin J. Krueger, and Louis S.Walter 1992. "Global Tracking of the SO2 Clouds from the June, 1991 Mount Pinatubo Eruptions." *Geophysical Research Letters* 19, no. 2: 151–154. https://doi.org/10.1029/91GL02792.

Bodansky, Daniel. 2012. "What's in a Concept? Global Public Goods, International Law, and Legitimacy." *The European Journal of International Law* 23, no. 3: 651–668. https://doi.org/10.1093/ejil/chs035.

Bonneuil, Christophe, and Jean-Baptiste Fressoz. 2016. *The Shock of the Anthropocene: The Earth, History and Us.* London and Brooklyn: Verso.

Bostrom, Nick, and Milan M. Ćirković. 2008. "Introduction." In *Global Catastrophic Risks*, edited by Nick Bostrom and Milan M. Ćirković, 1–29. New York: Oxford University Press.

Boucher, Olivier, Piers M. Forster, Nicolas Bruber, Minh Ha-Duong, Mark G. Lawrence, Timothy M. Lenton, Achim Maas, and Naomi E. Vaughan. 2014. "Rethinking Climate Engineering Categorization in the Context of Climate Change Mitigation and Adaptation." *WIREs Climate Change* 5, no. 1: 23–35. https://doi .org/10.1002/wcc.261.

Brander, Matthew, Francisco Ascui, Vivian Scott, and Simon Tett. 2021. "Carbon Accounting for Negative Emissions Technologies." *Climate Policy* 21, no. 5: 699–717. https://doi.org/10.1080/14693062.2021.1878009.

Briggs, R. A. 2019. "Normative Theories of Rational Choice: Expected Utility." In *The Stanford Encyclopedia of Philosophy*, edited by Edward N. Zalta, Fall. https:// plato.stanford.edu/archives/fall2019/entries/rationality-normative-utility/.

Brohé, Arnaud. 2016. *The Handbook of Carbon Accounting.* Saltaire: Greenleaf Publishing.

Brooke, John L. 2014. *Climate Change and the Course of Global History: A Rough Journey.* Cambridge: Cambridge University Press. https://doi.org/10.1093/ahr/120 .3.965.

Broome, John. 1999. *Ethics out of Economics.* Cambridge: Cambridge University Press. https://doi.org/10.1017/CBO9780511605888.

Buck, Holly Jean. 2017. "Climate Engineering." *Issues in Science and Technology* XXXIII, no. 4: 11.

Buck, Holly Jean. 2019. *After Geoengineering: Climate Tragedy, Repair and Restoration.* London and New York: Verso Books.

Budyko, Mikhail I. 1974. *Climate and Life.* New York and London: Academic Press (edited by David H. Miller).

Budyko, Mikhail I. 1977. *Climate Changes.* Washington, DC: American Geophysical Union.

Bunzl, Martin. 2009. "Researching Geoengineering: Should Not or Could Not?" *Environmental Research Letters* 4, no. 4: 045104. https://doi.org/10.1088/1748 -9326/4/4/045104.

Burger, Michael, and Justin Gundlach. 2018. "Research Governance." In *Climate Engineering and the Law: Regulation and Liability for Solar Radiation Management and Carbon Dioxide Removal*, edited by Michael B. Gerrard and Tracy Hester, 269–323. Cambridge: Cambridge University Press. https://doi.org /10.1017/9781316661864.006.

Burns, Elizabeth T., Jane A. Flegal, David W. Keith, Aseem Mahajan, Dustin Tingley, and Gernot Wagner. 2016. "What Do People Think When They Think About Solar Geoengineering? A Review of Empirical Social Science Literature, and Prospects for Future Research." *Earth's Future* 4, no. 11: 536–542. https://doi .org/10.1002/2016EF000461.

Burns, Williams C. G. 2016. *The Paris Agreement and Climate Geoengineering Governance: The Need for a Human Rights-Based Component.* Waterloo, ON: Center for International Governance Innovation.

Buylova, Alexandra, Mathias Fridahl, Naghmeh Nasiritousi, and Gunilla Reischl. 2021. "Cancel (Out) Emissions? The Envisaged Role of Carbon Dioxide Removal Technologies in Long-Term National Climate Strategies." *Frontiers in Climate* 3: 675499. https://doi.org/10.3389/fclim.2021.675499.

Cairns, Rose C. 2014. "Climate Geoengineering: Issues of Path-Dependence and Socio-Technical Lock-in." *WIREs Climate Change* 5, no. 5: 649–661. https://doi .org/10.1002/wcc.296.

Callendar, Guy S. 1938. "The Artificial Production of Carbon Dioxide and its Influence on Temperatures." *Quarterly Journal of the Royal Meteorological Society* 64, no. 275: 223–240. https://doi.org/10.1002/qj.49706427503.

Callies, Daniel Edwards. 2018. "Institutional Legitimacy and Geoengineering Governance." *Ethics, Policy & Environment* 21, no. 3: 324–340. https://doi.org/10 .1080/21550085.2018.1562523.

Callies, Daniel Edwards. 2019a. *Climate Engineering: A Normative Perspective.* London: Lexington Books.

Callies, Daniel Edwards. 2019b. "The Slippery Slope Argument against Geoengineering." *Journal of Applied Philosophy* 36, no. 4: 675–687. https://doi.org/10.1111/japp.12345.

Callies, Daniel Edwards, and Darrel Moellendorf. 2021. "Assessing Climate Policies: Catastrophe Avoidance and the Right to Sustainable Development." *Politics, Philosophy & Economics* 20, no. 2: 127–150. https://doi.org/10.1177/1470594X211003334.

Caney, Simon. 2015. "Climate Change." In *The Routledge Handbook of Global Ethics*, edited by Darrel Moellendorf and Heather Widdows, 372–386. London and New York: Routledge. https://doi.org/10.4324/9781315744520.

Cao, Long. 2018. "The Effects of Solar Radiation Management on the Carbon Cycle." *Current Climate Change Reports* 4, no. 1: 41–50. https://doi.org/10.1007/s40641-018-0088-z.

Cao, Long, Chao-Chao Gao, and Li-Yun Zhao. 2015. "Geoengineering: Basic Science and Ongoing Research Efforts in China." *Advances in Climate Change Research* 6, no. 3–4: 188–196. https://doi.org/10.1016/j.accre.2015.11.002.

Carlisle, Daniel P., Pamela M. Feetham, Malcolm J. Wright, and Damon A. H. Teagle. 2022. "Public Engagement with Emerging Technologies: Does Reflective Thinking Affect Survey Responses?" *Public Understanding of Science* 31, no. 5: 660–670. https://doi.org/10.1177/09636625211029438.

Carr, Wylie. 2019. "This is God's Stuff We're Messing with." In *Geoengineering our Climate? Ethics, Politics, and Governance*, edited by Jason J. Blackstock Sean Low, 66–70. London and New York: Routledge.

Carr, Wyle, and Christopher J. Preston. 2017. "Skewed Vulnerabilities and Moral Corruption in Global Perspectives on Climate Engineering." *Environmental Values*, 26, no. 6: 757–777. https://doi.org/10.3197/096327117X15046905490371.

Carr, Wylie A., and Laurie Yung. 2018. "Perceptions of Climate Engineering in the South Pacific, Sub-Saharan Africa, and North American Arctic." *Climatic Change*, 147, no. 1–2: 119–132. https://doi.org/10.1007/s10584-018-2138-x.

Carr, Wylie, Ashley Mercer, and Clare Palmer. 2012. "Public Concerns about the Ethics of Solar Radiation Management." In *Engineering the Climate: The Ethics of Solar Radiation Management*, edited by Christopher J. Preston, 169–186. Lanham: Lexington Books.

Carrer, Dominique, Gaétan Pique, Morgan Ferlicoq, Xavier Ceamanos, and Eric Ceschia. 2019. "What is the Potential of Cropland Albedo Management in the Fight against Global Warming?" *Environmental Research Letters*, 13, no. 4: 044030. https://doi.org/10.1088/1748-9326/aab650.

Carrico, Amanda R., Heather Barnes Truelove, Michael P. Vandenbergh, and David Dana. 2015. "Does Learning about Climate Change Adaptation Change Support for Mitigation?" *Journal of Environmental Psychology* 41, no. 1: 19–29. https://doi.org/10.1016/j.jenvp.2014.10.009.

Carton, Wim, Adeniyi Asiyanbi, Silke Beck, Holly J. Buck, and Jens F. Lund. 2020. "Negative Emissions and the Long History of Carbon Removal." *WIREs Climate Change*, 11, no. 6: e671. https://doi.org/10.1002/wcc.671.

Carvalho, António, and Mariana Riquito. 2022. "'It's Just a Band-Aid!': Public Engagement with Geoengineering and the Politics of the Climate Crisis." *Public Understanding of Science* 31, no. 7: 903–920. https://doi.org/10.1177/09636625221095353.

Chakrabarty, Dipesh. 2009. "The Climate of History: Four Theses." *Critical Inquiry* 35, no. 2: 197–222. https://doi.org/10.1086/596640.

Chavez, Anthony E. 2015. "Exclusive Rights to Saving the Planet: The Patenting of Geoengineering Inventions." *Northwestern Journal of Technology and Intellectual Property* 13, no. 1: 1–35.

Chazdon, Robin L. 2014. *Second Growth: The Promise of Tropical Forest Regeneration in an Age of Deforestation*. Chicago and London: The University of Chicago Press.

Chung, Ik Kyo, Calvyn F. A. Sondak, and John Beardall. 2017. "The Future of Seaweed Aquaculture in a Rapidly Changing World." *European Journal of Phycology* 52, no. 4: 495–505. https://doi.org/10.1080/09670262.2017.1359678.

Cicerone, Ralph J. 2006. "Geoengineering: Encouraging Research and Overseeing Implementation." *Climatic Change* 77, no. 3: 221–226. https://doi.org/10.1007/s10584-006-9102-x.

Clark, Peter U., Jeremy D. Shakun, Shaun A. Marcott, Alan C. Mix, Michael Eby, Scott Kulp, Anders Levermann, Glenn A. Milne, Patrik L. Pfister, Benjamin D. Santer, Daniel P. Schrag, Susan Solomon, Thomas F. Stocker, Benjamin H. Strauss, Andrew J. Weaver, Ricarda Winkelmann, David Archer, Edouard Bard, Aaron Goldner, Kurt Lambeck, Raymond T. Pierrehumbert, and Gian-Kasper Plattner. 2016. "Consequences of Twenty-first-Century Policy for Multi-Millennial Climate and Sea-Level Change." *Nature Climate Change* 6: 360–369. https://doi.org/10.1038/nclimate2923.

Clingerman, Forrest. 2012. "Between Babel and Pelagius: Religion, Theology, and Geoengineering." In *Engineering the Climate: The Ethics of Solar Radiation Management*, edited by Christopher J. Preston, 201–219. Lanham: Lexington Books.

Clingerman, Forrest, and Kevin J. O'Brien. 2014. "Playing God: Why Religion Belongs in the Climate Engineering Debate." *Bulletin of the Atomic Scientists* 70, no. 3: 27–37. https://doi.org/10.1177/0096340214531181.

Cohen, G. A. 1989. "On the Currency of Egalitarian Justice." *Ethics*, 99, no. 4: 906–44.

Cohen, G. A. 2001. *If You're an Egalitarian, How Come You're So Rich?* Cambridge, MA: Harvard University Press.

Collingridge, David. 1980. *The Social Control of Technology*. New York and London: St. Martin's Press & F. Pinter.

de Coninck, Heleen, Aromar Revi, Mustafa Babiker, Paolo Bertoldi, Marcos Buckeridge, Anton Cartwright, Wenjie Dong, James Ford, Sabine Fuss, Jean-Charles Hourcade, Debora Ley, Reinhard Mechler, Peter Newman, Anastasia

Revokatova, Seth Schultz, Linda Steg, and Taishi Sugiyama. 2018. "Strengthening and Implementing the Global Response." In *Global Warming of 1.5°C: IPCC Special Report on Impacts of Global Warming of 1.5°C above Pre-industrial Levels in Context of Strengthening Response to Climate Change, Sustainable Development, and Efforts to Eradicate Poverty, IPCC*, 313–443. Geneva: IPCC.

Consoli, Christopher. 2019. *Bioenergy and Carbon Capture and Storage: 2019 Perspective*. Melbourne: Global CCS Institute.

Corner, Adam, and Nick Pidgeon. 2014. "Geoengineering, Climate Change Scepticism and the 'Moral Hazard' Argument: An Experimental Study of UK Public Perceptions." *Philosophical Transactions of the Royal Society A* 372: 20140063. https://doi.org/10.1098/rsta.2014.0063.

Corner, Adam, and Nick Pidgeon, 2015. "Like Artificial Trees? The Effect of Framing by Natural Analogy on Public Perceptions of Geoengineering." *Climatic Change* 130, no. 3: 425–438. https://doi.org/10.1007/s10584-014-1148-6.

Corner, Adam, Nick Pidgeon, and Karen Parkhill. 2012. "Perceptions of Geoengineering: Public Attitudes, Stakeholder Perspectives, and the Challenge of 'Upstream' Engagement." *WIREs Climate Change* 3, no. 5: 451–466. https://doi.org/10.1002/wcc.176.

Corner, Adam, Karen Parkhill, Nick Pidgeon, and Naomi E. Vaughan. 2013. "Messing with Nature? Exploring Public Perceptions of Geoengineering." *Global Environmental Change* 23, no. 5: 938–947. https://doi.org/10.1016/j.gloenvcha.2013.06.002.

Corry, Olaf. 2017. "The International Politics of Geoengineering: The Feasibility of Plan B for Tackling Climate Change." *Security Dialogue* 48, no. 4: 297–315. https://doi.org/10.1177/0967010617704142.

Cox, Emily, and Neil Robert Edwards. 2019. "Beyond Carbon Pricing: Policy Levers for Negative Emissions Technologies." *Climate Policy* 19, no. 9: 1144–1156. https://doi.org/10.1080/14693062.2019.1634509.

Cox, Emily M., Nick Pidgeon, Elspeth Spence, and Gareth Thomas. 2018. "Blurred Lines: The Ethics and Policy of Greenhouse Removal at Scale." *Frontiers in Environmental Science* 6: 38. https://doi.org/10.3389/fenvs.2018.00038.

Cox, Emily, Elspeth Spence, and Nick Pidgeon. 2020. "Public Perceptions of Carbon Dioxide Removal in the United States and the United Kingdom." *Nature Climate change* 10: 744-749. https://doi.org/10.1038/s41558-020-0823-z

Cressey, Daniel. 2012. "Geoengineering Experiment Cancelled Amid Patent Row." *Nature,* May 15. https://www.nature.com/articles/nature.2012.10645.

Creutzig, Felix, N. H. Ravindranath, Göran Berndes, Simon Bolwig, Ryan Bright, Francesco Cherubini, Helena Chum, Esteve Corbera, Mark Delucchi, Andre Faaij, Joseph Fargione, Helmut Haberl, Garvin Heath, Oswaldo Lucon, Richard Plevin, Alexander Popp, Carmenza Robledo-Abad, Steven Rose, Pete Smith, Anders Stromman, Sangwon Suh, and Omar Masera. 2014. "Bioenergy and Climate Change Mitigation: An Assessment." *GCB – Bioenergy* 7, no. 5: 916–944. https://doi.org/10.1111/gcbb.12205.

Crosby, Alfred W. Jr. 1972. *The Columbian Exchange: Biological and Cultural Consequences of 1492*. Santa Barbara: Greenwood Publishing Group.

Crutzen, Paul J. 2002. "The Geology of Mankind." *Nature* 415: 23. https://doi.org/10 .1007/978-3-319-27460-7_10.

Crutzen, Paul J. 2006. "Albedo Enhancement by Stratospheric Sulfur Injections: A Contribution to Resolve a Policy Dilemma." *Climatic Change* 77, no. 3: 211–219. https://doi.org/10.1007/s10584-006-9101-y.

Crutzen, Paul J., and Eugene F. Stoermer. 2000. "The 'Anthropocene'." *Global Change Newsletter* 41 (May): 17–18.

Cui, Ryna Yiyun, Nathan Hultman, Morgan R. Edwards, Linlang He, Arijit Sen, Kavita Surana, Haewon McJeon, Gokul Iyer, Pralit Patel, Ted Nace, and Christine Shearer. 2019. "Quantifying Operational Lifetimes for Coal Power Plants under the Paris Goals." *Nature Communications,* 10: 4759. https://doi.org/10.1038/s41467 -019-12618-3.

Culver, Jennifer. 2019. "False Dilemma." In *Bad Arguments: 100 of the Most Important Fallacies of Western Philosophy,* edited by Robert Arp, Steven Barbone, and Michael Bruce, 346–347. Oxford: John Wiley & Sons Ltd.

Cummings, Christopher L., Sapphire H. Lin, and Benjamin D. Trump. 2017. "Public Perceptions of Climate Geoengineering: A Systematic Review of the Literature." *Climate Research* 73, no. 3: 247–264. https://doi.org/10.3354/cr01475.

Curry, Charles L., Jana Sillmann, David Bronaugh, Kari Alterskjaer, Jason N. S. Cole, Duoying Ji, Ben Kravitz, Jón Egill Kristjánsson, John C. Moore, Helene Muri, Ulrike Niemeier, Alan Robock, Simone Tilmes, Shuting Yang. 2014. "A Multimodel Examination of Climate Extremes in an Idealized Geoengineering Experiment." *JGR Atmospheres* 119, no. 7: 3900–3923. https://doi.org/10.1002 /2013JD020648.

Dahl, Robert A. 1970. *After the Revolution? Authority in a Good Society.* New Haven, MA: Yale University Press.

Davin, Edouard L., Sonia I. Seneviratnea, Philippe Ciaisb, Albert Oliosoc, and Tao Wang. 2014. "Preferential Cooling of Hot Extremes from Cropland Albedo Management." *Proceedings of the National Academy of Sciences* 111, no. 27: 9757–9761. https://doi.org/10.1073/pnas.1317323111.

Davis, Janae, Alex A. Moulton, Levi Van Sant, and Brian Williams. 2019. "Anthropocene, Capitalocene, … Plantationocen? A Manifesto for Ecological Justice in an Age of Global Crises." *Geography Compass* 13, no. 5: e12438. https:// doi.org/10.1111/gec3.12438.

Delina, Laurence L. 2021. "Southeast Asian Expert Perceptions of Solar Radiation Management Techniques and Carbon Dioxide Removal Approaches: Caution, Ambivalence, Risk Precaution, and Research Directions." *Environmental Research Communications* 3, no. 12: 125005. https://doi.org/10.1088/2515-7620 /ac3dc1.

Diamond, Michael S., Andrew Gettelman, Matthew D. Lebsock, Allison McComiskey, Lynn M. Russell, Robert Wood, and Graham Feingold. 2022. "To Assess Marine Cloud Brightening's Technical Feasibility, We Need to Know What to Study— and When to Stop." *Proceedings of the National Academy of Sciences of the United States of America* 119, no. 4: e2118379119. https://doi.org/10.1073/pnas .2118379119.

Diamond, Peter A., and Jerry A. Hausman. 1994. "Contingent Valuation: Is Some Number Better than No Number?" *Journal of Economic Perspectives* 8, no. 4: 45–64. https://doi.org/10.1257/jep.8.4.45.

Donner, Simon D. 2007. "Domain of the Gods: An Editorial Essay." *Climatic Change* 85, no. 3–4: 231–236. https://doi.org/10.1007/s10584-007-9307-7.

Dooley, Kate, Peter Christoff, and Kimberly A. Nicholas. 2018. "Co-producing Climate Policy and Negative Emissions: Trade-Offs for Sustainable Land-Use." *Global Sustainability* 1, e3: 1–10. https://doi.org/10.1017/sus.2018.6.

Dooley, Kate, Ellycia Harrould-Kolieb, and Anita Talberg. 2021. "Carbon-dioxide Removal and Biodiversity: A Threat Identification Framework." *Global Policy* 12, no. 1: 34–44. https://doi.org/10.1111/1758-5899.12828.

Drouet, Laurent, Valentina Bosetti, Simone A. Padoan, Lara Aleluia Reis, Christoph Bertram, Francesco Dalla Longa, Jacques Després, Johannes Emmerling, Florian Fosse, Kostas Fragkiadakis, Stefan Frank, Oliver Fricko, Shinichiro Fujimori, Mathijs Harmsen, Volker Krey, Ken Oshiro, Larissa P. Nogueira, Leonidas Paroussos, Franziska Piontek, Keywan Riahi, Pedro R. R. Rochedo, Roberto Schaeffer, Jun'ya Takakura, Kaj-Ivar van der Wijst, Bob van der Zwaan, Detlef van Vuuren, Zoi Vrontisi, Matthias Weitzel, Behnam Zakeri, and Massimo Tavoni. 2021. "Net Zero-Emission Pathways Reduce the Physical and Economic Risks of Climate Change." *Nature Climate Change* 11, no. 12: 1070–1076. https://doi.org/10.1038/s41558-021-01218-z.

Duan, Lei, Long Cao, Govindasamy Bala, and Ken Caldeira. 2018. "Comparison of the Fast and Slow Climate Response to Three Radiation Management Geoengineering Schemes." *Journal of Geophysical Research: Atmospheres* 123, no. 21: 11980–12001. https://doi.org/10.1029/2018JD029034.

Dworkin, Gerald, 1988. *The Theory and Practice of Autonomy*. Cambridge: Cambridge University Press.

Dworkin, Ronald. 1985. *A Matter of Principle*. Cambridge, MA: Harvard University Press. https://doi.org/10.2307/j.ctv1pncpxk.

Dworkin, Ronald. 2000. *Sovereign Virtue: The Theory and Practice of Equality*. Cambridge, MA: Harvard University Press. https://doi.org/10.2307/j.ctv1c3pd0r.

Early, James T. 1989. "Space-based solar shield to offset greenhouse effect." *Journal of the British Interplanetary Society* 42, 567–569.

EASAC. 2018. *Negative Emission Technologies: What Role in Meeting Paris Agreement Targets?* Brussels and Halle: EASAC.

EASAC. 2022. *Forest Bioenergy Update: BECCS and its Role in Integrated Assessment Models*. Brussels and Halle: EASAC.

Eastham, Sebastian D., Debra K. Weisenstein, David W. Keith, and Steven R. H. Barrett. 2018. "Quantifying the Impact of Sulfate Geoengineering on Mortality from Air Quality and UV-B Exposure." *Atmospheric Environment*, 187: 424–434. https://doi.org/10.1016/j.atmosenv.2018.05.047.

Ellery, Alex. 2016. "Low-Cost Space-Based Geoengineering: An Assessment Based on Self-Replicating Manufacturing of in-Situ Resources on the Moon." *International Journal of Environmental and Ecological Engineering* 10, no. 2: 278–285.

Ellis, Erle C. 2018. *Anthropocene: A Very Short Introduction*. Oxford: Oxford University Press. https://doi.org/10.1093/actrade/9780198792987.001.0001.

Ellis, Erle C., Mark Maslin, and Simon Lewis. 2020. "Planting Tress Won't Save the World." *The New York Times*, February 12, 2020. https://www.nytimes.com/2020/02/12/opinion/trump-climate-change-trees.html

Ellis, Erle C., Nicolas Gauthier, Kees Klein Goldewijk, Rebeccas Bliege Bird, Nicole Boivin, Sandra Díaz, Dorian Q. Fuller, Jacquelyn L. Gill, Jed O. Kaplan, Naomi Kingston, Harvey Locke, Crystal N. H. McMichael, Darren Ranco, Torben C. Rick, M. Rebecca Shaw, Lucas Stephens, Jens-Christian Svenning, and James E. M. Watson. 2021. "People Have Shaped Most of Terrestrial Nature for at Least 12,000 Years." *Proceedings of the National Academy of Sciences of the United States of America* 118, no. 17: e2023483118. https://doi.org/10.1073/pnas.2023483118.

Energy Future Initiatives. 2020. *Uncharted Waters: Expanding the Options for Carbon Dioxide Removal in Coastal and Ocean Environments*. Washington: Energy Futures Initiative.

Eysteinsson, Thröstur. 2017. *Forestry in a Treeless Land*. Egilsstaðir: Icelandic Forest Service.

Fabre, Adrien, and Gernot Wagner. 2020. "Availability of Risky Geoengineering Can Make an Ambitious Climate Mitigation Agreement More Likely." *Humanities and Social Sciences Communications* 7: 1. https://doi.org/10.1057/s41599-020-0492-6.

Fahlquist, Jessica Nihlén. 2019. *Moral Responsibility and Risk in Society: Examples from Emerging Technologies, Public Health and Environment*. Oxon and New York: Routledge.

Fajardy, Mathilde, Dr. Alexandre Köberle, Dr. Niall Mac Dowell, and Dr. Andrea Fantuzzi. 2019a. *BECCS Deployment: A Reality Check*. Graham Institute Briefing Paper No 28. London: Imperial College London. https://www.imperial.ac.uk/grantham/publications/briefing-papers/beccs-deployment-a-reality-check.php.

Fajardy, Mathilde, Piera Patrizio, Habiba Ahut Daggash, and Niall Mac Dowell. 2019b. "Negative Emissions: Priorities for Research and Policy Design." *Frontiers in Climate* 1: 6. https://doi.org/10.3389/fclim.2019.00006.

Fan, Yuanchao, Jerry Tjiputra, Helene Muri, Danica Lombardozzi, Chang-Eui Park, Shengjun Wu, and David Keith. 2021. "Solar Geoengineering Can Alleviate Climate Change Pressures on Crop Yields." *Nature Food* 2: 373–381. https://doi.org/10.1038/s43016-021-00278-w.

FAO. 2018. *The Global Status of Seaweed Production, Trade and Utilization*. Globefish Research Programme Volume 124. Rome: Food and Agriculture Organization of the United Nations.

Finkel, Adam M. 2018. "Demystifying Evidence-Based Policy Analysis by Revealing Hidden Value-Laden Constraints." *The Hastings Center Report* 48, S1: S21–S49. https://doi.org/10.1002/hast.818.

Finney, Stanley, and Lucy E. Edwards. 2016. "The 'Anthropocene' Epoch: Scientific Decision or Political Statement." *GSA Today* 26, no. 3–4: 4–10. https://doi.org/10.1130/GSATG270A.1.

Flegal, Jane, and Aarti Gupta. 2018. "Evoking Equity as a Rationale for Solar Geoengineering Research? Scrutinizing Emerging Expert Visions of Equity." *International Environmental Agreements: Politics, Law and Economics* 18, no. 1: 45–61. https://doi.org/10.1007/s10784-017-9377-6.

Flegal, Jane A., Anna-Maria Hubert, David R. Morrow, and Juan B. Moreno-Cruz. 2019. "Solar Geoengineering: Social Science, Legal, Ethical, and Economic Frameworks." *Annual Review of Environment and Resources* 44: 399–423. https://doi.org/10.1146/annurev-environ-102017-030032.

Fleming, James Rodger. 2007. "The Climate Engineers." *The Wilson Quarterly* 31, no. 2: 46–60.

Fleming, James Rodger. 2010. *Fixing the Sky: The Checkered History of Weather and Climate Control*. New York and Chichester: Columbia University Press.

Fletcher, Joseph O. 1968. *Changing Climate*. Santa Monica, CA: The RAND Corporation.

Fletcher, Joseph O. 1969. *Managing Climatic Resources*. Santa Monica, CA: The RAND Corporation.

Fofrich, Robert, Dan Tong, Katherine Calvin, Harmen Sytze de Boer, Johannes Emmerling, Oliver Fricko, Shinichiro Fujimori, Gunnar Luderer, Joeri Rogelj, and Steven J. Davis. 2020. "Early Retirement of Power Plants in Climate Mitigation Scenarios." *Environmental Research Letters* 15, no. 9: 094064. https://doi.org/10.1088/1748-9326/ab96d3.

Foote, Eunice. 1856. "Circumstances Affecting the Heat of the Sun's Rays." *The American Journal of Science and Arts* 22, no. 86: 383–384.

Fountain, Henry, and Christopher Flavelle. 2021. "Test Flight for Sunlight-Blocking Research is Canceled." *The New York Times*, April 2, 2021. https://www.nytimes.com/2021/04/02/climate/solar-geoengineering-block-sunlight.html.

Fragnière, Augustin and Stephen M. Gardiner. 2017. "Why Geoengineering Is Not 'Plan B'." In *Climate Justice and Geoengineering: Ethics and Policy in the Atmospheric Anthropocene*, edited by Christopher J. Preston, 15–32. London and New York: Rowman & Littlefield.

Frank, Robert H., Ben S. Bernanke, Kate Antonovics, and Ori Heffetz. 2022. *Principles of Microeconomics: A Streamlined Approach*. New York: McGraw Hill.

Fraser, Nancy, and Axel Honneth. 2003. *Redistribution or Recognition? A Political-Philosophical Exchange*. New York: Verso.

Fridahl, Mathias. 2017. "Socio-Political Prioritization of Bioenergy with Carbon Capture and Storage." *Energy Policy* 104: 89–99. https://doi.org/10.1016/j.enpol.2017.01.050.

Frisch, Mathias. 2018. "Modelling Climate Policies: The Social Cost of Carbon and Uncertainties in Climate Predictions." In *Climate Modelling: Philosophical and Conceptual Issues*, edited by Elisabeth A. Llyod and Eric Winsberg, 17–29. Cham: Palgrave Macmillan.

Frumhoff, Peter C., and Stephens Jennie C. 2018. "Towards Legitimacy of the Solar Geoengineering Research Enterprise." *Philosophical Transactions of the Royal Society A* 376: 20160459.20160459. https://doi.org/10.1098/rsta.2016.0459.

Fuglesang, Christer, and María García de Herreros Miciano. 2021. "Realistic Sunshade System at L1for Global Temperature Control." *Acta Astronautica* 186: 269–279. https://doi.org/10.1016/j.actaastro.2021.04.035.

Fujiwara, Masatomo, and Masahiro Sugiyama. 2016. *Public Perception of Climate Engineering in Japan: Results from Online and Classroom Surveys*. Tokyo: Policy Alternatives Research Institute. http://hdl.handle.net/2115/72566.

Fuss, Sabine, Josep G. Canadell, Glen P. Peters, Massimo Tavoni, Robbie M. Andrew, Philippe Ciais, Robert B. Jackson, Chris D. Jones, Florian Kraxner, Nebosja Nakicenovic, Corinne Le Quéré, Michael R. Raupach, Ayyoob Sharifi, Pete Smith, and Yoshiki Yamagata. 2014. "Betting on Negative Emissions." *Nature Climate Change* 4: 850–853. https://doi.org/10.1038/nclimate2392.

Fuss, Sabine, William F. Lamb, Max W. Callaghan, Jérôme Hilaire, Felix Creutzig, Thorben Amann, Tim Beringer, Wagner de Oliveira Garcia, Jens Hartmann, Tarun Khanna, Gunnar Luderer, Gregory F. Nemet, Joeri Rogelj, Pete Smith, José Luis Vicente Vicente, Jennifer Wilcox, Maria del Mar Zamora Dominguez, and Jan C. Minx. 2018. "Negative Emissions–Part 2: Costs, Potentials and Side Effects." *Environmental Research Letters* 13: 063002. https://doi.org/10.1088/1748-9326/aabf9f.

Garamone, Jim (2022). "Biden Signs National Defense Authorization Act into Law." December 23, 2022. Last accessed January 19, 2023. https://www.defense.gov/News/News-Stories/Article/Article/3252968/biden-signs-national-defense-authorization-act-into-law/.

Gardiner, Stephen M. 2006a. "A Core Precautionary Principle." *The Journal of Political Philosophy* 14, no. 1: 33–60. https://doi.org/10.1007/978-94-007-1433-5_38.

Gardiner, Stephen M. 2006b. "A Perfect Moral Storm: Climate Change, Intergenerational Ethics and the Problem of Moral Corruption." *Environmental Values* 15, no. 3: 397–413. https://doi.org/10.3197/096327106778226293.

Gardiner, Stephen M. 2010. "Is 'Arming the Future' with Geoengineering Really the Lesser Evil?" In *Climate Ethics: Essential Readings*, edited by Stephen M. Gardiner, Simon Caney, Dale Jamieson, and Henry Shue, 284–312. New York: Oxford University Press.

Gardiner, Stephen M. 2011. *A Perfect Moral Storm: The Ethical Tragedy of Climate Change*. New York: Oxford University Press.

Gardiner, Stephen M. 2013a. "The Desperation Argument for Geoengineering." *PS: Political Science and Politics* 46, no. 1: 28–33. https://doi.org/10.1017/S1049096512001424.

Gardiner, Stephen M. 2013b. "Why Geoengineering is Not a 'Global Public Good', and Why it is Ethically Misleading to Frame it as One." *Climatic Change* 121, no. 3: 513–525. https://doi.org/10.1007/s10584-013-0764-x.

Gardiner, Stephen M. 2014. "A Call for a Global Constitutional Convention Focused on Future Generations." *Ethics & International Affairs* 28, no. 3: 299–315. https://doi.org/10.1017/S0892679414000379.

Gardiner, Stephen M., and Augustin Fragnière. 2018. "The Tollgate Principles for the Governance of Geoengineering: Moving Beyond the Oxford Principles to an Ethically More Robust Approach." *Ethics, Policy & Environment* 21, no. 2: 143–174. https://doi.org/10.1080/21550085.2018.1509472.

Gasparini, Blaž, and Ulrike Lohmann. 2016. "Why Cirrus Cloud Seeding Cannot Substantially Cool the Planet." *Journal of Geophysical Research: Atmospheres* 121, no. 9: 4877–4893. https://doi.org/10.1002/2015JD024666.

Gasparini, Blaž, Zachary McGraw, Trude Storelvmo, and Ulrike Lohmann. 2020. "To What Extent Can Cirrus Cloud Seeding Counteract Global Warming?" *Environmental Research Letters* 15, no. 5: 054002. https://doi.org/10.1088/1748 -9326/ab71a3.

Geden, Oliver. 2018. "Politically Informed Advice for Climate Action." *Nature Geoscience* 11: 380–383. https://doi.org/10.1038/s41561-018-0143-3.

Geden, Oliver, Vivian Scott, and James Palmer. 2018. "Integrating Carbon Dioxide Removal into EU Climate Policy: Prospects for a Paradigm Shift." *WIREs Climate Change* 9, no. 4: e251. https://doi.org/10.1002/wcc.521.

Georgescu, Matei, Philip E. Morefield, Britta G. Bierwagen, and Christopher P. Weaver. 2014. "Urban adaptation Can Roll Back Warming of Emerging Megapolitan Regions." *Proceedings of the National Academy of Sciences of the United States* 111, no. 8: 2909–2914. https://doi.org/10.1073/pnas .1322280111.

Gerrard, Michael B., and Tracy Hester. 2018. *Climate Engineering and the Law: Regulation and Liability for Solar Radiation Management and Carbon Dioxide Removal.* Cambridge: Cambridge University Press. https://doi.org/10.1017 /9781316661864.

Glaberson, William. 1988. "Behind Du Pont's Shift on Loss of Ozone Layer." *The New York Times*, March 26, 1988. https://www.nytimes.com/1988/03/26/business/ behind-du-pont-s-shift-on-loss-of-ozone-layer.html.

Global CCS Institute. 2021. *Global Status of CCS 2021.* Melbourne: Global CCS Institute.

González, Miriam Ferrer, Tatiana Ilyina, Sebastian Sonntag, and Hauke Schmidt. 2018. "Enhanced Rates of Regional Warming and Ocean Acidification After Termination of Large-Scale Ocean Alkalinization." *Geophysical Research Letters* 45, no. 14: 7120–7129. https://doi.org/10.1029/2018GL077847.

Goodell, Jeff. 2010. *How to Cool the Planet: Geoengineering and the Audacious Quest to Fix Earth's Climate.* Boston and New York: Houghton Mifflin Harcourt.

Goodin, Robert E. 2007. "Enfranchising All Affected Interests, and Its Alternatives." *Philosophy & Public Affairs* 35, no. 1: 40–68. https://doi.org/10.1111/j.1088-4963 .2007.00098.x.

Gordon, Bart. 2010. "Plan B for the Climate." *Slate,* September 24, 2010. https:// slate.com/technology/2010/09/rep-bart-gordon-on-the-policy-implications-of -geoengineering.html.

Gough, Clair, and Sarah Mander. 2019. "Beyond Social Acceptability: Applying Lessons from CCS Social Science to Support Deployment of BECCS." *Current Sustainable/Renewable Energy Reports* 6, no. 4: 116–123. https://doi.org/10.1007 /s40518-019-00137-0.

Grasso, Marco. 2022. "Legitimacy and Procedural Justice: How Might Stratospheric Aerosol Injection Function in the Public Interest?" *Humanities and Social Sciences Communications* 9: 187. https://doi.org/10.1057/s41599-022-01213-5.

Griscom, Bronson W., Justin Adams, Peter W. Ellis, Richard A. Houghton, Guy Lomax, Daniela A. Miteva, William H. Schlesinger, David Shoch, Juha V. Siikamäki, Pete Smith, Peter Woodbury , Chris Zganjar, Allen Blackman, João Campari, Richard T. Conant, Christopher Delgado, Patricia Elias, Trisha Gopalakrishna, Marisa R. Hamsik, Mario Herrero, Joseph Kiesecker, Emily Landis, Lars Laestadius, Sara M. Leavitt, Susan Minnemeyer, Stephen Polasky, Peter Potapov, Francis E. Putz, Jonathan Sanderman, Marcel Silvius, Eva Wollenberg, and Joseph Fargione. 2017. "Natural Climate Solutions." *Proceedings of the National Academy of Sciences* 114, no. 44: 11645–11650. https://doi.org/10.1073/pnas.1710465114.

Grisé, Michelle, Emmi Yonekura, Jonathan S. Blake, David Desmet, Anusree Garg, and Benjamin Lee Preston. 2021. *Climate Control: International Legal Mechanisms for Managing the Geopolitical Risks of Geoengineering.* Santa Monica: RAND Corporation.

Gupta, Aarti, and Ina Möller. 2019. "De Facto Governance: How Authoritative Assessments Construct Climate Engineering as an Object of Governance." *Environmental Politics* 28, no. 3: 480–501. https://doi.org/10.1080/09644016.2018 .1452373.

Gupta, Aarti, Ina Möller, Frank Biermann, Sikina Jinnah, Prakash Kashwan, Vikrom Mathur, David R. Morrow, and Simon Nicholson. 2020. "Anticipatory Governance of Solar Geoengineering: Conflicting Visions of the Future and Their Links to Governance Proposals." *Current Opinion in Environmental Sustainability* 45: 10–19. https://doi.org/10.1016/j.cosust.2020.06.004.

Haddaway, Neal R., Katarina Hedlund, Louise E. Jackson, Thomas Kätterer, Emanuele Lugato, Ingrid K. Thomsen, Helene B. Jørgensen, and Per-Erik Isberg. 2017. "How Does Tillage Intensity Affect Soil Organic Carbon? A Systematic Review." *Environmental Evidence* 6: 30. https://doi.org/10.1186/s13750-017-0108 -9.

Haikola, Simon, Anders Hansson, and Matthias Fridahl. 2018. "Views of BECCS among Modellers and Policymakers." In *Bioenergy with Carbon Capture and Storage: From Global Potentials to Domestic Realities*, edited by Mathias Fridahl, 17–29. Brussels: Liberal European Forum.

Hale, Benjamin. 2012. "The World That Would Have Been: Moral Hazard Arguments against Geoengineering." In *Engineering the Climate: The Ethics of Solar Radiation Management*, edited by Christopher Preston, 113–131. Lahman: Lexington Books.

Hale, Benjamin, and Lisa Dilling. 2011. "Geoengineering, Ocean Fertilization, and the Problem of Permissible Pollution." *Science, Technology, & Human Values* 36, no. 2: 190–212. https://doi.org/10.1177/0162243910366150.

Halon, Michelle L. D. 2022. "Lunar Mining and Moon Land Claims Fall into a Gray Area of International Law, But Negotiations Are Underway to Avoid Conflict and Damage to Spacecraft." *The Conversation,* August 23, 2022. https:// theconversation.com/lunar-mining-and-moon-land-claims-fall-into-a-gray-area-of

-international-law-but-negotiations-are-underway-to-avoid-conflict-and-damage
-to-spacecraft-188426.

Halstead, John. 2018. "Stratospheric Aerosol Injection Research and Existential
Risk." *Futures* 102: 63–77. https://doi.org/10.1016/j.futures.2018.03.004.

Hamilton, Clive. 2011. "Ethical Anxieties about Geoengineering: Moral hazard,
Slippery Slope and Playing God." Paper presented to a conference of the
Australian Academy of Science Canberra, 27 September 2011. http://www
.homepages.ed.ac.uk/shs/Climatechange/Geo-politics/ethical_anxieties_about
_geoengineering.pdf.

Hamilton, Clive. 2013a. *Earthmasters: The Dawn of the Age of Climate Engineering*.
New Haven: Yale University Press.

Hamilton, Clive. 2013b. "No, We Should Not Just 'At Least Do the Research'."
Nature 496: 139. https://doi.org/10.1038/496139a.

Hamilton, Clive. 2014. "Geoengineering and the Politics of Science." *Bulletin of the
Atomic Scientists* 70, no. 3: 17–26. https://doi.org/10.1177/0096340214531173.

Hardin, Garrett. 1968. "The Tragedy of the Commons." *Science* 162, no. 3859:
1243–1248.

Hardin, Russell. 1982. *Collective Action*. Washington: Resources for the Future.
https://doi.org/10.4324/9781315044330.

Hartman, Laura M. 2017. "Climate Engineering and the Playing God Critique."
Ethics & International Affairs 31, no. 3: 313–333. https://doi.org/10.1017/
S0892679417000223.

Hartzell-Nichols, Lauren. 2012. "Precaution and Solar Radiation Management."
Ethics, Policy & Environment 15, no. 2: 158–171. https://doi.org/10.1080
/21550085.2012.685561.

Hausfather, Zeke and Glen P. Peters. 2020. "Emissions-the 'Business as Usual' Story
is Misleading." *Nature* 577: 618–620. https://doi.org/10.1038/d41586-020-00177
-3.

Hausman, Daniel M. 2021. "Philosophy of Economics." In *The Stanford Encyclopedia
of Philosophy*, edited by Edward N. Zalta, Winter. https://plato.stanford.edu/
archives/win2021/entries/economics/.

Heath, Joseph. 2020. *The Machinery of Government: Public Administration and the
Liberal State*. New York: Oxford University Press. https://doi.org/10.1093/oso
/9780197509616.001.0001.

Heath, Joseph. 2021. *Philosophical Foundations of Climate Change Policy*. New
York: Oxford University Press. https://doi.org/10.1093/oso/9780197567982.001
.0001.

Heck, Vera, Dieter Gerten, Wolfgang Lucht, and Lena R. Boysen. 2016. "Is Extensive
Terrestrial Carbon Dioxide Removal a 'Green' Form of Geoengineering? A Global
Modelling Study." *Global and Planetary Change* 137: 123–130. https://doi.org/10
.1016/j.gloplacha.2015.12.008.

Hepburn, Cameron, Ella Adlen, John Beddington, Emily A. Carter, Sabine Fuss, Niall
Mac Dowell, Jan C. Minx, Pete Smith, and Charlotte K. Williams. 2019. "The
Technological and Economic Prospects for CO_2 Utilization and Removal." *Nature*
575: 87–97. https://doi.org/10.1038/s41586-019-1681-6.

Hermansson, Hélène, and Sven Ove Hansson. 2007. "A Three-Party Model Tool for Ethical Risk Analysis." *Risk Management* 9, no. 3: 129–144. https://doi.org/10.1057/palgrave.rm.8250028.

Heyen, Daniel, Joshua Horton, and Juan Moreno-Cruz. 2019. "Strategic Implications of Counter-Engineering: Clash or Cooperation." *Journal of Environmental Economics and Management* 95: 153–177. https://doi.org/10.1016/j.jeem.2019.03.005.

Heymann, Matthias, and Dania Achermann. 2018. "From Climatology to Climate Science in the Twentieth Century." In *The Palgrave Handbook of Climate History*, edited by Sam White, Christian Pfister, and Franz Mauelshagen, 605–632. London: Palgrave Macmillan. https://doi.org/10.1057/978-1-137-43020-5_38.

Heyward, Clare. 2013. "Situating and Abandoning Geoengineering: A Typology of Five Responses to Dangerous Climate Change." *Political Science & Politics* 46, no. 1: 23–27. https://doi.org/10.1017/S1049096512001436.

Heyward, Clare. 2014. "Benefiting from Climate Geoengineering and Corresponding Remedial Duties: The Case of Unforeseeable Harms." *Journal of Applied Philosophy* 31, no. 4: 405–419. https://doi.org/10.1111/japp.12075.

Heyward, Clare. 2015. "Is There Anything New Under the Sun? Exceptionalism, Novelty and Debating Geoengineering Governance." In *The Ethics of Climate Governance*, edited by Aaron Maltais and Catriona McKinnon, 135–154. London and New York: Rowman & Littlefield.

Heyward, Clare, and Dominic Roser. 2016. *Climate Justice in a Non-Ideal World*. Oxford: Oxford University Press.

Himmelsbach, Raffael. 2018. "How Scientists Advising the European Commission on Research Priorities View Climate Engineering Proposals." *Science and Public Policy* 45, no. 1: 124–133. https://doi.org/10.1093/scipol/scx053.

Honegger, Matthias, Wil Burns, and David R. Morrow. 2021. "Is Carbon Dioxide Removal 'Mitigation of Climate Change'?" *Review of European, Comparative & International Environmental Law* 30, no. 3: 327–335. https://doi.org/10.1111/reel.12401.

Horton, Joshua B. 2013. "Geoengineering and the Myth of Unilateralism: Pressures and Prospects for international Cooperation." In *Climate Change Geoengineering: Philosophical Perspectives, Legal Issues, and Governance Frameworks*, edited by Wil C. G. Burns and Andrew L. Strauss, 168–181. New York: Cambridge University Press. https://doi.org/10.1017/CBO9781139161824.010.

Horton, Joshua B. 2015. "The Emergency Framing of Solar Geoengineering: Time for a Different Approach." *The Anthropocene Review* 2, no. 2: 147–151. https://doi.org/10.1177/2053019615579922.

Horton, Joshua B., and David W. Keith. 2016. "Solar Geoengineering and Obligations to the Global Poor." In *Climate Justice and Geoengineering: Ethics and Policy in the Atmospheric Anthropocene*, edited by Christopher J. Preston, 79–92. London and New York: Rowman & Littlefield.

Horton, Joshua B., David W. Keith, and Matthias Honegger. 2016. *Implications of the Paris Agreement for Carbon Dioxide Removal and Solar Geoengineering* (Policy Brief). Belfer Center: Harvard Project on Climate Agreements.

Horton Joshua B., Jessie L. Reynolds, Holly Jean Buck, Daniel Callies, Stefan Schäfer, David W. Keith, and Steve Rayner. 2018. "Solar Geoengineering and Democracy." *Global Environment Politics* 18, no. 3: 5–24. https://doi.org/10.1162/glep_a_00466.

Houlton, Benjamin Z. 2020. "An Effective Climate Change Solution May Lie in Rocks Beneath Our Feet." *The Conversation*, July 16, 2020. https://theconversation.com/an-effective-climate-change-solution-may-lie-in-rocks-beneath-our-feet-142462.

Hourdequin, Marion. 2018. "Climate Change, Climate Engineering, and the 'Global Poor': What Does Justice Require?" *Ethics, Politics, & Environment* 21, no. 3: 270–288. https://doi.org/10.1080/21550085.2018.1562525.

Hourdequin, Marion. 2019. "Geoengineering Justice: The Role of Recognition." *Science, Technology, & Human Values* 44, no. 3: 448–477. https://doi.org/10.1177/0162243918802893.

Hubert, Anna-Maria. 2021. "Code of Conduct for Responsible Geoengineering Research." *Global Policy* 12, S1: 82–96. https://doi.org/10.1111/1758-5899.12845.

Hulme, Mike. 2014. *Can Science Fix Climate Change?* Cambridge: Polity Press.

Huntingford, Chris and Jason Lowe. 2007. "'Overshoot' Scenarios and Climate Change." *Science* 316, no. 5826: 829b. https://doi.org/10.1126/science.316.5826.829b.

Hursthouse, Rosalind, and Glen Pettigrove. 2022. "Virtue Ethics." *The Stanford Encyclopedia of Philosophy*, edited by Edward N. Zalta and Uri Nodelman, Winter. https://plato.stanford.edu/archives/win2022/entries/ethics-virtue/.

IEA. 2021a. *Direct Air Capture*. Paris: IEA. https://www.iea.org/reports/direct-air-capture.

IEA. 2021b. *World Energy Balances: Overview*. Paris: IEA. https://www.iea.org/reports/world-energy-balances-overview

IEA. 2021c. *Is Carbon Capture Too Expensive?* Paris: IEA. https://www.iea.org/commentaries/is-carbon-capture-too-expensive.

IEA. 2022a. *Global CO2 Emissions Rebounded To Their Highest Level in History in 2021*. Paris: IEA. https://www.iea.org/news/global-co2-emissions-rebounded-to-their-highest-level-in-history-in-2021.

IEA. 2022b. *Greenhouse Gas Emissions from Energy Data Explorer*. Paris: IEA. https://www.iea.org/data-and-statistics/data-tools/greenhouse-gas-emissions-from-energy-data-explorer.

IEA. 2023. *Credible Pathways to 1.5°C Four Pillars for Action in the 2020s*. Paris: IEA. https://www.iea.org/reports/credible-pathways-to-150c.

Incropera, Frank P. 2016. *Climate Change: A Wicked Problem*. New York: Cambridge University Press. https://doi.org/10.1017/CBO9781316266274.

Institute of Medicine, National Academy of Sciences, and National Academy of Engineering. 1992. *Policy Implications of Greenhouse Warming: Mitigation, Adaptation, and the Science Base*. Washington, DC: The National Academies Press. https://doi.org/10.17226/1605.

IPCC. 2005. *IPCC Special Report on Carbon Dioxide Capture and Storage*. New York: Cambridge University Press.

IPCC. 2015. *Climate Change 2014: Synthesis Report. Contribution of Working Groups I, II and III to the Fifth Assessment Report of the Intergovernmental Panel on Climate Change.* Geneva: IPCC.

IPCC. 2018. *Global Warming of 1.5°C: IPCC Special Report on Impacts of Global Warming of 1.5°C above Pre-industrial Levels in Context of Strengthening Response to Climate Change, Sustainable Development, and Efforts to Eradicate Poverty.* Geneva: IPCC.

IPCC. 2019. *Summary for Policymakers. In: Climate Change and Land: An IPCC Special Report on Climate Change, Desertification, Land Degradation, Sustainable Land Management, Food Security, and Greenhouse Gas Fluxes in Terrestrial Ecosystems.* Geneva: IPCC.

IPCC. 2021. *Summary for Policymakers.* Cambridge, UK and New York, NY, USA: Cambridge University Press.

IPCC. 2023. *Climate Change 2023: Synthesis Report.* Geneva: IPCC.

Irvine, Peter J., and David W. Keith. 2020. "Halving Warming with Stratospheric Aerosol Geoengineering Moderates Policy-Relevant Climate Hazards." *Environmental Research Letters* 15, no. 4: 044011. https://doi.org/10.1088/1748 -9326/ab76de.

Irvine, Peter J., Andy Ridgwell, and Daniel J. Lunt. 2010. "Assessing the Regional Disparities in Geoengineering Impacts." *Geophysical Research Letters* 37, no. 18: L18702. https://doi.org/10.1029/2010GL044447.

Irvine, Peter J., Andy Ridgwell, and Daniel J. Lunt. 2011. "Climatic Effects of Surface Albedo Geoengineering." *Journal of Geophysical Research – Atmospheres* 116, no. D24: D24112. https://doi.org/10.1029/2011JD016281.

Irvine, Peter J., Ryan L. Sriver, and Klaus Keller. 2012. "Tension between Reducing Sea-Level Rise and Global Warming Through Solar Radiation Management." *Nature Climate Change* 2, no. 2: 97–100. https://doi.org/10.1038/nclimate1351.

Irvine, Peter J., Kerry Emanuel, Jie He, Larry W. Horowitze, Gabriel Vecchi, and David Keith. 2019. "Halving Warming with Idealized Solar Geoengineering Moderates Key Climate Hazards." *Nature Climate Change,* 9: 295–299. https://doi .org/10.1038/s41558-019-0398-8.

Irvine, Peter J., Ben Kravitz, Mark G. Lawrence, Dieter Gerten, Cyril Caminade, Simon N. Gosling, Erica J. Hendy, Belay T. Kassie, W. Daniel Kissling, Helene Muri, Andreas Oschlies, and Steven J. Smith. 2017. "Towards a Comprehensive Climate Impacts Assessment of Solar Geoengineering." *Earth's Future* 5, no. 1: 93–106. https://doi.org/10.1002/2016EF000389.

Jacobson, Mark Z., and John E. Ten Hoeve. 2012. "Effects of Urban Surfaces and White Roofs on Global and Regional Climate." *Journal of Climate* 25, no. 3: 1028–1044. https://doi.org/10.1175/JCLI-D-11-00032.1.

Jamieson, Dale. 1996. "Ethics and Intentional Climate Change." *Climatic Change* 33, no. 3: 323–336. https://doi.org/10.1007/BF00142580.

Jamieson, Dale. 2013. "Some Whats, Whys and Worries of Geoengineering." *Climatic Change* 121, no. 3: 527–537. https://doi.org/10.1007/s10584-013-0862-9.

Jinnah, Sikina, Simon Nicholson, David R. Morrow, Zachary Dove, Paul Wapner, Walter Valdivia, Leslie Paul Thiele, Catriona McKinnon, Andrew Light, Myanna

Lahsen, Prakash Kashwan, Aarti Gupta, Alexander Gillespie, Richard Falk, Ken Conca, Dan Chong, and Netra Chhetri. 2019. "Governing Climate Engineering: A Proposal for Immediate Governance of Solar Radiation Management." *Sustainability* 11, no. 14. https://doi.org/10.3390/su11143954.

Johannessen, Sophia C., and Robbie W. Macdonald. 2016. "Geoengineering with Seagrasses: Is Credit Due Where Credit Is Given?" *Environmental Research Letters* 11:113001. https://doi.org/10.1088/1748-9326/11/11/113001.

Jones, Andy, Jim M. Haywood, Kari Alterskjær, Olivier Boucher, Jason N. S. Cole, Charles L. Curry, Peter J. Irvine, Duoying Ji, Ben Kravitz, Jón Egill Kristjánsson, John C. Moore, Ulrike Niemeier, Alan Robock, Hauke Schmidt, Balwinder Singh, Simone Tilmes, Shingo Watanabe, and Jin-Ho Yoon. 2013. "The Impact of Abrupt Suspension of Solar Radiation Management (Termination Effect) in Experiment G2 of the Geoengineering Model Intercomparison Project (GeoMIP)." *Journal of Geophysical Research: Atmospheres* 118, no. 17: 9743–9752. https://doi.org/10.1002/jgrd.50762.

Kahneman, Daniel, and Amos Tversky. 2000. "Choices, Values, and Frames." *American Psychologist* 39, no. 4: 582–591. https://psycnet.apa.org/doi/10.1037/0003-066X.39.4.341.

Katz, Eric. 2015. "Geoengineering, Restoration, and the Construction of Nature: Oobleck and the Meaning of Solar Radiation Management." *Environmental Ethics* 37, no. 4: 485–498. https://doi.org/10.5840/enviroethics201537444.

Keith, David W. 2000. "Geoengineering the Climate: History and Prospect." *Annual Review of Energy and the Environment* 25: 245–284. https://doi.org/10.1146/annurev.energy.25.1.245.

Keith, David W. 2013. *A Case for Climate Engineering*. Cambridge, MA and London: The MIT Press.

Keith, David W. 2017. "Toward a Responsible Solar Geoengineering Research Program." *Issues in Science & Technology* 33, no. 3: 71–77.

Keith, David W., and Peter J. Irvine. 2016. "Solar Geoengineering Could Substantially Reduce Climate Risks – A Research Hypothesis for the Next Decade." *Earth's Future* 4, no. 11: 549–559. https://doi.org/10.1002/2016EF000465.

Keith, David W., Edward A. Parson, and M. Granger Morgan. 2010. "Research on Global Sun Block Needed Now." *Nature,* 463: 426–427. https://doi.org/10.1038/463426a.

Keith, David W., Riley Duren, and Douglas G. MacMartin. 2014. "Field Experiments on Solar Geoengineering: Report of a Workshop Exploring a Representative Research Portfolio." *Philosophical Transactions of the Royal Society A* 372, no. 2031: 20140175. https://doi.org/10.1098/rsta.2014.0175.

Keith, David W., Debra K. Weisenstein, John A. Dykema, and Frank N. Keutsch. 2016. "Stratospheric Solar Geoengineering without Ozone Loss." *Proceedings of the National Academy of Sciences* 113, no. 52: 14910–14914. https://doi.org/10.1073/pnas.1615572113.

Kellogg, Louise H., Donald L. Turcotte, and Harsha Lokavarapu. 2019. "On the Role of the Urey Reaction in Extracting Carbon from the Earth's Atmosphere and

Adding It to the Continental Crust." *Frontiers in Astronomy and Space Sciences* 6: 62. https://doi.org/10.3389/fspas.2019.00062.

Kellogg, William W., and Stephen H. Schneider. 1974. "Climate Stabilization: For Better or for Worse?" *Science* 186, no. 4170: 1163–1172. https://doi.org/10.1126/science.186.4170.1163.

Khadka, Navin Singh. 2022. "How Phantom Forests Are Used for Greenwashing." *BBC*, May 3, 2022. https://www.bbc.com/news/science-environment-61300708.

Kintisch, Eli. 2010. *Hack the Planet*. Hoboken NJ: John Wiley & Sons.

Knight, Frank. 1921. *Risk, Uncertainty, and Profit*. Boston and New York: Houghton Mifflin Company.

Köhler, Peter, Jens Hartmann, and Dieter A. Wolf-Gladrow. 2010. "Geoengineering Potential of Artificially Enhanced Silicate Weathering of Olivine." *Proceedings of the National Academy of Sciences of the United States of America* 107, no. 47: 20228–20233. https://doi.org/10.1073/pnas.1000545107.

Köhler, Peter, Jesse F. Abrams, Christoph Völker, Judith Hauck and Dieter A. Wolff-Gladrow. 2013. "Geoengineering Impact of Open Ocean Dissolution of Olivine on Atmospheric CO_2, Surface Ocean pH and Marine Biology." *Environmental Research Letters* 8, no. 1: 014009. https://doi.org/10.1088/1748-9326/8/1/014009.

Kolb, Leigh. 2019. "Guilt by Association Fallacy." In *Bad Arguments: 100 of the Most Important Fallacies of Western Philosophy,* edited by Robert Arp, Steven Barbone, and Michael Bruce, 351–353. Oxford: John Wiley & Sons Ltd.

Kolodny, Niko, and John Brunero. 2023. "Instrumental Rationality." In *Stanford Encyclopedia of Philosophy*, edited by Edward N. Zalta and Uri Nodelman, Summer. https://plato.stanford.edu/archives/sum2023/entries/rationality-instrumental/.

Komorita, Samuel S., and Craig D. Parks. 1996. *Social Dilemmas*. Boulder, CO and Oxford: Westview Press.

Koop, Christel, and Martin Lodge. 2017. "What is Regulation? An Interdisciplinary Concept Analysis." *Regulation & Governance* 11, no. 1: 95–108. https://doi.org/10.1111/rego.12094.

Krause-Jensen, Dorte, and Carlos M. Duarte. 2016. "Substantial Role of Macroalgae in Marine Carbon Sequestration." *Nature Geoscience* 9: 737–742. https://doi.org/10.1038/ngeo2790.

Kravitz, Ben, Douglas G. MacMartin, and Ken Caldeira. 2012. "Geoengineering: Whiter Skies?" *Geophysical Research Letters* 39, no. 11: L11801. https://doi.org/10.1029/2012GL051652.

Kravitz, Ben, Douglas G. MacMartin, Hailong Wang, and Philip J. Rasch. 2016. "Geoengineering as a Design Problem." *Earth System Dynamics* 7, no. 2: 469–497. https://doi.org/10.5194/esd-7-469-2016.

Kuebbeler, Miriam, Ulrike Lohmann, and Johann Feitcher. 2012. "Effects of Stratospheric Aerosol Geo-Engineering on Cirrus Clouds." *Geophysical Research Letters* 39, no. 23: L23803. https://doi.org/10.1029/2012GL053797.

Kwiatkowski, Lester, Katharine L. Ricke, and Ken Caldeira. 2015. "Atmospheric Consequences of Disruption of the Ocean Thermocline." *Environmental Research Letters* 10: 034016. https://doi.org/10.1088/1748-9326/10/3/034016.

Lampitt, Richard S., Eric P. Achterberg, Thomas R. Anderson, Hughes J.A, Debora M. Iglesias-Rodriguez, Boris A. Kelly-Gerreyn, Mary Lucas, Ekaterina E. Popova, Richard Sanders, John G. Shepherd, Denise Smythe-Wright, and Andrew Yool. 2008. "Ocean Fertilization: A Potential Means of Geoengineering?" *Philosophical Transactions of the Royals Society A* 366, no. 1882: 3919–3945. https://doi.org/10 .1098/rsta.2008.0139.

Landes, Xavier. 2013. "Insurance." In *Encyclopedia of Corporate Social Responsibility*, edited by Samuel Idowu, Nicholas Capaldi, Liangrong Zu, Ananda Das Gupta, 1433–1440. New York, Springer. https://doi.org/10.1007/978-3-642-28036-8_75.

Landes, Xavier. 2015. "Moral Hazard." In *Dictionary of Corporate Social Responsibility,* edited by Samuel Idowu, Nicholas Capaldi, Liangrong Zu, and Ananda Das Gupta, 389. New York: Springer.

Landes, Xavier. 2017. "Consensus and Liberal Legitimacy: From First to Second Best?" *The Ethics Forum* 12, no. 1: 84–106. https://doi.org/10.7202/1042279ar.

Latham, John, Alan Gadian, Jim Fournier, Ben Parkes, Peter Wadhams, and Jack Chen. 2014. "Marine Cloud Brightening: Regional Applications." *Philosophical Transactions of the Royal Society A* 372, no. 2031: 20140053. https://doi.org/10 .1098/rsta.2014.0053.

Latham, John, Joan Kleypas, Rachel Hauser, Ben Parkes, and Alan Gadian. 2013. "Can Marine Cloud Brightening Reduce Coral Bleaching?" *Atmospheric Science Letters* 14, no. 4: 214–219. https://doi.org/10.1098%2Frsta.2014.0053.

Latham, John, Ben Parkes, Alan Gadian, and Stephen Salter. 2012. "Weakening of Hurricanes Via Marine Cloud Brightening (MCB)." *Atmospheric Science Letters* 13, no. 4: 231–237. https://doi.org/10.1002/asl.402.

Latham, John, Philip Rasch, Chih-Chieh Chen, Laura Kettles, Alan Gadian, Andrew Gettelman, Hugh Morrison, Keith Bower, and Tom Choularton. 2008. "Global Temperature Stabilization Via Controlled Albedo Enhancement of Low-Level Maritime Clouds." *Philosophical Transactions of the Royal Society A* 336, no. 1882: 1–19. https://doi.org/10.1098/rsta.2008.0137.

Lauderdale, Jonathan Maitland, Rogier Braakman, Gaël Forget, and Michael J. Follows. 2020. "Microbial Feedbacks Optimize Ocean Iron Availability." *Proceedings of the National Academy of Sciences of the United States of America* 117, no. 9: 4842–4849. https://doi.org/10.1073/pnas.1917277117.

Lawford-Smith, Holly, and Adrian Currie. 2017. "Accelerating the Carbon Cycle: The Ethics of Enhanced Weathering." *Biology Letters* 13, no. 4. https://doi.org/10 .1098/rsbl.2016.0859.

Lee, Walker R., Douglas G. MacMartin, Daniele Visioni, and Ben Kravitz. 2021. "High-Latitude Stratospheric Aerosol Geoengineering Can Be More Effective if Injection Is Limited to Spring." *Geophysical Research Letters* 48, no. 9: e2021GL092696. https://doi.org/10.1029/2021GL092696.

Lempert, Robert J., Steven W. Popper, and Steven C. Bankes. 2003. *Shaping the Next One Hundred Years: New Methods for Quantitative, Long-Term Policy Analysis.* Santa Monica: RAND.

Lenferna, Georges Alexandre, Rick D. Russotto, Amanda Tan, Stephen Gardiner, and Thomas P. Ackerman. 2017. "Relevant Climate Response Tests for Stratospheric

Aerosol Injection: A Combined Ethical and Scientific Analysis." *Earth's Future* 5, no. 6: 577–591. https://doi.org/10.1002/2016EF000504.

Lenton, Timothy M. 2019. "Can Emergency Geoengineering Really Prevent Climate Tipping Points?" In *Geoengineering our Climate? Ethics, Politics, and Governance*, edited by Jason J. Blackstock and Sean Low, 43–46. London and New York: Rowman & Littlefield.

Lenton, Timothy M. 2021. "Tipping Points in the Climate System." *Weather* 76, no. 10: 325–326. https://doi.org/10.1002/wea.4058.

Lenton, Timothy M., Johan Rockström, Owen Gaffney, Stefan Rahmstorf, Katherine Richardson, Will Steffen and Hans Joachim Schellnhuber. 2019. "Climate Tipping Points — Too Risky to Bet Against." *Nature* 575: 592–595. https://doi.org/10.1038/d41586-019-03595-0.

Lenzi, Dominic, William F. Lamb, Jérôme Hilaire, Martin Kowarsch, and Jan C. Minx. 2018. "Weigh the Ethics of Plans to Mop Up Carbon Dioxide." *Nature* 561: 303–305. https://doi.org/10.1038/d41586-018-06695-5.

LeRoy, Stephen F., and Larry D. Singell Jr. 1987. "Knight on Risk and Uncertainty." *Journal of Political Economy* 95, no. 2: 394–406. https://doi.org/10.1086/261461.

Lewis, Simon L., and Mark A. Maslin. 2015. "Defining the Anthropocene." *Nature* 519: 171–180. https://doi.org/10.1038/nature14258.

Lewis, Simon L., and Mark A. Maslin. 2018. *The Human Planet: How We Created the Anthropocene*. London: A Pelican Book.

Lieberman, Benjamin, and Elizabeth Gordon. *Climate Change in Human History: Prehistory to the Present*. London: Bloomsbury Academic.

Lin, Albert C. 2013. "Does Geoengineering Presents a Moral Hazard?" *Ecology Law Quarterly* 40, no. 3: 673–712. https://doi.org/10.15779/Z38JP1J.

Lin, Albert C. 2019. "Carbon Dioxide Removal after Paris." *Ecology Law Quarterly* 45, no. 3: 533–582. https://doi.org/10.15779/Z386M3340F.

Lior, Noam. 2013. "Mirrors in the Sky: Status, Sustainability, and Some Supporting Material Experiments." *Renewable and Sustainable Energy Reviews* 18: 401–415. https://doi.org/10.1016/j.rser.2012.09.008.

Lockley, Andrew, Michael Wolovick, Bowie Keefer, Rupert Gladstone, Li-Yun Zhao, and John C. Moore. 2020. "Glacier Geoengineering to Address Sea-Level Rise: A Geotechnical Approach." *Advances in Climate Change Research* 11, no. 4: 401–414. https://doi.org/10.1016/j.accre.2020.11.008.

Logan Mitchell, Ed Brook, James E. Lee, Christo Buizert, and Todd Sowers. 2013. "Constraints on the Late Holocene Anthropogenic Contribution to the Atmospheric Methane Budget." *Science* 342, no. 6161: 964–966. https://doi.org/10.1126/science.1238920.

Lohmann, Ulrike, and Blaž Gasparini. 2017. "A Cirrus Cloud Climate Dial?" *Science* 357, no. 6348: 248–249. https://doi.org/10.1126/science.aan3325.

Lomborg, Bjorn. 2020. *False Alarm: How Climate Change Panic Costs Us Trillions, Hurts the Poor, and Fails to Fix the Planet*. New York: Basic Books.

Long, Jane C. S. 2017. "Coordinated Action against Climate Change: A New World Symphony." *Issues in Science and Technology* 33, no. 3: 78–82.

Long, Jane C. S., Frank Loy, and M. Granger Morgan. 2015. "Start Research on Climate Engineering." *Nature* 518: 29–31. https://doi.org/10.1038/518029a.

L'Orange Seigo, Selma, Simone Dohle, and Michael Siegrist. 2014. "Public Perception of Carbon Capture and Storage (CCS): A Review." *Renewable and Sustainable Energy Reviews* 38: 848–863. https://doi.org/10.1016/j.rser.2014.07.017.

Lovelock, James. 1987. *Gaia: A New Look at Life on Earth.* Oxford: Oxford University Press.

Low, Sean, Chad M. Baum, and Benjamin K. Sovacool. 2022. "Taking it Outside: Exploring Social Opposition to 21 Early-Stage Experiments in Radical Climate Interventions." *Energy Research & Social Science* 90: 102594. https://doi.org/10.1016/j.erss.2022.102594.

Lowe, Scott A. 2016. "An Energy and Mortality Impact Assessment of the Urban Heat Island in the US." *Environmental Impact Assessment Review* 56: 139–144. https://doi.org/10.1016/j.eiar.2015.10.004.

Lunt Daniel J., Andy Ridgwell, Paul J. Valdes, and A. Seale. 2008. "'Sunshade World': A Fully Coupled GCM Evaluation of the Climatic Impacts of Geoengineering." *Geophysical Research Letters* 35, no. 12: L12710. https://doi.org/10.1029/2008GL033674.

Mace, M. J., Claire L. Fyson, Michiel Schaeffer, and William L. Hare. 2021. "Large-Scale Carbon Dioxide Removal to Meet the 1.5°C Limit: Key Governance Gaps, Challenges and Priority Responses." *Global Policy* 12, S1: 67–81. https://doi.org/10.1111/1758-5899.12921.

MacMartin, Douglas G., and Ben Kravitz. 2019. "Mission-Driven Research for Stratospheric Aerosol Geoengineering." *Proceedings of the National Academy of Sciences of the United States of America* 116, no. 4: 1089–1094. https://doi.org/10.1073/pnas.1811022116.

MacMartin, Douglas G., Katharine L. Ricke, and David W. Keith. 2018. "Solar Geoengineering as Part of An Overall Strategy for Meeting the 1.5°C Paris Target." *Philosophical Transactions of the Royal Society A* 376, no. 2119: 20160454. https://doi.org/10.1098/rsta.2016.0454.

MacMartin, Douglas G., David W. Keith, Ben Kravitz, and Ken Caldeira. 2021. "Management of Trade-offs in Geoengineering Through Optimal Choice of Non-Uniform Radiative Forcing." *Nature Climate Change* 3: 365–368. https://doi.org/10.1038/nclimate1722.

MacMartin, Douglas G., Ben Kravitz, Jane C. S. Long, and Philip J. Rasch. 2016. "Geoengineering with Stratospheric Aerosols: What Do We Not Know after a Decade of Research?" *Earth's Future* 4, no. 11: 543–548. https://doi.org/10.1002/2016EF000418.

MacMartin, Douglas G., Daniele Visioni, Ben Kravitz, Jadwiga H. Richter, Tyler Felgenhauer, Walker R. Lee, David R. Morrow, Edward A. Parson, and Masahiro Sugiyama. 2022. "Scenarios for Modeling Solar Radiation Modification." *Proceedings of the National Academy of Sciences of the United States of America* 119, no. 33: e2202230119. https://doi.org/10.1073/pnas.2202230119.

MacMartin, Douglas G., Wenli Wang, Ben Kravitz, Simone Tilmes, Jadwiga H. Richter, and Michael J. Mills. 2019. "Timescale for Detecting the Climate

Response to Stratospheric Aerosol Geoengineering." *JGR Atmospheres* 124, no. 3: 1233–1247. https://doi.org/10.1029/2018JD028906.

Macreadie, Peter I., Daniel A. Nielsen, Jeffrey J. Kelleway, Trisha B. Atwood, Justin R. Seymour, Katherina Petrou, Rod M. Connolly, Alexandra C. G. Thomson, Stacey M. Trevathan-Tackett, and Peter J. Ralph. 2017. "Can We Manage Coastal Ecosystems to Sequester More Blue Carbon." *Frontiers in Ecology and the Environment* 15, no. 4: 206–213. https://doi.org/10.1002/fee .1484.

Mahajan, Aseem, Dusting Tingley, and Gernot Wagner. 2018. "Fast, Cheap, and Imperfect? US Public Opinion about Solar Geoengineering." *Environmental Politics* 28, no. 3: 523–543. https://doi.org/10.1080/09644016.2018.1479101.

Makido, Yasuyo, Dana Hellman, and Vivek Shandas. 2019. "Nature-Based Designs to Mitigate Urban Heat: The Efficacy of Green Infrastructure Treatments in Portland, Oregon." *Atmosphere* 10, no. 5: 282. https://doi.org/10.3390/atmos10050282.

Manabe, Syukuro, and Richard T. Wetherald. 1975. "The Effects of Doubling the CO2 Concentration on the Climate of a General Circulation Model." *Journal of the Atmospheric Sciences* 31, no. 1: 3–15. https://doi.org/10.1175/1520 -0469(1975)032%3C0003:TEODTC%3E2.0.CO;2.

Mankiw, Gregory. 2021. *Principles of Macroeconomics*. 9th ed. Boston: Cengage.

Mann, Michael E. 2012. *The Hockey Stick and Climate Wars*. New York: Columbia University Press.

Mann, Michael E. 2021. *The New Climate War: The Fight to Take Back Our Planet*. Melbourne and London: Scribe.

Marchetti, Cesare. 1977. "On Geoengineering and the CO_2 Problem." *Climate Change* 1, 59–68. https://doi.org/10.1007/BF00162777.

Martin, Josh H. 1990. "Glacial-Interglacial CO2 Change: The Iron Hypothesis." *Paleoceanography and Paleoclimatology* 5, no. 1: 1–13. https://doi.org/10.5194/ cp-15-981-2019.

Martin, Josh H., R. Michael Gordon, and Steve E. Fitzwater. 1990. "Iron in Antarctic Waters." *Nature* 345: 156–158. https://doi.org/10.1038/345156a0.

Matloff, Greg, C. Bangs, and Les Johnson. 2014. *Harvesting Space for a Greener Earth*. New York: Springer. https://doi.org/10.1007/978-1-4614-9426-3.

Matthews, Damon H., and Ken Caldeira. 2007. "Transient Climate-Carbon Simulations of Planetary Geoengineering." *Proceedings of the National Academy of Sciences of the United States of America* 104, no. 24: 9949–9954. https://doi.org /10.1073/pnas.0700419104.

McGinn, Robert. 2018. *The Ethical Engineer: Contemporary Concepts and Cases*. Princeton and Oxford: Princeton University Press.

McKibben, Bill. 2006. *The End of Nature*. New York: Random House.

McKinnon, Catriona. 2019. "The Panglossian Politics of the Geoclique." *Critical Review of International Social and Political Philosophy* 23, no. 5: 584–599. https:// doi.org/10.1073/pnas.0700419104.

McLaren, Duncan P. 2018. "Whose Climate and Whose Ethics? Conceptions of Justice in Solar Geoengineering Modelling." *Energy Research & Social Science* 44: 209–221. https://doi.org/10.1016/j.erss.2018.05.021.

McLaren, Duncan P. 2021. "Recognizing the Injustice in Geoengineering: Negotiating a Path to Restorative Justice through a Political Account of Justice as Recognition." In *Has It Come to This? The Promises and Perils of Geoengineering on the Brink*, edited by J. P. Sapinski, Holly Jean Buck, and Andreas Malm, 82–98. New Brunswick, Camden, and Newark, New Jersey, and London: Rutgers University Press.

McLaren, Duncan P., and Olaf Corry. 2021. "The Politics and Governance of Research into Solar Geoengineering." *WIREs Climate Change* 12, no. 3: e707. https://doi.org/10.1002/wcc.707.

McLaren, Duncan P., David P. Tyfield, Rebecca Willis, Bronislaw Szerszynski, and Nils O. Markusson. 2019. "Beyond 'Net-Zero': A Case for Separate Targets for Emissions Reduction and Negative Emissions." *Frontiers in Climate* 1: 4. https://doi.org/10.3389/fclim.2019.00004.

McMichael, Anthony J. (with Alistair Woodward and Cameron Muir). 2017. *Climate Change and the Health of Nations: Famines, Fevers, and the Fate of Populations.* New York: Oxford University Press. https://doi.org/10.1093/oso/9780190262952.001.0001.

Merk, Christine, Gert Pönitzch, and Katrin Rehdanz. 2016. "Knowledge about Aerosol Injection Does Not Reduce Individual Mitigation Efforts." *Environmental Research Letters* 11, no. 5: 054009. https://doi.org/10.1088/1748-9326/11/5/054009.

Meyer, Kristen, and Christian Uhle. 2015. *Geoengineering and the Accusation of Hubris.* THESys Discussion Paper No. 20153. Berlin, Germany: Humboldt-Universität zu Berlin.

Meyerson, Denise, and Catriona Mackenzie. 2018. "Procedural Justice and the Law." *Philosophical Compass* 13, no. 12: e12548. https://doi.org/10.1111/phc3.12548.

Michaelson, Jay. 2013. "Geoengineering and Climate Management: From Marginality to Inevitability." In *Climate Change Geoengineering: Philosophical Perspectives, Legal Issues, and Governance Frameworks*, edited by Wil C. G. Burns and Andrew L. Strauss, 81–114. New York: Cambridge University Press. https://doi.org/10.1017/CBO9781139161824.010

Milman, Oliver, and Dominic Rushe. 2021. "The Latest Must-Have among US Billionaires? A Plan to End the Climate Crisis." *The Guardian*, March 25, 2021. https://www.theguardian.com/us-news/2021/mar/25/elon-musk-climate-plan-reward-jeff-bezos-gates-investments.

Minx, Jan C., William F. Lamb, Max W. Callaghan, Sabine Fuss, Jérôme Hilaire, Felix Creutzig, Thorben Amann, Tim Beringer, Wagner de Oliveira Garcia, Jens Hartmann, Tarun Khanna, Dominic Lenzi, Gunnar Luderer, Gregory F. Nemet, Joeri Rogelj, Pete Smith, Jose Luis Vicente Vicente, Jennifer Wilcox, and Maria del Mar Zamora Dominguez. 2018. "Negative Emissions-Part 1: Research Landscape and Synthesis." *Environmental Research Letters* 13, no. 6: 063001. https://doi.org/10.1088/1748-9326/aabf9b.

Miocic, Johannes M., Stuart M. V. Gilfillan, Norbert Frank, Andrea Schroeder-Ritzrau, Neil M. Burnside, and R. Stuart Haszeldine. 2019. "420,000-Year

Assessment of Fault Leakage Rates Shows Geological Carbon Storage Is Secure." *Scientific Reports* 9: 769. https://doi.org/10.1038/s41598-018-36974-0.

Mitchell, David L., and William Finnegan. 2009. "Modification of Cirrus Clouds to Reduce Global Warming." *Environmental Research Letters,* 4, no. 4: 045102. https://doi.org/10.1088/1748-9326/4/4/045102.

Mitchell, David L., Subhashree Mishra, and R. Paul Lawson. 2011. "Cirrus Clouds and Climate Engineering: New Findings on Ice Nucleation and Theoretical Basis." In *Planet Earth 2011: Global Warming Challenges and Opportunities for Policy and Practice,* edited by Elias Carayannis, 257–288. London: IntechOpen.

Moellendorf, Darrel. 2015. "Can Dangerous Climate Change Be Avoided?" *Global Justice: Theory Practice Rhetoric* 8, no. 2: 66–85. https://doi.org/10.21248/gjn.8 .2.94.

Möller, Ina. 2021. "Winning Hearts and Minds? Explaining the Rise of the Geoengineering Idea." In *Has It Come to This? The Promises and Perils of Geoengineering on the Brink,* edited by J. P. Sapinski, Holly Jean Buck, and Andreas Malm, 21–33. New Brunswick, Camden, and Newark, New Jersey, and London: Rutgers University Press.

Moore, Bryan L. 2017. *Ecological Literature and the Critique of Anthropocentrism.* Cham: Palgrave Macmillan.

Moore, Jason W. 2017. "The Capitalocene, Part I: On the Nature and Origins of Our Ecological Crisis." *The Journal of Peasant Studies* 44, no. 3: 594–630. https://doi .org/10.1080/03066150.2016.1235036.

Moore, John C., Ilona Mettiäinen, Michael Wolovick, Liyun Zhao, Rupert Gladstone, Ying Chen, Stefan Kirchner, and Timo Koivurova. 2021. "Targeted Geoengineering: Local Interventions with Global Implications." *Global Policy* 12, S1: 108–118. https://doi.org/10.1111/1758-5899.12867.

Morrow, David R. 2014a. "Starting a Flood to Stop a Fire? Some Moral Constraints on Solar Radiation Management." *Ethics, Policy, & Environment* 17, no. 2:123– 138. https://doi.org/10.1080/21550085.2014.926056.

Morrow, David R. 2014b. "Why Geoengineering is a Public Good, Even if it is Bad." *Climatic Change* 123, no. 2: 95–100. https://doi.org/10.1007/s10584-013-0764-x

Morrow, David R. 2014c. "Ethical Aspects of the Mitigation Obstruction Argument against Climate Engineering Research." *Philosophical Transactions of the Royal Society A* 372, no. 2031: 20140062. https://doi.org/10.1098/rsta.2014.0062.

Morrow David R. 2020a. "A Mission-Driven Research Program on Solar Geoengineering Could Promote Justice and Legitimacy." *Critical Review of International Social and Political Philosophy* 23, no. 5: 618–640. https://doi.org /10.1080/13698230.2020.1694220.

Morrow, David R. 2020b. *Values in Climate Policy.* London and New York: Rowman & Littlefield.

Morrow, David R., and Toby Svoboda. 2016. "Geoengineering and Non-Ideal Theory." *Public Affairs Quarterly* 30, no. 1: 83–102.

Morrow, David R., Robert E. Kopp, and Michael Oppenheimer. 2009. "Toward Ethical Norms and Institutions for Climate Engineering Research." *Environmental Research Letters* 4, no. 4: 045106. https://doi.org/10.1088/1748-9326/4/4/045106.

Morton, Oliver. 2015. *The Planet Remade: How Geoengineering Could Change the World*. London: Granta.

Moss, David A. 2002. *When All Else Fails: Government as the Ultimate Risk Manager*. Cambridge, MA: Harvard University Press. https://doi.org/10.2307/j.ctv23985qn.

Muri, Helene, Jón Egill Kristjánsson, Trude Storelvmo, and Melissa Anne Pfeffer. 2014. "The Climatic Effects of Modifying Cirrus Clouds in a Climate Engineering Framework." *Journal of Geophysical Research – Atmospheres* 119, no. 7: 4174–4191. https://doi.org/10.1002/2013JD021063.

Myhre, Gunnar, Drew Shindell, François-Marie Bréon, William Collins, Jan Fuglestvedt, Jianping Huang, Dorothy Koch, Jean-François Lamarque, David Lee, Blanca Mendoza, Teruyuki Nakajima, Alan Robock, Graeme Stephens, Toshihiko Takemura and Huan Zhang. 2013. "Anthropogenic and Natural Radiative Forcing." In: *Climate Change 2013: The Physical Science Basis*, edited by IPCC, 659–740. Cambridge and New York, NY: Cambridge University Press. https://doi.org/10.1017/CBO9781107415324.018.

Nalam, Aditya, Govindasamy Bala, and Angshuman Modak. 2018. "Effects of Arctic Geoengineering on Precipitation in the Tropical Monsoon Regions." *Climate Dynamics* 50, no. 9–10: 3375–3395. https://doi.org/10.1007/s00382-017-3810-y.

NASA. 2016. *Systems Engineering Handbook. NASA SP-2016-6105 Rev2*. Washington: National Aeronautics and Space Administration.

National Academies of Sciences, Engineering, and Medicine. 2019. *Negative Emissions Technologies and Reliable Sequestration: A Research Agenda*. Washington, DC: The National Academies Press. https://doi.org/10.17226/25259.

National Academies of Sciences, Engineering, and Medicine. 2021. *Reflecting Sunlight: Recommendations for Solar Geoengineering Research and Research Governance*. Washington, DC: The National Academies Press. https://doi.org/10.17226/25762.

National Research Council. 1983. *Changing Climate: Report of the Carbon Dioxide Assessment Committee*. Washington, DC: The National Academies Press. Assessment Committee. Washington, DC: The National Academies Press. https://doi.org/10.17226/18714.

National Research Council. 2015a. *Climate Intervention: Carbon Dioxide Removal and Reliable Sequestration*. Washington, DC: The National Academies Press. https://doi.org/10.17226/18805.

National Research Council. 2015b. *Climate Intervention: Reflecting Sunlight to Cool Earth*. Washington, DC: The National Academies Press. https://doi.org/10.17226/18988.

Nature. 2021. "Give Research into Solar Geoengineering a Chance." *Nature* 593: 167. https://doi.org/10.1038/d41586-021-01243-0.

Nehrbass-Ahles, Christoph, Jinhwa Shin, Jochen Schmitt, Bernhard Bereiter, Fortunat Joos, Adrian Schilt, Loïc Schmidely, Lucas Silva, Gregory Teste, Roberto Grilli, Jérôme Chappellaz, David Hodell, Hubertus Fischer, and Thomas F. Stocker. 2020. "Abrupt CO_2 release to the atmosphere under glacial and early interglacial climate conditions." *Science* 369, no. 6506: 1000–1005. https://doi.org/10.1126/science .aay8178.

Neimark, Benjamin. 2018. "Greenwashing: Corporate Tree Planting Generates Goodwill But May Sometimes Harm the Planet." *The Conversation,* September 25, 2018. https://theconversation.com/greenwashing-corporate-tree-planting-generates -goodwill-but-may-sometimes-harm-the-planet-103457.

Nemet, Gregory F., Max W. Callaghan, Felix Creutzig, Sabine Fuss, Jens Hartmann, Jérôme Hilaire, William F. Lamb, Jan C. Minx, Sophia Rogers, and Pete Smith. 2018. "Negative Emissions—Part 3: Innovation and Upscaling." *Environmental Research Letters* 13: 063003. https://doi.org/10.1088/1748-9326/aabff4.

Nordhaus, Ted. 2018. "THE TWO-DEGREE DELUSION." *Foreign Affairs*, February 8, 2018. https://www.foreignaffairs.com/world/two-degree-delusion.

Nordhaus, William. 2007. "A Review of the *Stern Review on the Economics of Climate Change.*" *Journal of Economic Literature* 45, no. 3: 686–702. https://doi .org/10.1257/jel.45.3.686.

Nordhaus, William. 2008. *A Question of Balance: Weighing Options on Global Warming Policies.* New Haven and London: Yale University Press.

Nordhaus, William. 2019. "Climate Change: The Ultimate Challenge for Economics." *American Economic Review* 109, no. 6: 1991–2014. https://doi.org/10.1257/aer .109.6.1991.

Nozick, Robert. 1974. *Anarchy, State, and Utopia.* New York: Basic Books.

Nunn, Nathan, and Nancy Quian. 2010. "The Columbian Exchange: A History of Disease, Food, and Ideas." *Journal of Economic Perspectives* 24, no. 2: 163–188. https://doi.org/10.1257/jep.24.2.163.

Nusbaumer, Jesse, and Katsumi Matsumoto. 2008. "Climate and Carbon Cycle Changes under the Overshoot Scenario." *Global and Planetary Change* 62, no. 1–2: 164–172. https://doi.org/10.1016/j.gloplacha.2008.01.002.

Oakes, William C., and Les L. Leone. 2018. *Engineering Your Future: A Comprehensive Introduction to Engineering.* 9th ed. New York and Oxford: Oxford University Press.

OECD/IEA. 2016. *20 Years of Carbon Capture and Storage: Accelerating Future Deployment.* Paris: International Energy Agency.

Ogle, Stephen M., Cody Alsaker, Jeff Baldock, Martial Bernoux, F. Jay Breidt, Brian McConkey, Kristiina Regina, and Gabriel G. Vazquez-Amabile. 2019. "Climate and Soil Characteristics Determine Where No-Till Management Can Store Carbon in Soils and Mitigate Greenhouse Gas Emissions." *Scientific Reports* 9: 11665. https://doi.org/10.1038/s41598-019-47861-7.

Oksanen, Aslak-Antti. 2023. "Dimming the Midnight Sun? Implications of the Sámi Council's Intervention against the SCoPEx Project." *Frontiers in Climate* 5: 994193. https://doi.org/10.3389/fclim.2023.994193.

Olechowski, Alison L., Steven D. Eppinger, Nitin Joglekar, and Katharina Tomaschek. 2020. "Technological Readiness Levels: Shortcomings and Improvement Opportunities." *Systems Engineering* 23, no. 4: 395–408. https://doi.org/10.1002 /sys.21533.

Oleson, Keith W., Gordon B. Bonan, and Johannes J. Feddema. 2010. "Effects of White Roofs on Urban Temperature in a Global Climate Model." *Geophysical Research Letters* 37, no. 3. https://doi.org/10.1029/2009GL042194.

Olsaretti, Serena. 2018. *The Oxford Handbook of Distributive Justice*. Oxford: Oxford University Press. https://doi.org/10.1093/oxfordhb/9780199645121.001 .0001.

Olson, Robert L. 2012. "Soft Geoengineering: A Gentler Approach to Addressing Climate Change." *Environment: Science and Policy for Sustainable Development* 54, no. 5: 29–39. https://doi.org/10.1080/00139157.2012.711672.

Oreskes, Naomi, and Erik M. Conway. 2010. *Merchants of Doubt: How a Handful of Scientists Obscured the Truth on Issues from Tobacco Smoke to Global Warming*. London: Bloomsbury.

Osaka, Shannon. 2023. "This Firm Is Working to Control the Climate. Should the World Let it?" *The Washington Post*, January 9, 2023. https://www.washingtonpost.com/ climate-environment/2023/01/09/make-sunsets-solar-geoengineering-climate/.

Oschlies, Andreas, Markus Pahlow, Andrew Yool, and Richard J. Matear. 2010. "Climate Engineering by Artificial Ocean Upwelling: Channelling the Sorcerer's Apprentice." *Geophysical Research Letters* 37, no. 4: L04701. https://doi.org/10 .1029/2009GL041961.

Ott, Konrad. 2012. "Might Solar Radiation Management Constitute a Dilemma?" In *Engineering the Climate: The Ethics of Solar Radiation Management*, edited by Christopher J. Preston, 33–42. Lanham: Lexington Books.

Otto, Danny, Terese Thoni, Felix Wittstock, and Silke Beck. 2021. "Exploring Narratives on Negative Emissions Technologies in the Post-Paris Era." *Frontiers in Climate* 3. https://doi.org/10.3389/fclim.2021.684135.

Palmgren Michael G., Anna Kristina Edenbrandt, Suzanne Elizabeth Vedel, Martin Marchman Andersen, Xavier Landes, Jeppe Thulin Østerberg, Janus Falhof, Lene Irene Olsen, Søren Brøgger Christensen, Peter Sandøe, Christian Gamborg, Klemens Kappel, Bo Jellesmark Thorsen, and Peter Pagh. 2015. "Are We Ready for Back-to-Nature Crop Breeding?" *Trends in Plant Science* 20, no. 3: 155–164. https://doi.org/10.1016/j.tplants.2014.11.003.

Pamplany, Augustine, Bert Gordijn, and Patrick Brereton. 2020. "The Ethics of Geoengineering: A Literature Review." *Science and Engineering* 26, no. 6: 3069– 3119. https://doi.org/10.1007/s11948-020-00258-6.

Parker, Andy, and Peter J. Irvine. 2018. "The Risk of Termination Shock from Solar Geoengineering." Earth's Future 6, no. 3: 456-467. https://doi.org/10.1002 /2017EF000735

Parker, Andy, Joshua B. Horton, and David W. Keith. 2018. "Stopping Solar Geo-engineering Through Technical Means: A Preliminary Assessment of Counter-Geoengineering." *Earth's Future* 6, no. 8: 1058–1065. https://doi.org/10.1029 /2018EF000864.

Peters, Glen P., and Oliver Geden. 2017. "Catalysing a Political Shift from Low to Negative Carbon." *Nature Climate Change* 7: 619–621. https://doi.org/10.1038/ nclimate3369.

Pettit, Philip. 1997. *Republicanism: A Theory of Freedom and Government.* New York: Oxford University Press. https://doi.org/10.1093/0198296428.001.0001.

Pidgeon, Nick F., and Elspeth Spence. 2017. "Perceptions of Enhanced Weathering as a Biological Negative Emissions Option." *Biology Letters* 13, no. 4. https://doi.org/10.1098/rsbl.2017.0024.

Pidgeon, Nick F., Karen Parkhill, Adam Corner, and Naomi Vaughan. 2013. "Deliberating Stratospheric Aerosols for Climate Geoengineering and the SPICE Project." *Nature Climate Change* 3: 451–457. https://doi.org/10.1038/nclimate1807.

Pierrehumbert, Raymond. 2019. "There Is no Plan B for Dealing with the Climate Crisis." *Bulletin of the Atomic Scientists* 75, no. 5: 215–221. https://doi.org/10.1080/00963402.2019.1654255.

Popp, József, Zoltán Lakner, Mónika Harangi-Rákos, and Miklós Gábor Fári. 2014. "The Effect of Bioenergy Expansion: Food, Energy, and Environment." *Renewable and Sustainable Energy Reviews* 32: 559–578. https://doi.org/10.1016/j.rser.2014.01.056.

Preston, Christopher J. 2012. "Solar Radiation Management and Vulnerable Populations: The Moral Deficit and its Prospects." In *Engineering the Climate: The Ethics of Solar Radiation Management*, edited by Christopher J. Preston, 77–93. Plymouth: Lexington Books.

Preston, Christopher J. 2013. "Ethics and Geoengineering: Reviewing the Moral Issues Raised by Solar Radiation Management and Carbon Dioxide Removal." *WIREs Climate Change* 41, no. 1: 23–37. https://doi.org/10.1002/wcc.198.

Preston, Christopher J., and Wylie Carr. 2018. "Recognitional Justice, Climate Engineering, and the Care Approach." *Ethics, Policy, & Environment* 21, no. 3: 308–323. https://doi.org/10.1080/21550085.2018.1562527.

Proctor, Jonathan, Solomon Hsiang, Jennifer Burney, Marshall Burke, and Wolfram Schlenker. 2018. "Estimating Global Agricultural Effects of Geoengineering Using Volcanic Eruptions." *Nature* 560: 480–483. https://doi.org/10.1038/s41586-018-0417-3.

Quaas, Martin F., Johannes Quaas, Wilfried Rickels, and Olivier Boucher. 2017. "Are There Reasons against Open-Ended Research into Solar Radiation Management? A Model of Intergenerational Decision-Making under Uncertainty." *Journal of Environmental Economics and Management* 84: 1–17. https://doi.org/10.1016/j.jeem.2017.02.002.

Rahman, Atiq A, Paulo Artaxo, Asfawossen Asrat, and Andy Parker. 2018. "Developing Countries Must Lead on Solar Geoengineering Research." *Nature* 556: 22–24. https://doi.org/10.1038/d41586-018-03917-8.

Raimi, Kaitlin T. 2021. "Public Perceptions of Geoengineering." *Current Opinion in Psychology* 42: 66–70. https://doi.org/10.1016/j.copsyc.2021.03.012.

Raimi, Kaitlin T., Alexander Maki, David Dana, and Michael Vandenbergh. 2019. "Framing of Geoengineering Affects Support for Climate Change Mitigation." *Environmental Communication* 13, no. 3: 300–319. https://doi.org/10.1080/17524032.2019.1575258.

Rawls, John. 1971. *A Theory of Justice.* Cambridge, MA: Belknap Press.

Rawls, John. 2005. *Political Liberalism: Expanded Edition*. New York: Columbia University.

Rayner, Steve, Clare Heyward, Tim Kruger, Nick Pidgeon, Catherine Redgwell, and Julian Savulescu. 2013. "The Oxford Principles." *Climatic Change* 121, no. 3: 499–512. https://doi.org/10.1007/s10584-012-0675-2.

Renforth, Phil. 2012. "The Potential of Enhanced Weathering in the UK." *International Journal of Greenhouse Gas Control* 10: 229–243. https://doi.org/10.1016/j.ijggc .2012.06.011.

Reynolds, Jesse. 2015. "A Critical Examination of the Climate Engineering Moral Hazard and Risk Compensation Concern." *The Anthropocene Review* 2, no. 2: 174–191. https://doi.org/10.1177/2053019614554304

Reynolds, Jesse. 2018. "International Law." In *Climate Engineering and the Law: Regulation and Liability for Solar Radiation Management and Carbon Dioxide Removal*, edited by Michael B. Gerrard and Tracy Hester, 57–153. Cambridge: Cambridge University Press. https://doi.org/10.1017/9781316661864.003.

Reynolds, Jesse. 2020. "Is Solar Geoengineering Ungovernable? A Critical Assessment of Governance Challenges Identified by the Intergovernmental Panel on Climate Change." *WIREs Climate Change* 12, no. 2: e690. https://doi.org/10 .1002/wcc.690.

Riahi, Keywan, Christoph Bertram, Daniel Huppmann, Joeri Rogelj, Valentina Bosetti, Anique-Marie Cabardos, Andre Deppermann, Laurent Drouet, Stefan Frank, Oliver Fricko, Shinichiro Fujimori, Mathijs Harmsen, Tomoko Hasegawa, Volker Krey, Gunnar Luderer, Leonidas Paroussos, Roberto Schaeffer, Matthias Weitzel, Bob van der Zwann, Zoi Vrontisi, Francesco Dalla Longa, Jacques Després, Florian Fosse, Kostas Meinshausen, Larissa P. Nogeira, Ken Oshiro, Alexander Popp, Pedro R. R. Rochedo, Gamze Ünlü, Bas van Rujiven, Junya Takkura, Massimo Tavoni, Detlef van Vuuren, and Behnam Zakeri. 2021. "Cost and Attainability of Meeting Stringent Climate Targets without Overshoot." *Nature Climate Change* 11, no. 12: 1063–1069. https://doi.org/10.1038/s41558-021-01215 -2.

Ridgwell, Andy, Joy S. Singarayer, Allstair M. Hetherington, and Paul J. Valdes. 2009. "Tackling Regional Climate Change by Leaf Albedo Bio-geoengineering." *Current Biology* 19, no. 2: 146–150. https://doi.org/10.1016/j.cub.2008.12.025.

Robock, Alan. 2008a. "20 Reasons Why Geoengineering May Be a Bad Idea." *Bulletin of the Atomic Scientists* 64, no. 2: 14–18. https://doi.org/10.2968 /064002006.

Robock, Alan. 2008b. "Geoengineering Shouldn't Distract from Investing in Emissions Reduction." *Bulletin of the Atomic Scientists,* May 29, 2008. https:// thebulletin.org/roundtable_entry/geoengineering-shouldnt-distract-from-investing -in-emissions-reduction/.

Robock, Alan. 2016. "Albedo Enhancement by Stratospheric Sulfur Injections: More Research Needed." *Earth's Future* 4, no. 12: 644–648. https://doi.org/10.1002 /2016EF000407.

Robock, Alan. 2020. "Benefits and Risks of Stratospheric Solar Radiation Management for Climate Intervention (Geoengineering)." *Bridge* 50, no. 1: 59–67.

Robock, Alan, Luke Oman, and Georgiy L. Stenchikov. 2008. "Regional Climate Responses to Geoengineering with Tropical and Arctic SO2 Injections." *Journal of Geophysical Research* 113, no. D16. https://doi.org/10.1029/2008JD010050.

Robock, Alan, Martin Bunzl, Ben Kravitz, and Georgiy L. Stenchikov. 2010. "A Test for Geoengineering?" *Science* 327, no. 5965: 530–531. https://doi.org/10.1126/science.1186237.

Rockström, Johan, Owen Gaffney, Joeri Rogelj, Malte Meinshausen, Nebojsa Nakicenovic, Hans Joachim Schellnhuber. 2017. "A Roadmap for Rapid Decarbonization." *Science* 355, no. 6331: 1269–1271. https://doi.org/10.1126/science.aah3443.

Rogelj, Joeri, Taryn Fransen, Michel G. J. Den Elzen, Robin D. Lamboll, Clea Schumer, Takeshi Kuramochi, Frederic Hans, Silke Mooldijk, and Joana Portugal-Pereira. 2023. "Credibility Gap in Net-Zero Climate Targets Leaves World at High Risk." *Science* 380, no. 6649: 1014–1016. https://doi.org/10.1126/science.adg6248.

Rogelj, Joeri, Daniel Huppmann, Volker Krey, Keywan Riahi, Leon Clarke, Matthew Gidden, Zebedee Nicholls, and Malte Meinshausen. 2019. "A New Scenario Logic for Paris Agreement Long-Term Temperature Goal." *Nature* 573: 357–363. https://doi.org/10.1038/s41586-019-1541-4.

Roy, Kenneth. 2022. "The Solar Shield Concept: Current Status and Future Possibilities." *Acta Astronautica* 197: 368–374. https://doi.org/10.1016/j.actaastro.2022.02.022.

Royal Society and Royal Academy of Engineering. 2018. "Greenhouse Gas Removal." https://royalsociety.org/topics-policy/projects/greenhouse-gas-removal/.

Ruddiman, William F. 2003. "The Anthropogenic Greenhouse Era Began Thousands of Years Ago." *Climate Change* 61: 261–293. https://doi.org/10.1023/B:CLIM.0000004577.17928.fa.

Russell, Lynn M., Armin Sorooshian, John H. Seinfeld, Bruce A. Albrecht, Athanasios Nenes, Lars Ahlm, Yi-Chun Chen, Matthew Coggon, Jill S. Craven, Richard C. Flagan, Amanda A. Frossard, Haflidi Jonsson, Eunsil Jung, Jack J. Lin, Andrew R. Metcalfe, Robin Modini, Johannes Mülmenstädt, Greg C. Roberts, Taylor Shingler, Siwon Song, Zhen Wang, and Anna Wonaschütz. 2013. "Eastern Pacific Emitted Aerosol Cloud Experiment." *Bulletin of the American Meteorological Society* 94, no. 5: 709–729. https://doi.org/10.1175/BAMS-D-12-00015.1.

Salter, Stephen, Graham Sortino, and John Latham. 2008. "Sea-going Hardware for the Cloud Albedo Method of Reversing Global Warming." *Philosophical Transactions of the Royal Society A* 366, no. 1882: 3989–4006. https://doi.org/10.1098/rsta.2008.0136.

Samiraddi. 2021. "Open Letter Requesting Cancellation of Plans for Geoengineering Related Test Flights in Kiruna." March 2, 2021. https://www.saamicouncil.net/news-archive/open-letter-requesting-cancellation-of-plans-for-geoengineering.

Sánchez, Joan-Pau, and Colin R. McInnes. 2015. "Optimal Sunshade Configurations for Space-Based Geoengineering near the Sun-Earth L_1 Point." *PLoS One* 10, no. 8: e0136648. https://doi.org/10.1371/journal.pone.0136648.

Santana, Carlos. 2019. "Waiting for the Anthropocene." *British Journal of Philosophy of Science* 70, no. 4: 1073–1096. https://doi.org/10.1093/bjps/axy022.

Sarewitz, Daniel, and Richard Nelson. 2008. "Three Rules for Technological Fixes." *Nature* 456: 871–872. https://doi.org/10.1038/456871a.

Searchinger, Tim, Richard Waite, Craig Hanson, Janet Ranganathan, and Emily Matthews. 2019. *World Resources Report: Creating a Sustainable Food Future.* Washington, DC: World Resources Institute.

Selznick, Philip 1985. "Focusing Organisational Research on Regulation." In *Regulatory Policy and the Social Sciences*, edited by Roger G. Noll, 363–368. Berkeley: University of California Press.

Sen, Amartya. 1980. "Equality of What?" In *Tanner Lectures on Human Values, Volume 1*, edited by Sterling M. McMurrin, 197–220. Cambridge: Cambridge University Press.

Seymour, Frances. 2020. "Seeing the Forests as Well as the (Trillion) Trees in Corporate Climate Strategies." *One Earth* 2, no. 5: 390–393. https://doi.org/10.1016/j.oneear.2020.05.006.

Schäfer, Stefan, Mark Lawrence, Harald Stelzer, Wanda Born, Sean Low, Asbjørn Aaheim, Paola Adriázola, Gregor Betz, Olivier Boucher, Alexander Carius, Patrick Devine-Right, Anne Therese Gullberg, Stuart Haszeldine, Jim Haywood, Katherine Houghton, Rodrigo Ibarrola, Peter Irvine, Jon-Egill Kristjansson, Tim Lenton, Jasmin S. A. Link, Achim Maas, Lukas Meyer, Helene Muri, Andreas Oschlies, Alexander Proelß, Tim Rayner, Wilfried Rickels, Lena Ruthner, Jürgen Scheffran, Hauke Schmidt, Michael Schulz, Vivian Scott, Simon Shackley, Dennis Tänzler, Matt Watson, and Naomi Vaughan. 2015. *The European Transdisciplinary Assessment of Climate Engineering (EuTRACE): Removing Greenhouse Gases from the Atmosphere and Reflecting Sunlight Away from Earth.* Potsdam: Institute for Advanced Sustainability Studies Potsdam.

Scheer, Dirk, and Ortwin Renn. 2014. "Public Perception of Geoengineering and its Consequences for Public Debate." *Climatic Change* 125, no. 3–4: 305–318. https://doi.org/10.1007/s10584-014-1177-1.

Schenuit, Felix, Rebecca Colvin, Mathias Fridahl, Barry McMullin, Andy Reisinger, Daniel L. Sanchez, Stephen M. Smith, Asbjørn Torvanger, Anita Wreford, and Oliver Geden. 2021. "Carbon Dioxide Removal Policy in the Making: Assessing Developments in 9 OECD Cases." *Frontiers in Climate* 3: 638805. https://doi.org/10.3389/fclim.2021.638805.

Schneider, Stephen H. 1996. "Geoengineering: Could – or Should – We Do It?" *Climatic Change* 33, no. 3: 291–302. https://doi.org/10.1007/BF00142577.

Schneider, Tapio, Colleen M. Kaul, and Kyle G. Pressel. 2020. "Solar Geoengineering May Not Prevent Strong Warming from Direct Effects of CO_2 on Stratocumulus Cloud Cover." *Proceedings of the National Academy of Sciences of the United States of America* 117, no. 48: 30179–30185. https://doi.org/10.1073/pnas.2003730117.

Schrag, Daniel P. 2009. "Storage of Carbon Dioxide in Offshore Sediments." *Science* 325, no. 5948: 1658–1659. https://doi.org/10.1126/science.1175750.

Scott, Dane. 2012. "Insurance Policy or Technological Fix? The Ethical Implications of Framing Solar Radiation Management." In *Engineering the Climate: The Ethics of Solar Radiation Management*, edited by Christopher J. Preston, 151–168. Plymouth: Lexington Books.

Seneviratne, Sonia I., Steven J. Phipps, Andrew J. Pitman, Annette L. Hirsch, Edouard L. Davin, Markus G. Donat, Martin Hirschi, Andrew Lenton, Micah Wilhelm, and Ben Kravitz. 2018. "Land Radiative Management as Contributor to Regional-Scale Climate Adaptation and Mitigation." *Nature Geoscience* 11: 88–96. https://doi.org/10.1038/s41561-017-0057-5.

Shackley, Simon, and Michael Thompson. 2012. "Lost in the Mix: Will the Technologies of Carbon Dioxide Capture and Storage Provide Us with a Breathing Space as We Strive to Make the Transition from Fossil Fuels to Renewables?" *Climatic Change* 110, no. 1–2: 101–121. https://doi.org/10.1007/s10584-011-0071-3.

Shemaloff, Victor, Cayne Layton, Masayuki Tatsumi, Matthew J. Cameron, Jeffrey T. Wright, Graham J. Edgar, and Craig R. Johnson. 2020. "High kelp Density Attracts Fishes Except for Recruiting Cryptobenthic Species." *Marine Environmental Research* 161: 105127. https://doi.org/10.1016/j.marenvres.2020.105127.

Shepherd, John, Ken Caldeira, Peter Cox, Joanna Haigh, David Keith, Brian Launder, Georgina Mace, Gordon MacKerron, John Pyle, Steve Rayner, Catherine Redgwell, and Andrew Watson. 2009. *Geoengineering the Climate: Science, Governance and Uncertainty*. London: Royal Society.

Shue, Henry. 2010. "Deadly Delays, Saving Opportunities Creating a More Dangerous World?" In *Climate Ethics: Essential Readings*, edited by Stephen M. Gardiner, Simon Caney, Dale Jamieson, and Henry Shue, 146–162. New York: Oxford University Press. https://doi.org/10.1093/oso/9780195399622.003.0017.

Shue, Henry. 2017. "Climate Dreaming: Negative Emissions, Risk Transfer, and Irreversibility." *Journal of Human Rights and the Environment* 8, no. 2: 203–216. https://doi.org/10.4337/jhre.2017.02.02.

Shrum, Trisha R., Ezra Markowitz, Holly Buck, Robin Gregory, Sander van der Linden, Shahzeen Z. Attari, and Leaf Van Boven. 2020. "Behavioural Frameworks to Understand Public Perceptions of and Risk Response to Carbon Dioxide Removal." *Interface Focus* 10, no. 5: 20200002. https://doi.org/10.1098/rsfs.2020.0002.

Siegel, David A., Tim DeVries, Scott C. Doney, and Tom Bell. 2021. "Assessing the Sequestration Time Scales of Some Ocean-Based Carbon Dioxide Reduction Strategies." *Environmental Research Letters* 16, no. 10: 105127. https://doi.org/10.1088/1748-9326/ac0be0.

Singarayer, Joy S., Andy Ridgwell, and Peter Irvine. 2009. "Assessing the Benefits of Crop Albedo Bio-Geoengineering." *Environmental Research Letters* 4, no. 4: 045110. https://doi.org/10.1088/1748-9326/4/4/045110.

Sinnott-Armstrong, Walter. 2022. "Consequentialism." In *The Stanford Encyclopedia of Philosophy*, edited by Edward N. Zalta and Uri Nodelman, Winter. https://plato.stanford.edu/archives/win2022/entries/consequentialism/.

Smith, Patrick Taylor. 2018. "Legitimacy and Non-Domination in Solar Radiation Management Research." *Ethics, Policy & Environment* 21, no. 3: 341–361. https://doi.org/10.1080/21550085.2018.1562528.

Smith, Pete, Steven J. Davis, Felix Creutzig, Sabine Fuss, Jan Minx, Benoit Gabrielle, Etsushi Kato, Robert B. Jackson, Annette Cowie, Elmar Kriegler, Detlef P. van Vuuren, Joeri Rogelj, Philippe Ciais, Jennifer Milne, Josep G. Canadell, David McCollum, Glen Peters, Robbie Andrew, Volker Krey, Gyami Shrestha, Pierre Friedlingstein, Thomas Gasser, Arnulf Grübler, Wolfgang K. Heidug, Matthias Jonas, Chris D. Jones, Florian Kraxner, Emma Littleton, Jason Lowe, José Roberto Moreira, Nebojsa Nakicenovic, Michael Obersteiner, Anand Patwardhan, Mathis Rogner, Ed Rubin, Ayyoob Sharifi, Asbjørn Torvanger, Yoshiki Yamagata, Jae Edmonds, and Cho Yongsung. 2016. "Biophysical and Economic Limits to Negative CO2 Emissions." *Nature Climate Change* 6: 42–50. https://doi.org/10.1038/nclimate2870.

Smith, Stephen M., Oliver Geden, Gregory F. Nemet, Matthew Gidden, William F. Lamb, Carter Powis, Rob Bellamy, Max Callaghan, Annette Cowie, Emily Cox, Sabine Fuss, Thomas Gasser, Giacomo Grassi, Jenna Greene, Sarah Lück, Aniruddh Mohan, Finn Müller-Hansen, Glen Peters, Yoga Pratama, Tim Repke, Keywan Riahi, Felix Schenuit, Jan Steinhauser, Jessica Strefler, Jose Maria Valenzuela, and Jan C. Minx. 2023. *The State of Carbon Dioxide Removal.* 1st ed. https://www.stateofcdr.org.

Smith, Wake. 2020. "The Cost of Stratospheric Aerosol Injection through 2100." *Environmental Research Letters* 15, no. 11: 114004. https://doi.org/10.1088/1748-9326/aba7e7.

Smith, Wake, and Gernot Wagner. 2018. "Stratospheric Aerosol Injection Tactics and Costs in the First 15 Years of Deployment." *Environmental Research Letters* 13, no. 12: 124001. https://doi.org/10.1088/1748-9326/aae98d.

Smith, Wake, Umang Bhattarai, Donald C. Bingaman, James I. Mace, and Christian C. Rice. 2022a. "Review of Possible Very High-Altitude Platforms for Stratospheric Aerosol Injection." *Environmental Research Communications* 4, no. 3: 031002. https://doi.org/10.1088/2515-7620/ac4f5d.

Smith, Wake, Umang Bhattarai, Douglas G. MacMartin, Walker R. Lee, Daniele Visioni, Ben Kravitz, and Christian V. Rice. 2022b. "A Subpolar-Focused Stratospheric Aerosol Injection Deployment Scenario." *Environmental Research Communications* 4, no. 9: 095009. https://doi.org/10.1088/2515-7620/ac8cd3.

Soden, Brian J., Richard T. Wetherald, Georgiy L. Stenchikov, and Alan Robock. 2022. "Global Cooling After the Eruption of Mount Pinatubo: A Test of Climate Feedback by Water Vapor." *Science,* 296, no. 5568: 727–730. https://doi.org/10.1126/science.296.5568.727.

Spielthenner, Georg. 2010. "Lesser Evil Reasoning and its Pitfalls." *Argumentation* 24: 139–152. https://doi.org/10.1007/s10503-009-9158-7.

SRMGI (Solar Radiation Management Governance Initiative). 2013. *Solar Radiation Management: The Governance of Research.* London: The Royal Society.

Stanovich, Keith E. 2010. *Decision Making and Rationality in the Modern World.* New York and Oxford: Oxford University Press.

Staw, Barry M. 1976. "Knee-Deep in the Big Muddy: A Study of Escalating Commitment to a Chosen Course of Action." *Organizational Behavior and Human Performance* 16, no. 1: 27–44. https://doi.org/10.1016/0030-5073(76)90005-2.

Steel, Daniel. 2015. *Philosophy and the Precautionary Principle: Science, Evidence, and Environmental Policy.* Cambridge: Cambridge University Press. https://doi .org/10.1017/CBO9781139939652.

Steiner, Gary. 2005. *Anthropocentrism and its Discontents: The Moral Status of Animals in the History of Western Philosophy.* Pittsburgh: University of Pittsburgh Press.

Stephens, Jennie C., and Kevin Surprise. 2020. "The Hidden Injustices of Advancing Solar Geoengineering Research." *Global Sustainability* 3: e2. https://doi.org/10 .1017/sus.2019.28.

Stern, Nicholas. 2007. *The Economics of Climate Change: The Stern Review.* Cambridge University Press. https://doi.org/10.1017/CBO9780511817434.

Stilgoe, Jack. 2015. *Experiment Earth: Responsible Innovation in Geoengineering.* London and New York: Routledge.

Storelvmo, Trude, William R. Boos, and Nadja Herger. 2014. "Cirrus Cloud Seeding: A Climate Engineering Mechanism with Reduced Side Effects?" *Philosophical Transactions of the Royal Society A* 372, no. 2031: 20140116. https://doi.org/10 .1098/rsta.2014.0116.

Storelvmo, Trude, Jón Egill Kristjánsson, Helene Muri, Melissa Anne Pfeffer, Donifan Barahona, and Athanasios Nenes. 2013. "Cirrus Cloud Seeding Has Potential to Cool Climate." *Geophysical Research Letters* 40, no. 1: 178–182. https://doi.org/10.1029/2012GL054201.

Suarez, Pablo, and Maarten K. van Aalst. 2017. "Geoengineering: A Humanitarian Concern." *Earth's Future* 5, no. 2: 183–195. https://doi.org/10.1002/2016EF000464.

Sugiyama, Masahiro, Shinichiro Asayama, and Takanobu Kosugi. 2020. "The North–South Divide on Public Perceptions of Stratospheric Aerosol Geoengineering? A Survey in Six Asia-Pacific Countries." *Environmental Communication* 14, no. 5: 641–656. https://doi.org/10.1080/17524032.2019.1699137.

Sun, Weiyo, Bin Wang, Deliang Chen, Chaochao Gao, Guonian Lu, and Jian Liu. 2020. "Global Monsoon Response to Tropical and Arctic Stratospheric Aerosol Injection." *Climate Dynamics* 55, no. 7–8: 2107–2121. https://doi.org/10.1007/ s00382-020-05371-7.

Sunstein, Cass R. 2005. *Laws of Fear: Beyond the Precautionary Principle.* New York: Cambridge University Press. https://doi.org/10.1017/CBO9780511790850.

Supekar, Sarang D., and Steven J. Skerlos. 2015. "Reassessing the Efficiency Penalty from Carbon Capture in Coal-Fired Power Plants." *Environmental Science & Technology* 49, no. 20: 12576–12584. https://doi.org/10.1021/acs.est.5b03052.

Svoboda, Toby. 2017. *The Ethics of Climate Engineering: Solar Radiation Management and Non-Ideal Justice.* London and New York: Routledge.

Svoboda, Toby, Klaus Keller, Marlos Goes, and Nancy Tuana. 2011. "Sulfate Aerosol Geoengineering: The Question of Justice." *Public Affairs Quarterly* 25, no. 3: 157–179.

Sweet, Shannan K., Jonathon P. Schuldt, Johannes Lehmann, Deborah A. Bossio, and Dominic Woolf. 2021. "Perceptions of Naturalness Predict US Public Support for Soil Carbon Storage as a Climate Solution." *Climatic Change* 166, no. 1–2. https:// doi.org/10.1007/s10584-021-03121-0.

Swoboda, Philipp, Thomas F. Döring, and Martin Hamer. 2022. "Remineralizing Soils? The Agricultural Usage of Silicate Powders: A Review." *Science of the Total Environment* 807, no. 3: 150976. https://doi.org/10.1016/j.scitotenv.2021 .150976.

Szerszynski, Bronislaw, Matthew Kearnes, Phil Macnaghten, Richard Owen, and Jack Stilgoe. 2013. "Why Solar Radiation Management Geoengineering and Democracy Won't Mix." *Environment and Planning A: Economy and Space* 45, no. 12: 2809–2816. https://doi.org/10.1068/a45649.

Talberg, Anita, Peter Christoff, Sebastian Thomas, and David Karoly. 2018a. "Geoengineering Governance-by-Default: An Earth System Governance Perspective." *International Environmental Agreements: Politics, Law and Economics* 18, no. 2: 229–253. https://doi.org/10.1007/s10784-017-9374-9.

Talberg, Anita, Sebastian Thomas, Peter Christoff, and David Karoly. 2018b. "How Geoengineering Scenarios Frame Assumptions and Create Expectations." *Sustainability Science* 13, no. 4: 1093–1104. https://doi.org/10.1007/s11625-018 -0527-8.

Taleb, Nassim Nicholas. 2007. *The Black Swan: The Impact of the Highly Improbable.* New York: Random House.

Tamme, Eva, and Larissa Beck. 2021. "European Carbon Dioxide Removal Policy: Current Status and Future Opportunities." *Frontiers in Climate* 3: 682882. https:// doi.org/10.3389/fclim.2021.682882.

Tang, Aaron, and Luke Kemp. 2021. "A Fate Worse Than Warming? Stratospheric Aerosol Injection and Global Catastrophic Risk." *Frontiers in Climate* 3: 720312. https://doi.org/10.3389/fclim.2021.720312.

Taylor, Paul W. 1986. *Respect for Nature: A Theory of Environmental Ethics.* Princeton and Oxford: Princeton University Press.

Teller, Edward, Lowell Wood, and Roderick Hyde. 1997. "Global Warming and Ice Ages: I. Prospects for Physics-Based Modulation of Global Age." *Lawrence Livermore National Laboratory.*

Teller Edward, Roderick Hyde, and Lowell Wood. 2002. "Active Climate Stabilization: Practical Physics-Based Approaches to Prevention of Climate Change." *Lawrence Livermore National Laboratory.*

Temple, James. 2022. "A startup Says It's Begun Releasing Particles into the Atmosphere, in An Effort to Tweak the Climate." *MIT Technological Review,* December 24, 2022. https://www.technologyreview.com/2022/12/24/1066041 /a-startup-says-its-begun-releasing-particles-into-the-atmosphere-in-an-effort-to -tweak-the-climate/.

Thiele, Leslie Paul. 2019. "Geoengineering and Sustainability." *Environmental Politics* 28, no. 3: 460–479. https://doi.org/10.1080/09644016.2018.1449602.

Tilmes, Simone, Benjamin M. Sanderson, and Brian C. O'Neill. 2016. "Climate impacts of geoengineering in a delayed mitigation scenario." *Geophysical Research Letters* 43, no. 15: 8222–8229. https://doi.org/10.1002/2016GL070122.

Tilmes, Simone, John Fasullo, Jean-Francois Lamarque, Daniel R. Marsh, Michael Mills, Kari Alterskjær, Helene Muri, Jón E. Kristjánsson, Olivier Boucher, Michael Schulz, Jason N. S. Cole, Charles L. Curry, Andy Jones, Jim Haywood, Peter J.

Irvine, Duoying Ji, John C. Moore, Diana B. Karam, Ben Kravitz, Philip J. Rasch, Balwinder Singh, Jin-Ho Yoon, Ulrike Niemeier, Hauke Schmidt, Alan Robock, Shuting Yang, and Shingo Watanabe. 2013. "The Hydrological Impact of Geoengineering in the Geoengineering Model Intercomparison Project (GeoMIP)." *Journal of Geophysical Research: Atmospheres* 118, no. 19: 11036–11058. https://doi.org/10.1002/jgrd.50868.

Tilmes, Simone, Jadwiga H. Richter, Michael J. Mills, Ben Kravitz, Douglas G. MacMartin, Francis Vitt, Joseph J. Tribbia, and Jean-Francois Lamarque. 2017. "Sensitivity of Aerosol Distribution and Climate Response to Stratospheric SO_2 Injection Locations." *Journal of Geophysical Research: Atmospheres* 122, no. 23: 12591–12615. https://doi.org/10.1002/2017JD026888.

Tol, Richard S. J. 2016. "Distributional Implications of Geoengineering." In *Climate Justice and Geoengineering: Ethics and Policy in the Atmospheric Anthropocene*, edited by Christopher J. Preston, 189–200. London and New York: Rowman & Littlefield.

Tollefson, Jeff. 2008. "UN Decision Puts Brakes on Ocean Fertilization." *Nature* 453: 704. https://doi.org/10.1038/453704b.

Tollefson, Jeff. 2012. "Ocean-Fertilization Project off Canada Sparks Furore." *Nature* 490: 458–459. https://doi.org/10.1038/490458a.

Tollefson, Jeff. 2017. "Plankton-Boosting Project in Chile Sparks Controversy." *Nature* 545: 393–394. https://doi.org/10.1038/545393a.

Tollefson, Jeff. 2018. "Clock Ticking on Climate Action." *Nature* 562: 172–173. https://doi.org/10.1038/d41586-018-06876-2.

Tong, Dan, Qiang Zhang, Yixuan Zheng, Ken Caldeira, Christine Shearer, Chaopeng Hong, Yue Quin, and Steven J. Davis. 2019. "Committed Emissions from Existing Energy Infrastructure Jeopardize 1.5 °C Climate Target." *Nature* 572: 373–377. https://doi.org/10.1038/s41586-019-1364-3.

Tracy, Samantha M., Jonathan M. Moch, Sebastian D. Eastham, and Jonathan J. Buonocore. 2022. "Stratospheric Aerosol Injection May Impact Global Systems and Human Health Outcomes." *Elementa: Science of the Anthropocene* 10, no. 1: 00047. https://doi.org/10.1525/elementa.2022.00047.

Transparency International. 2022. "Climate Geoengineering Technologies: Corruption and Integrity Gaps." June 03, 2022. https://www.transparency.org/en/publications/climate-geoengineering-technologies-corruption-integrity-gaps.

Trencher, Gregory, Adrian Rinscheid, Mert Duygan, Nhi Truong, and Jusen Asuka. 2020. "Revisiting Carbon Lock-in in Energy Systems: Explaining the Perpetuation of Coal Power in Japan." *Energy Research & Social Science,* 69: 101770. https://doi.org/10.1016/j.erss.2020.101770.

Trisos, Christopher H., Giuseppe Amatulli, Jessica Gurevitch, Alan Robock, Lili Xia, and Brian Zambri. 2018. "Potentially Dangerous Consequences for Biodiversity of Solar Geoengineering Implementation and Termination." *Nature Ecology & Evolution* 2: 475–482. https://doi.org/10.1038/s41559-017-0431-0.

Tyka, Michael D., Christopher Van Arsdale, and John C. Platt. 2022. "CO_2 Capture by Pumping Surface Acidity to the Deep Ocean." *Energy & Environmental Science* 15: 786–798. https://doi.org/10.1039/D1EE01532J.

United Nations. 2019. *World Population Prospects 2019: Highlights (ST/ESA/ SER.A/423)*. New York: United Nations, Department of Economic and Social Affairs, Population Division.

United Nations Environment Programme. 2022. *The Closing Window: Climate Crisis Calls for Rapid Transformation of Societies*. Kenya, Nairobi. https://www.unep .org/resources/emissions-gap-report-2022.

United Nations Environment Programme. 2023. *One Atmosphere: An Independent Expert Review on Solar Radiation Modification Research and Deployment*. Kenya, Nairobi. https://www.unep.org/resources/report/Solar-Radiation-Modification -research-deployment.

Unruh, Gregory C. 2000. "Understanding Carbon Lock-in." *Energy Policy* 28, no. 12: 817–830. https://doi.org/10.1016/S0301-4215(00)00070-7.

Van Straaten, Peter. 2007. *Agroecology: The Use of Rocks for Crops*. Cambridge ON: Enviroquest Ltd.

Velbel, Michael A. 2009. "Dissolution of Olivine during Natural Weathering." *Geochimica et Cosmochimica Acta* 73, no. 20: 6098–6113. https://doi.org/10.1016 /j.gca.2009.07.024.

Vergragt, Philip J., Nils Markusson, and Henrik Karlsson. 2011. "Carbon Capture and Storage, Bio-Energy with Carbon Capture and Storage, and the Escape from the Fossil-Fuel Lock-in." *Global Environmental Change* 21, no. 2: 282–292. https:// doi.org/10.1016/j.gloenvcha.2011.01.020.

Victor, David G. 2008. "On the Regulation of Geoengineering." *Oxford Review of Economic Policy* 24, no. 2: 322–336. https://doi.org/10.1093/oxrep/grn018.

Victor, David G., M. Granger Morgan, Jay Apt, John Steinbruner, and Katharine Ricke. 2009. "The Geoengineering Option: A Last Resort against Global Warming?" *Foreign Affairs* 88, no. 2: 64–76.

Visioni, Daniele, Douglas G. MacMartin, Ben Kravitz, Simone Tilmes, Michael J. Mills, Jadwiga H. Richter, and Matthew P. Boudreau. 2019. "Seasonal Injection Strategies for Stratospheric Aerosol Geoengineering." *Geophysical Research Letters* 46, no. 13: 7790–7799. https://doi.org/10.1029/2019GL083680.

Visioni, Daniele, Giovanni Pitari, Paolo Tuccella, and Gabriele Curci. 2018. "Sulfur Deposition Changes under Sulfate Geoengineering Conditions: Quasi-Biennial Oscillation Effects on the Transport and Lifetime of Stratospheric Aerosols." *Atmospheric Chemistry and Physics* 18, no. 4: 2787–2808. https://doi.org/10.5194 /acp-18-2787-2018.

Visioni, Daniele, Eric Slessarev, Douglas G. MacMartin, Natalie M. Mahowald, Christine L. Goodale, and Lili Xia. 2020. "What Goes Up Must Come Down: Impacts of Deposition in a Sulfate Geoengineering Scenario." *Environmental Research Letters* 15, no. 9: 094063. https://doi.org/10.1088/1748-9326/ab94eb.

Visschers, Vivianne H. M., Jing Shi, Michael Siegrist, and Joseph Arvai. 2017. "Beliefs and Values Explain International Differences in Perception of Solar Radiation Management: Insights from a Cross-Country Survey." *Climatic Change* 142, no. 3–4: 531–544. https://doi.org/10.1007/s10584-017-1970-8.

Wagner, Gernot. 2021. *Geoengineering: The Gamble*. Cambridge: Polity Press.

Wagner, Gernot, and Martin L. Weitzman. 2012. "Playing God." *Foreign Policy*, October 24, 2020. https://foreignpolicy.com/2012/10/24/playing-god/.

Wagner, Gernot, and Martin L. Weitzman. 2015. *Climate Shock: The Economic Consequences of a Hotter Planet*. Princeton and Oxford: Princeton University Press.

Walzer, Michael. 1983. *Spheres of Justice*. New York: Basic Books.

Wang, Ye, Xiadong Yan, and Zhaomin Wang. "The Biogeophysical Effects of Extreme Afforestation in Modeling Future Climate." *Theoretical and Applied Climatology* 118, no. 3: 511–521. https://doi.org/10.1007/s00704-013-1085-8.

Warren, Mark E. 2017. *The All Affected Interests Principle in Democratic Theory and Principle*. IHS Political Science Series Working Paper No. 145, July 2017. http://aei.pitt.edu/93142/1/Warren_-_All_Affected_Interests_Principle.pdf.

Weinberg, Alvin. 1967. *Reflections on Big Science*. Cambridge MA: MIT Press.

Weitzman, Martin L. 2015. "A Voting Architecture for the Governance of Free-Driver Externalities, with Application to Geoengineering." *Scandinavian Journal of Economics* 117, no. 4: 1049–1068. https://doi.org/10.1111/sjoe.12120.

Whyte, Kyle Powys. 2012. "Now This! Indigenous Sovereignty, Political Obliviousness and Governance Models for SRM Research." *Ethics, Politics, & Environment* 15, no. 2: 172–187. https://doi.org/10.1080/21550085.2012.685570.

Whyte, Kyle Powys. 2018. "Indigeneity in Geoengineering Discourses: Some Considerations." *Ethics, Policy & Environment* 21, no. 3: 289–307. https://doi.org/10.1080/21550085.2018.1562529.

Wibeck, Victoria, Anders Hansson, and Jonas Anshelm. 2015. "Questioning the Technological Fix to Climate Change – Lay Sense-Making of Geoengineering in Sweden." *Energy Research & Social Science* 7: 23–30. https://doi.org/10.1016/j.erss.2015.03.001.

Wiener, Jonathan. 2015. "Towards an Effective System of Monitoring, Reporting and Verification." In *Towards a Workable and Effective Climate Regime*, edited by Scott Barrett, Carlo Carraro and Jaime de Melo, 183–200. London and Clermont-Ferrand: CEPR Press and FERDI.

Wigley, Tom L. 2006. "A Combined Mitigation/Geoengineering Approach to Climate Stabilization." *Science* 314, no. 5798: 452–454. https://doi.org/10.1126/science.1131728.

Wigley, Tom L. 2014. "Why Nuclear Power May Be the Only Way to Avoid Geoengineering." *Bulletin of the Atomic Scientists* 70, no. 3: 10–16. https://doi.org/10.1177/0096340214531174.

Wilby, Robert L. 2017. *Climate Change in Practice*. Cambridge: Cambridge University Press. https://doi.org/10.1017/9781316534588.

Wilkinson, Sara, and Tim Dixon. 2016. *Green Roof Retrofit: Building Urban Resilience*. Oxford: Wiley Blackwell. https://doi.org/10.1002/9781119055587.

Winickoff, David E., Jane A. Flegal, and Asfawossen Asrat. 2015. "Engaging the Global South on Climate Engineering Research." *Nature Climate Change* 5, no. 7: 627–634. https://doi.org/10.1038/nclimate2632.

Winsberg, Eric. 2021. "A Modest Defense of Geoengineering Research: A Case Study in the Cost of Learning." *Philosophy & Technology* 34: 1109–1134. https://doi.org/10.1007/s13347-021-00452-9.

Wolff, Jonathan. 2019. "Fighting Risk with Risk: Solar Radiation Management, Regulatory Drift, and Minimal Justice." *Critical Review of International Social and Political Philosophy* 23, no. 5: 564–583. https://doi.org/10.1080/13698230.2020.1694214.

Wolske, Kimberly S., Kaitlin T. Raimi, Victoria Campbell-Arvai, and P. Sol Hart. 2019. "Public Support for Carbon Dioxide Removal Strategies: The Role of Tampering with Nature Perceptions." *Climatic Change* 152, no. 3–4: 345–361. https://doi.org/10.1007/s10584-019-02375-z.

Wong, Pak-Hang. 2016. "Consenting to Geoengineering." *Philosophy and Technology* 29, no. 2: 173–188. https://doi.org/10.1007/s13347-015-0203-1.

Wong Pak-Hang. 2017. "Maintenance Required: The Ethics of Geoengineering and Post-Implementation Scenarios." *Ethics, Policy, & Environment* 17, no. 2: 186–191. https://doi.org/10.1080/21550085.2014.926090.

World Bank. 2022. "What You Need to Know about the Measurement, Reporting, and Verification (MRV) of Carbon Credits." https://www.worldbank.org/en/news/feature/2022/07/27/what-you-need-to-know-about-the-measurement-reporting-and-verification-mrv-of-carbon-credits.

Zalasiewicz, Jan, Mark Williams, Alan Smith, Tiffany L. Barry, Angela L. Coe, Paul R. Bown, Patrick Brenchley, David Cantrill, Andrew Gale, Philip Gibbard, F. John Gregory, Mark W. Hounslow, Andrew C. Kerr, Paul Pearson, Robert Knox, John Powell, Colin Waters, John Marshall, Michael Oates, Peter Rawson, and Philip Stone. 2008. "Are We Now Living in the Anthropocene?" *GSA Today* 18, no. 2: 4–8. https://doi.org/10.1130/GSAT01802A.1.

Zarnetske, Phoebe L., Jessica Gurevitch, Janet Franklin, Peter M. Groffman, Cheryl S. Harrison, Jessica J. Hellmann, Forrest M. Hoffman, Shan Kothari, Alan Robock, Simone Tilmes, Daniele Visioni, Jin Wu, Lili Xia, and Cheng-En Yang. 2021. "Potential Ecological Impacts of Climate Intervention by Reflecting Sunlight to Cool Earth." *Proceedings of the National Academy of Sciences of the United States of America* 118, no. 5: e1921854118. https://doi.org/10.1073/pnas.1921854118.

Index

About the Author

Xavier Landes (Ph.D. Université de Montréal) is an associate professor at the Stockholm School of Economics in Riga after having held positions at Université de Montréal, University of Toronto, and Copenhagen University. His areas of competence are political and moral philosophy. He has published on topics such as multiculturalism, the welfare state, public insurance, and happiness. Climate change and climate alteration are currently among his focuses of research and teaching.